T0296117

LONDON MATHEMATICAL SOCIETY LECTURE NOTE SERIES

Managing Editor:
Professor M. Reid, Mathematics Institute, University of Warwick, Coventry CV4 7AL, United Kingdom

The titles below are available from booksellers, or from Cambridge University Press at
http://www.cambridge.org/mathematics

287 Topics on Riemann surfaces and Fuchsian groups, E. BUJALANCE, A.F. COSTA & E. MARTÍNEZ (eds)
288 Surveys in combinatorics, 2001, J.W.P. HIRSCHFELD (ed)
289 Aspects of Sobolev-type inequalities, L. SALOFF-COSTE
290 Quantum groups and Lie theory, A. PRESSLEY (ed)
291 Tits buildings and the model theory of groups, K. TENT (ed)
292 A quantum groups primer, S. MAJID
293 Second order partial differential equations in Hilbert spaces, G. DA PRATO & J. ZABCZYK
294 Introduction to operator space theory, G. PISIER
295 Geometry and integrability, L. MASON & Y. NUTKU (eds)
296 Lectures on invariant theory, I. DOLGACHEV
297 The homotopy category of simply connected 4-manifolds, H.-J. BAUES
298 Higher operads, higher categories, T. LEINSTER (ed)
299 Kleinian groups and hyperbolic 3-manifolds, Y. KOMORI, V. MARKOVIC & C. SERIES (eds)
300 Introduction to Möbius differential geometry, U. HERTRICH-JEROMIN
301 Stable modules and the D(2)-problem, F.E.A. JOHNSON
302 Discrete and continuous nonlinear Schrödinger systems, M.J. ABLOWITZ, B. PRINARI & A.D. TRUBATCH
303 Number theory and algebraic geometry, M. REID & A. SKOROBOGATOV (eds)
304 Groups St Andrews 2001 in Oxford I, C.M. CAMPBELL, E.F. ROBERTSON & G.C. SMITH (eds)
305 Groups St Andrews 2001 in Oxford II, C.M. CAMPBELL, E.F. ROBERTSON & G.C. SMITH (eds)
306 Geometric mechanics and symmetry, J. MONTALDI & T. RATIU (eds)
307 Surveys in combinatorics 2003, C.D. WENSLEY (ed.)
308 Topology, geometry and quantum field theory, U.L. TILLMANN (ed)
309 Corings and comodules, T. BRZEZINSKI & R. WISBAUER
310 Topics in dynamics and ergodic theory, S. BEZUGLYI & S. KOLYADA (eds)
311 Groups: topological, combinatorial and arithmetic aspects, T.W. MÜLLER (ed)
312 Foundations of computational mathematics, Minneapolis 2002, F. CUCKER *et al.* (eds)
313 Transcendental aspects of algebraic cycles, S. MÜLLER-STACH & C. PETERS (eds)
314 Spectral generalizations of line graphs, D. CVETKOVIC, P. ROWLINSON & S. SIMIC
315 Structured ring spectra, A. BAKER & B. RICHTER (eds)
316 Linear logic in computer science, T. EHRHARD, P. RUET, J.-Y. GIRARD & P. SCOTT (eds)
317 Advances in elliptic curve cryptography, I.F. BLAKE, G. SEROUSSI & N.P. SMART (eds)
318 Perturbation of the boundary in boundary-value problems of partial differential equations, D. HENRY
319 Double affine Hecke algebras, I. CHEREDNIK
320 L-functions and Galois representations, D. BURNS, K. BUZZARD & J. NEKOVÁR (eds)
321 Surveys in modern mathematics, V. PRASOLOV & Y. ILYASHENKO (eds)
322 Recent perspectives in random matrix theory and number theory, F. MEZZADRI & N.C. SNAITH (eds)
323 Poisson geometry, deformation quantisation and group representations, S. GUTT *et al* (eds)
324 Singularities and computer algebra, C. LOSSEN & G. PFISTER (eds)
325 Lectures on the Ricci flow, P. TOPPING
326 Modular representations of finite groups of Lie type, J.E. HUMPHREYS
327 Surveys in combinatorics 2005, B.S. WEBB (ed)
328 Fundamentals of hyperbolic manifolds, R. CANARY, D. EPSTEIN & A. MARDEN (eds)
329 Spaces of Kleinian groups, Y. MINSKY, M. SAKUMA & C. SERIES (eds)
330 Noncommutative localization in algebra and topology, A. RANICKI (ed)
331 Foundations of computational mathematics, Santander 2005, L.M PARDO, A. PINKUS, E. SÜLI & M.J. TODD (eds)
332 Handbook of tilting theory, L. ANGELERI HÜGEL, D. HAPPEL & H. KRAUSE (eds)
333 Synthetic differential geometry (2nd Edition), A. KOCK
334 The Navier–Stokes equations, N. RILEY & P. DRAZIN
335 Lectures on the combinatorics of free probability, A. NICA & R. SPEICHER
336 Integral closure of ideals, rings, and modules, I. SWANSON & C. HUNEKE
337 Methods in Banach space theory, J.M.F. CASTILLO & W.B. JOHNSON (eds)
338 Surveys in geometry and number theory, N. YOUNG (ed)
339 Groups St Andrews 2005 I, C.M. CAMPBELL, M.R. QUICK, E.F. ROBERTSON & G.C. SMITH (eds)
340 Groups St Andrews 2005 II, C.M. CAMPBELL, M.R. QUICK, E.F. ROBERTSON & G.C. SMITH (eds)
341 Ranks of elliptic curves and random matrix theory, J.B. CONREY, D.W. FARMER, F. MEZZADRI & N.C. SNAITH (eds)
342 Elliptic cohomology, H.R. MILLER & D.C. RAVENEL (eds)
343 Algebraic cycles and motives I, J. NAGEL & C. PETERS (eds)
344 Algebraic cycles and motives II, J. NAGEL & C. PETERS (eds)
345 Algebraic and analytic geometry, A. NEEMAN
346 Surveys in combinatorics 2007, A. HILTON & J. TALBOT (eds)

347 Surveys in contemporary mathematics, N. YOUNG & Y. CHOI (eds)
348 Transcendental dynamics and complex analysis, P.J. RIPPON & G.M. STALLARD (eds)
349 Model theory with applications to algebra and analysis I, Z. CHATZIDAKIS, D. MACPHERSON, A. PILLAY & A. WILKIE (eds)
350 Model theory with applications to algebra and analysis II, Z. CHATZIDAKIS, D. MACPHERSON, A. PILLAY & A. WILKIE (eds)
351 Finite von Neumann algebras and masas, A.M. SINCLAIR & R.R. SMITH
352 Number theory and polynomials, J. MCKEE & C. SMYTH (eds)
353 Trends in stochastic analysis, J. BLATH, P. MÖRTERS & M. SCHEUTZOW (eds)
354 Groups and analysis, K. TENT (ed)
355 Non-equilibrium statistical mechanics and turbulence, J. CARDY, G. FALKOVICH & K. GAWEDZKI
356 Elliptic curves and big Galois representations, D. DELBOURGO
357 Algebraic theory of differential equations, M.A.H. MACCALLUM & A.V. MIKHAILOV (eds)
358 Geometric and cohomological methods in group theory, M.R. BRIDSON, P.H. KROPHOLLER & I.J. LEARY (eds)
359 Moduli spaces and vector bundles, L. BRAMBILA-PAZ, S.B. BRADLOW, O. GARCÍA-PRADA & S. RAMANAN (eds)
360 Zariski geometries, B. ZILBER
361 Words: Notes on verbal width in groups, D. SEGAL
362 Differential tensor algebras and their module categories, R. BAUTISTA, L. SALMERÓN & R. ZUAZUA
363 Foundations of computational mathematics, Hong Kong 2008, F. CUCKER, A. PINKUS & M.J. TODD (eds)
364 Partial differential equations and fluid mechanics, J.C. ROBINSON & J.L. RODRIGO (eds)
365 Surveys in combinatorics 2009, S. HUCZYNSKA, J.D. MITCHELL & C.M. RONEY-DOUGAL (eds)
366 Highly oscillatory problems, B. ENGQUIST, A. FOKAS, E. HAIRER & A. ISERLES (eds)
367 Random matrices: High dimensional phenomena, G. BLOWER
368 Geometry of Riemann surfaces, F.P. GARDINER, G. GONZÁLEZ-DIEZ & C. KOUROUNIOTIS (eds)
369 Epidemics and rumours in complex networks, M. DRAIEF & L. MASSOULIÉ
370 Theory of *p*-adic distributions, S. ALBEVERIO, A.YU. KHRENNIKOV & V.M. SHELKOVICH
371 Conformal fractals, F. PRZYTYCKI & M. URBANSKI
372 Moonshine: The first quarter century and beyond, J. LEPOWSKY, J. MCKAY & M.P. TUITE (eds)
373 Smoothness, regularity and complete intersection, J. MAJADAS & A. G. RODICIO
374 Geometric analysis of hyperbolic differential equations: An introduction, S. ALINHAC
375 Triangulated categories, T. HOLM, P. JØRGENSEN & R. ROUQUIER (eds)
376 Permutation patterns, S. LINTON, N. RUŠKUC & V. VATTER (eds)
377 An introduction to Galois cohomology and its applications, G. BERHUY
378 Probability and mathematical genetics, N. H. BINGHAM & C. M. GOLDIE (eds)
379 Finite and algorithmic model theory, J. ESPARZA, C. MICHAUX & C. STEINHORN (eds)
380 Real and complex singularities, M. MANOEL, M.C. ROMERO FUSTER & C.T.C WALL (eds)
381 Symmetries and integrability of difference equations, D. LEVI, P. OLVER, Z. THOMOVA & P. WINTERNITZ (eds)
382 Forcing with random variables and proof complexity, J. KRAJÍCEK
383 Motivic integration and its interactions with model theory and non-Archimedean geometry I, R. CLUCKERS, J. NICAISE & J. SEBAG (eds)
384 Motivic integration and its interactions with model theory and non-Archimedean geometry II, R. CLUCKERS, J. NICAISE & J. SEBAG (eds)
385 Entropy of hidden Markov processes and connections to dynamical systems, B. MARCUS, K. PETERSEN & T. WEISSMAN (eds)
386 Independence-friendly logic, A.L. MANN, G. SANDU & M. SEVENSTER
387 Groups St Andrews 2009 in Bath I, C.M. CAMPBELL *et al.* (eds)
388 Groups St Andrews 2009 in Bath II, C.M. CAMPBELL *et al.* (eds)
389 Random fields on the sphere, D. MARINUCCI & G. PECCATI
390 Localization in periodic potentials, D.E. PELINOVSKY
391 Fusion systems in algebra and topology, M. ASCHBACHER, R. KESSAR & B. OLIVER
392 Surveys in combinatorics 2011, R. CHAPMAN (ed)
393 Non-abelian fundamental groups and Iwasawa theory, J. COATES *et al.* (eds)
394 Variational roblems in differential geometry, R. BIELAWSKI, K. HOUSTON & M. SPEIGHT (eds)
395 How groups grow, A. MANN
396 Arithmetic dfferential operators over the *p*-adic Integers, C.C. RALPH & S.R. SIMANCA
397 Hyperbolic geometry and applications in quantum Chaos and cosmology, J. BOLTE & F. STEINER (eds)
398 Mathematical models in contact mechanics, M. SOFONEA & A. MATEI
399 Circuit double cover of graphs, C.-Q. ZHANG
400 Dense sphere packings: a blueprint for formal proofs, T. HALES
401 A double Hall algebra approach to affine quantum Schur–Weyl theory, B. DENG, J. DU & Q. FU
402 Mathematical aspects of fluid mechanics, J. ROBINSON, J.L. RODRIGO & W. SADOWSKI (eds)
403 Foundations of computational mathematics: Budapest 2011, F. CUCKER, T. KRICK, A. SZANTO & A. PINKUS (eds)
404 Operator methods for boundary value problems, S. HASSI, H.S.V. DE SNOO & F.H. SZAFRANIEC (eds)
405 Torsors, étale homotopy and applications to rational points, A.N. SKOROBOGATOV (ed)
406 Appalachian set theory, J. CUMMINGS & E. SCHIMMERLING (eds)
407 The maximal subgroups of the low-dimensional finite classical groups, J.N. BRAY, D.F. HOLT & C.M. RONEY-DOUGAL
408 Complexity science: the Warwick master's course, R. BALL, R.S. MACKAY & V. KOLOKOLTSOV (eds)
409 Surveys in combinatorics 2013, S. BLACKBURN, S. GERKE & M. WILDON (eds)
410 Representation theory and harmonic analysis of wreath products of finite groups, T. CECCHERINI SILBERSTEIN, F. SCARABOTTI & F. TOLLI
411 Moduli Spaces, L. BRAMBILA-PAZ, O. GARCIA-PRADA, P. NEWSTEAD & R. THOMAS (eds)

London Mathematical Society Lecture Note Series: 412

Automorphisms and Equivalence Relations in Topological Dynamics

DAVID B. ELLIS
Beloit College, Wisconsin

ROBERT ELLIS
University of Minnesota

CAMBRIDGE
UNIVERSITY PRESS

University Printing House, Cambridge CB2 8BS, United Kingdom

One Liberty Plaza, 20th Floor, New York, NY 10006, USA

477 Williamstown Road, Port Melbourne, VIC 3207, Australia

314-321, 3rd Floor, Plot 3, Splendor Forum, Jasola District Centre, New Delhi - 110025, India

103 Penang Road, #05-06/07, Visioncrest Commercial, Singapore 238467

Cambridge University Press is part of the University of Cambridge.

It furthers the University's mission by disseminating knowledge in the pursuit of education, learning and research at the highest international levels of excellence.

www.cambridge.org
Information on this title: www.cambridge.org/9781107633223

© D. B. Ellis and R. Ellis 2014

This publication is in copyright. Subject to statutory exception and to the provisions of relevant collective licensing agreements, no reproduction of any part may take place without the written permission of Cambridge University Press.

First published 2014

A catalogue record for this publication is available from the British Library

Library of Congress Cataloging in Publication data
Ellis, D. (David), 1958– Automorphisms and equivalence relations in topological dynamics / David B. Ellis, Beloit College, Wisconsin, Robert Ellis, University of Minnesota.
pages cm. – (London Mathematical Society lecture note series ; 412)
ISBN 978-1-107-63322-3 (pbk.)
1. Topological dynamics. 2. Algebraic topology. 3. Automorphisms.
I. Ellis, Robert, 1926–2013 II. Title.
QA611.5.E394 2014
512'.55–dc23
2013043992

ISBN 978-1-107-63322-3 Paperback

Cambridge University Press has no responsibility for the persistence or accuracy of URLs for external or third-party internet websites referred to in this publication, and does not guarantee that any content on such websites is, or will remain, accurate or appropriate.

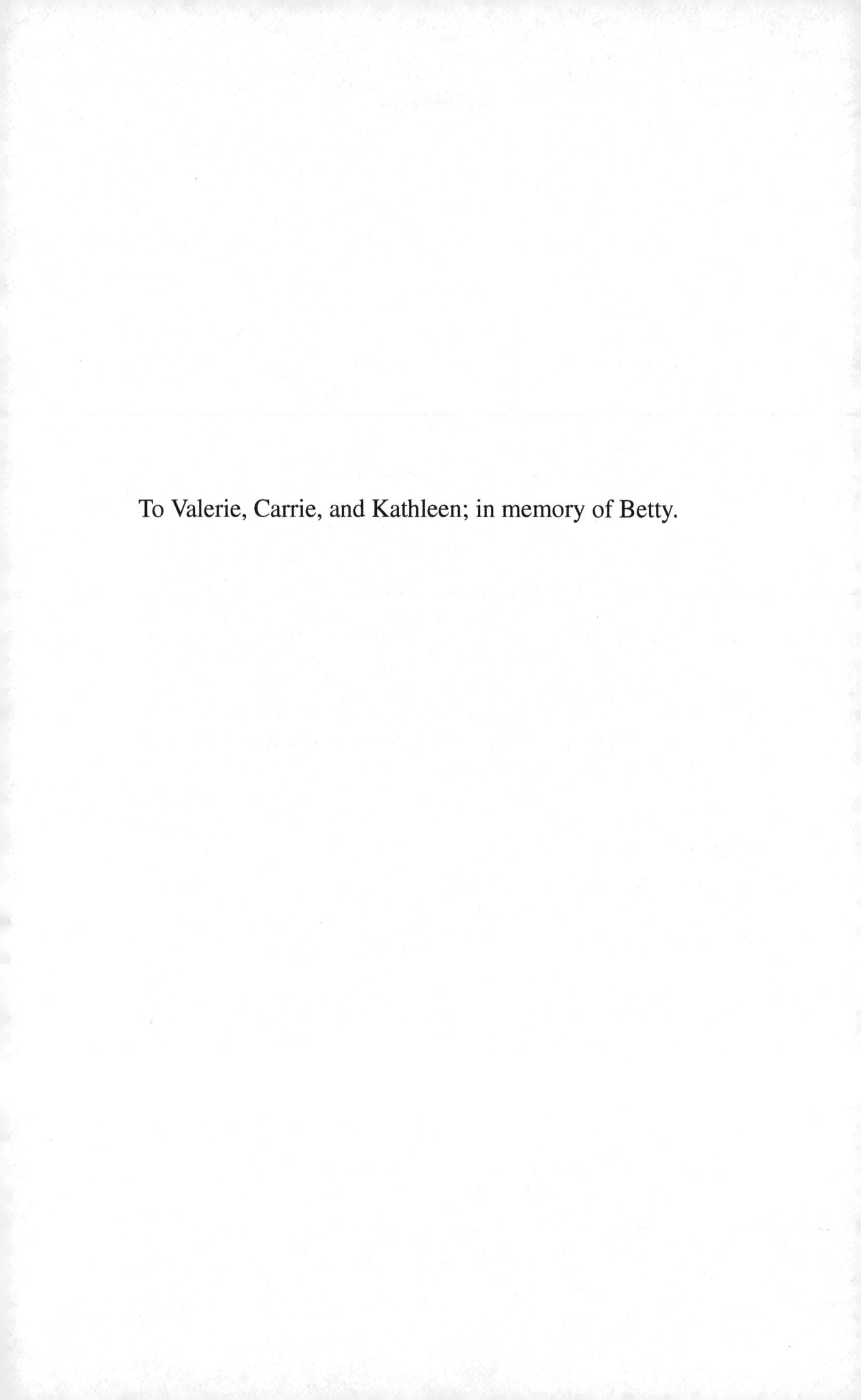

To Valerie, Carrie, and Kathleen; in memory of Betty.

Contents

Introduction

To a large extent this book is an updated version of *Lectures on Topological Dynamics* by Robert Ellis [Ellis, R., (1969)]. That book gave an exposition of what might be called an algebraic theory of minimal sets. Our goal here is to give a clear, self contained exposition of a new approach to the theory which allows for more straightforward proofs and develops a clearer language for expressing many of the fundamental ideas. We have included a treatment of many of the results in the aforementioned exposition, in addition to more recent developments in the theory; we have not attempted, however, to give a complete or exhaustive treatment of all the known results in the algebraic theory of minimal sets. Our hope is that the reader will be motivated to use the language and techniques to study related topics not touched on here. Some of these are mentioned either in the exercises or notes given at the end of various sections. This book should be suitable for a graduate course in topological dynamics whose prerequisites need only include some background in topology. We assume the reader is familiar with compact Hausdorff spaces, convergence of nets, etc., and perhaps has had some exposure to uniform structures and pseudo metrics which play a limited role in our exposition.

A *flow* is a triple (X, T, π) where X is a compact Hausdorff space, T is a topological group, and $\pi : X \times T \to X$ is a continuous action of T on X, so that $xe = x$ and $(xt)s = x(ts)$ for all $x \in X$, $s, t \in T$. Here we write $xt = \pi(x, t)$ for all $x \in X$ and $t \in T$, and e is the identity of the group T. Usually the symbol π will be omitted and the flow (X, T, π) denoted by (X, T) or simply by X. In the situations considered here there is no loss of generality if T is given the discrete topology. The assumptions made thus far do not suffice to produce an interesting theory. The group T may be too "small" in its action on X. Thus for example, the trivial case where $xt = x_0$, a fixed element of X, for all $x \in X$ and $t \in T$, is not ruled out. To eliminate such degenerate behavior it is convenient to assume that the flow (X, T) is *point transitive*, i.e. that there exists $x_0 \in X$ such that its orbit $x_0T \equiv \{x_0t \mid t \in T\}$ is dense in X.

The category $\mathcal{P}$ of point transitive flows has the remarkable property that it possesses a universal object; i,e, there exists a point transitive flow $(\beta T, T)$ such that any flow in $\mathcal{P}$ is a homomorphic image of βT. (See section 1 for a description of $(\beta T, T)$ and the proof of its universality.) Moreover, one may associate in a canonical fashion with any flow (X, T) a point transitive flow $E(X, T)$. The latter, called the *enveloping semigroup* has proved extremely useful in the study of the dynamical properties of the original flow (X, T). The enveloping semigroup is defined and studied in section 2, and examples of its use are scattered throughout the subsequent sections.

This exposition focuses, however, on the category, $\mathcal{M}$, of *minimal* flows. These are flows for which the orbit of every point $x \in X$ is dense; that is $\overline{xT} = \overline{\{xt \mid t \in T\}} = X$ for all $x \in X$. Again there exists a (unique up to isomorphism) universal object M in $\mathcal{M}$. This fact was exploited in several papers to develop an "algebraic theory" of minimal flows. In particular a group was associated with each such flow and various relations among minimal flows studied by means of these groups. One purpose of this volume is to collect in one place the techniques which have proved useful in this study; another goal is to provide an exposition of a new approach to this material.

The account of this algebraic theory of minimal flows given in *Lectures on Topological Dynamics* depends heavily on an algebraic point of view derived by studying the collection $\mathcal{C}(X)$ of continuous functions on X rather than X itself. In this volume we instead exploit the fact that X, as a homomorphic image of M, is of the form M/R for some *icer* (invariant closed equivalence relation) on M. We study the flow (X, T) via the icer R rather than the algebra $\mathcal{C}(X)$.

Another change is that the role of the group of automorphisms of a flow is emphasized. In particular the group G of automorphisms of M plays a crucial role. It is used both to codify the algebraic structure of M, and to define the groups associated to the minimal flows in $\mathcal{M}$. In the earlier approach G was viewed as a subset Mu of M, where $u \in M$ was a fixed idempotent. The new approach eliminates the asymmetrical treatment of the idempotents. Instead we view $M = \biguplus G(u)$ as a disjoint union (taken over all the idempotents in $u \in M$) of the images of the idempotents under the group G. Thus we explicitly take advantage of the fact that every $p \in M$ can be written uniquely in the form $\alpha(u)$ with $\alpha \in G$ and u an idempotent in M. This approach also makes reliance on the concept of a pointed flow unnecessary. Previously the concept of a pointed flow was used to define, up to conjugacy, the group of a minimal flow; a different choice of base point corresponding to a conjugate subgroup of Mu. From the point of view of icers on M, the group of the flow M/R is the subgroup:

$$G(R) = \{\alpha \in G \mid gr(\alpha) \subset R\},$$

of G. Here $gr(\alpha) = \{(p, \alpha(p)) \mid p \in M\}$ is the graph of the automorphism α of M. Again if S is an icer with $M/R \cong M/S$, then $G(S)$ is conjugate to $G(R)$.

One of the important tools for the study of minimal flows is the so-called τ-topology on G. In section 10 we show how one can define a topology on the automorphism group $Aut(X)$ of any regular flow (X, T). Since $G = Aut(M)$ and (M, T) is regular, this allows one to define a topology on G. This topology on G coincides with the original definition of the τ-topology. (The idea for this viewpoint stems from J. Auslander's approach to the τ-topology–private communication.)

We would now like to make a few comments on some of the results which have been included herein. In part I we lay the foundation for what follows by treating the universal constructions upon which much of the later material is based. This includes an introduction to βT, the enveloping semigroup, and the universal minimal flow. The flow $(2^X, T)$ whose minimal subflows are the so-called quasi-factors of the minimal flow (X, T) is discussed in section 5. Here 2^X is the collection of non-empty closed subsets of X. The space 2^X is given the Vietoris topology detailed in the appendix to section 5. The extension of the action of T on 2^X to an action of βT on 2^X via the circle operator is also discussed in section 5, and used later in sections 12 and 17.

Part II develops many of the techniques and language critical to our approach. As mentioned above, this approach hinges on identifying minimal flows as quotients of M by icers. We need not only to associate to any minimal flow an icer on M, but to any icer on M a minimal flow. The basic topological result needed is that the quotient of any compact Hausdorff space by a closed equivalence relation is again a compact Hausdorff space. Section 6 includes a proof of this result and a discussion of the relative product of two relations, a useful tool for constructing equivalence relations.

The fundamental result concerning icers on M is proven in section 7. We show that any icer R on M can be written as a relative product

$$R = (R \cap P_0) \circ gr(G(R))$$

where $P_0 = \{(\alpha(u), \alpha(v)) \mid \alpha \in G \text{ and } u, v \text{ are idempotents in } M\}$ (see **7.21**). Regular flows, whose original definition is motivated in terms of automorphisms, are those flows whose representation as a quotient M/R is unique. The flow (M, T) is of course regular, and its structure serves as a prototype for the algebraic structure of regular flows outlined in section 8. In particular, if (X, T) is a regular flow, then the pair $\{X, Aut(X)\}$ has properties analogous to those of the pair $\{M, G\}$, some of which were alluded to above.

In part III we give a detailed exposition of the approach to the τ-topology mentioned earlier. When applied to the group $Aut(X)$, for any regular minimal flow (X, T), we obtain a topology which is compact and T_1 but not Hausdorff. The construction of a *derived group* F' for any closed subgroup $F \subset Aut(X)$ is given in section 11. F' is a normal subgroup of F which measures the extent to which F fails to be Hausdorff; in fact for any closed subgroup $H \subset F$, the quotient space F/H is Hausdorff if and only if $F' \subset H$ (see **11.10**). In section 12 we give a proof of the fact that there exists a minimal flow X whose group $G(X) = A$ if and only if A is a τ-closed subgroup of G. One example of such a flow is M/R where

$$R = \overline{gr(A)} = \overline{\bigcup\{gr(\alpha) \mid \alpha \in A\}}.$$

The basic idea of the proof, which uses the material on quasi-factors, is the same as in *Lectures on Topological Dynamics* but the language of the current approach allows a more efficient treatment.

Part IV is motivated by the questions: How are the various subgroups of G related to one another, and what do they tell us about the dynamics of minimal flows? It has long been known that the subcategories $\mathcal{D}$ and $\mathcal{E}$ of minimal distal and minimal equicontinuous flows respectively also possess universal objects $X_{\mathcal{D}}$ and $X_{\mathcal{E}}$. Heretofore the groups D and E have been defined as the groups associated to these flows, i.e. $D = G(X_{\mathcal{D}})$ and $E = G(X_{\mathcal{E}})$. In sections 14 and 15 we obtain intrinsic characterizations of D (see **14.6**), and E (see **15.23**) respectively. This gives content to the statements: if X is distal, then $D \subset G(X)$, and if X is equicontinuous, then $E \subset G(X)$. In fact, emphasizing the language of icers, M/R is distal (respectively equicontinuous) if and only if

$$R = P_0 \circ gr(A)$$

with $D \subset A$ (respectively $E \subset A$). For proofs see **14.10** and **15.14** respectively. In particular distal and equicontinuous flows are completely determined by their groups. We show in **15.21** that $G'D = E$, from which it follows immediately that (X, T) is equicontinuous if and only if (X, T) is distal and $G' \subset G(X)$. In section 13 we discuss the proximal relation $P(X)$ on a minimal flow X. In analogy with the distal and equicontinuous cases, we give a description of a subgroup $P \subset G$ and show that $P(X)$ is an equivalence relation if and only if $P \subset G(X)$. Another subgroup $G^J \subset G$ is introduced and we show that $P(X)$ is an equivalence relation with closed cells if and only if $PG^J \subset G(X)$. In fact $PG^J \subsetneq D$ which is consistent with the well-known result that $P(X)$ is a closed invariant equivalence relation on X if and only if $D \subset G(X)$. (see **14.8**) In section 15 the regionally proximal relation, $Q(X)$ of a minimal flow (X, T) is introduced to facilitate the study of equicontinuous flows. (Recall that (X, T) is equicontinuous if and only if $Q(X) = \Delta_X$ the diagonal in $X \times X$.) The case $Q \equiv Q(M)$ is also used to define the group E. Equicontinuous minimal flows are discussed from the point of view of icers on M in the body of section 15, while the approach to the same material via the enveloping semigroup is treated in the appendix. $Q(X)$ is discussed in further detail in section 15 where we give a new proof of the fact that if $E \subset G'G(X)$, then $Q(X)$ is an equivalence relation.

To a large extent part V is concerned with generalizing the results of part IV to homomorphisms (extensions) of minimal flows. For instance for icers $R \subset S$ on M, the canonical projection $M/R \to M/S$ is a distal homomorphism if and only if

$$S = (R \cap P_0) \circ gr(G(S)),$$

moreover the extension is equicontinuous if and only if $G(S)' \subset G(R)$. We close with a section devoted to four theorems all of which are equivalent to the Furstenberg structure theorem for distal extensions; this section uses the language of icers and the techniques developed in the earlier sections to give proofs that all four theorems are equivalent. This fact does not seem to have been emphasized in the literature, and provides a good opportunity to illustrate the language and techniques developed in the book. This analysis also illustrates an interesting twist to the icer approach. Here not only does the structure of the icers R and S come into play in understanding the extension $M/R \to M/S$, the *dynamics* of the *icer* on M/R whose quotient gives M/S also plays an important role. The construction of the so-called Furstenberg tower provides another nice illustration of the language of icers; the stages in the tower are explicitly constructed using icers which are themselves constructed from the groups involved.

Section 20 itself does not contain the proof of the Furstenberg theorem. Instead we give a chart describing where proofs of various special cases appear in the text. On the other hand a complete proof for compact Hausdorff spaces, of the fact that any icer on a minimal flow which is both topologically transitive and pointwise almost periodic must be trivial (one of the equivalents of the Furstenberg structure theorem) appears in **9.13**. This is because our proof relies on the concept of the quasi-relative product developed in section 9. Indeed the quasi-relative product arose during our attempt to give a proof of the Furstenberg theorem in its full generality. The metric case of the theorem follows immediately from the fact that for metric flows the notions of point-transitivity and topological transitivity coincide. Our proof in the general case proceeds by reducing it to the metric case; the key tool in the construction which enables this is the quasi-relative product. While the quasi-relative product is only necessary for the most general version of the Furstenberg theorem, it turns out to be closely connected to quasi-factors (hence the name) and RIC extensions. We detail these connections in sections 9 and 17 respectively.

A word about format

We have written this book using a theorem-proof format. All the proofs are given using a sequence of numbered steps for which reasons are given at each stage. There are two main reasons for this approach. The first is to make sure that the arguments are as clear and accessible as possible. We found that insisting on numbering our steps and giving reasons forced a rigor, clarity, and attention to detail we hope the reader will appreciate. We have attempted to avoid situations where as the material becomes more complex the reader is expected to fill in more gaps in the arguments.

In addition to a better understanding of the details of the individual arguments, we hope that the format adds to the clarity of the overall exposition. The assumptions and conclusions of each of the lemmas, propositions, and theorems are stated carefully and precisely in a consistent format. These items are all numbered so that they can be referred to in a precise and unambiguous way as the exposition proceeds. We have tried to keep the proofs reasonably short and have divided the material into short sections, typically ten to fifteen pages long. In addition, we begin each section with an introduction designed to give an informal outline and motivation for the material in that section. The reader who wishes to go lightly on the intricate details, may wish to follow the train of thought by focusing on the introductions to each section and skipping the proofs. In this case, if a specific result attracts the reader's interest, then the numbering system should facilitate a more careful reading of the details. This format is designed especially for the student who is not yet an expert; it assures that careful attention is paid to the details and that the train of thought is readily accessible.

PART I

Universal constructions

Our focus in the first part of this book is on the construction of certain universal objects that are crucial to the algebraic approach to the study of the asymptotic behavior of dynamical systems (flows). For the purposes of this exposition a flow is a pair (X, T), where X is a compact Hausdorff space, and T is a group which acts on X (on the right). A homomorphism of flows is a continuous mapping which preserves the actions. When the orbit closure of some point $x_0 \in X$ is all of X, that is $\overline{x_0T} = X$, we say that the flow (X, T) is point transitive. If $\overline{xT} = X$ for all $x \in X$ we say that (X, T) is minimal. The collection of point transitive flows has a universal object, $(\beta T, T)$, in the sense that every point transitive flow is a homomorphic image of $(\beta T, T)$ (see **2.5**). The action of T on βT extends in a natural way to a semigroup structure on βT which plays an important role in the study of flows. In section 1, we give an exposition of the structure of βT, relegating its construction via ultrafilters on T to an appendix.

We exploit the properties of βT in section 2 to give a treatment of the enveloping semigroup $E(X, T)$ of a flow (X, T). In Section 4 βT and $E(X, T)$ are used to introduce many of the fundamental notions which will be studied throughout the book. Of particular importance is the structure of the minimal ideals in $E(X, T)$ discussed in section 3. The fact that βT is its own enveloping semigroup allows us to apply these ideas to a minimal right ideal $M \subset \beta T$. On the other hand for such a minimal ideal, the flow (M, T) is a universal object for the collection of all minimal flows (see **3.16**). Our approach to the study of minimal flows involves exploiting the structure of M and the group of automorphisms of M to gain an understanding of the structure of the icers (closed invariant equivalence relations) on M. These ideas are pursued further in section 7 of Part II.

Another construction which will play a significant role in our exposition is that of a quasi-factor of the flow (X, T); this is by definition a subflow of the flow $(2^X, T)$. Here by 2^X we mean the space whose elements are closed non-empty subsets of X. In the appendix to section 5, we give an outline of the construction of the Vietoris topology, a compact Hausdorff topology on 2^X. In the body of section 5 we develop some of the properties of the flow $(2^X, T)$, including the extension of the natural action of T on 2^X to an action of βT on 2^X given by the so-called circle operator.

1

The Stone-cech compactification βT

The Stone-Cech compactification βT, is a compact Hausdorff space containing the discrete group T as a dense subset. Of course one can construct the Stone-Cech compactification of any discrete set; a construction via ultrafilters is outlined in the appendix to this section. On the other hand βT is characterized by certain properties which we take as its definition for the purposes of this section. When T is a group there is a natural semigroup structure on βT, for which left multiplication by all elements, and right multiplication by elements of T are continuous. This semigroup structure plays a fundamental role in our study. In proposition **1.3** we deduce this structure as a consequence of the characterizing properties of βT; in the appendix the semigroup structure is defined directly in terms of ultrafilters.

Definition 1.1 Let T be a set with the discrete topology. The *Stone-Cech compactification* βT *of* T is determined up to homeomorphism by the following properties:

(i) $T \subset \beta T$ with $\overline{T} = \beta T$,
(ii) βT is a compact Hausdorff space, and
(iii) if X is a compact Hausdorff space and $f{:}T \to X$, then there exists a unique continuous extension $\hat{f}\colon \beta T \to X$.

The uniqueness of the extension in (iii) above is crucial. For instance it has as a consequence the fact that βT is unique up to homeomorphism. Indeed if Y is any space satisfying (i), (ii), and (iii), then the inclusions $T \subset Y$ and $T \subset \beta T$ extend to continuous maps $\varphi\colon \beta T \to Y$ and $\psi\colon Y \to \beta T$. The composition $\varphi \circ \psi$ is thus a continuous extension of the inclusion $T \subset Y$ to Y, and hence by uniqueness must be the identity. Similarly $\psi \circ \varphi$ is the identity on βT, and therefore φ is a homeomorphism with inverse ψ. This shows that as a topological space βT is completely determined by the conditions in **1.1**.

The following theorem confirms this by exhibiting a base for the topology on βT. It is interesting to note that this base consists of sets which are both open and closed in βT.

Theorem 1.2 Let:

(i) T be a set with the discrete topology and βT be its Stone-Cech compactification,
(ii) $A \subset T$, and
(iii) $V \subset \beta T$ be an open set.

Then:

(a) $\beta T = \overline{A} \cup \overline{T \setminus A}$ is a disjoint union, and thus $\overline{A}$ is both open and closed (clopen) in βT,
(b) $\overline{V} = \overline{V \cap T}$, and hence $\overline{V}$ is both open and closed, and
(c) $\{\overline{A} \mid A \subset T\}$ is a base for the topology on βT.

PROOF: (a) 1. Let $\emptyset \neq A \subset T$.

2. Let $\chi_A : T \to \{0, 1\}$ be defined by $\chi_A(t) = \begin{cases} 1 & \text{if } t \in A \\ 0 & \text{otherwise} \end{cases}$.

3. There exists a continuous extension $\hat{\chi}_A : \beta T \to \{0, 1\}$. (by 2, **1.1**(iii))

4. $\hat{\chi}_A^{-1}(1)$ and $\hat{\chi}_A^{-1}(0)$ are clopen with $\overline{A} \subset \hat{\chi}_A^{-1}(1)$ and $\overline{T \setminus A} \subset \hat{\chi}_A^{-1}(0)$. (by 2, 3)

5. Let $p \in \hat{\chi}_A^{-1}(1)$ and $W \subset \beta T$ be open with $p \in W$.

6. There exists $t \in T$ with $t \in W \cap \hat{\chi}_A^{-1}(1)$. (by 4, 5, **1.1**(i))

7. $t \in A \cap W$. (by 2, 3, 6)

8. $p \in \overline{A}$. (by 5, 7)

9. $\hat{\chi}_A^{-1}(1) \subset \overline{A}$. (by 5, 8)

10. $\hat{\chi}_A^{-1}(0) \subset \overline{T \setminus A}$. (similar argument)

11. $\hat{\chi}_A^{-1}(1) = \overline{A}$ and $\hat{\chi}_A^{-1}(0) = \overline{T \setminus A}$. (by 4, 9, 10)

(b) 1. Clearly $\overline{V \cap T} \subset \overline{V}$.

2. Let $W \subset \beta T$ be open and $p \in \overline{V} \cap W$.

3. There exists $t \in T$ with $t \in V \cap W$. (by 2, **1.1**(i))

4. $t \in (V \cap T) \cap W$. (by 3)

5. $p \in \overline{V \cap T}$. (by 2, 4)

(c) 1. Let $\emptyset \neq V \subset \beta T$ be open and $p \in V$.

2. There exists W open with $p \in W \subset \overline{W} \subset V$. ($\beta T$ is compact Hausdorff)

3. $p \in \overline{W} = \overline{W \cap T} \subset V$. (by 2, part (b))

4. $\{\overline{A} \mid A \subset T\}$ is a base for the topology on βT. (by 1, 3)

We will be most interested in the space βT when T is a group. In this case, and in fact whenever T is a semigroup, the semigroup structure on T induces a semigroup structure on βT. Once again the uniqueness of the extension in **1.1** (iii) is crucial. The following proposition details the construction.

Proposition 1.3 Let T be a semigroup, so that T is provided with an associative binary operation:

$$T \times T \to T$$
$$(s, t) \to st.$$

Then the semigroup structure on T extends to one on βT,

$$\beta T \times \beta T \to \beta T$$
$$(p, q) \to pq$$

such that:

(a) the right multiplication map $R_t : \beta T \to \beta T$ is continuous for all $t \in T$,
$$p \to pt$$
and
(b) the left multiplication map $L_p : \beta T \to \beta T$ is continuous for all $p \in \beta T$.
$$q \to pq$$

PROOF: 1. Let $m^t(s) = st$ for all $s, t \in T$.
2. There exists a continuous extension $R_t : \beta T \to \beta T$ of m^t for every $t \in T$. (by (iii) of **1.1**)
3. There exists a continuous extension $L_p : \beta T \to \beta T$ of the map
$$\begin{array}{ccc} T & \to & \beta T \\ t & \to & R_t(p) \end{array}.$$
(by (iii) of **1.1**)
4. For $p, q \in \beta T$ we define $pq \equiv L_p(q)$.
5. Let $t, s \in T$. Then the maps $\begin{array}{ccc} \beta T & \to & \beta T \\ p & \to & (ps)t \end{array}$ and $\begin{array}{ccc} \beta T & \to & \beta T \\ p & \to & p(st) \end{array}$
are both continuous extensions of the map

$$T \to \beta T$$
$$t' \to (t's)t = t'(st).$$

6. $p(st) = (ps)t$ for all $p \in \beta T$ and $s, t \in T$.
(by 5 and uniqueness in (iii) of **1.1**)
7. Let $p, q \in \beta T$. Then the maps $\begin{array}{ccc} \beta T & \to & \beta T \\ q & \to & (pq)t \end{array}$ and $\begin{array}{ccc} \beta T & \to & \beta T \\ q & \to & p(qt) \end{array}$
are both continuous extensions of the map

$$T \to \beta T$$
$$s \to (ps)t = p(st).$$
(by 3, 6)

8. $p(qt) = (pq)t$ for all $p \in \beta T$ and $t \in T$.
(by 7 and uniqueness in (iii) of **1.1**)

9. The maps $\begin{array}{rcl} \beta T & \to & \beta T \\ r & \to & (pq)r \end{array}$ and $\begin{array}{rcl} \beta T & \to & \beta T \\ r & \to & p(qr) \end{array}$ are both continuous extensions of the map

$$\begin{array}{l} T \to \beta T \\ t \to (pq)t = p(qt). \end{array} \qquad \text{(by 3, 8)}$$

10. $p(qr) = (pq)r$ for all $p, q, r \in \beta T$. (by 9 and uniqueness in (iii) of **1.1**)

The space βT can be provided with a (different) semigroup structure in which left multiplication is continuous for all $t \in T$, and right multiplication is continuous for all $p \in \beta T$. Merely mimic the proof of **1.2** starting with the map $\begin{array}{rcl} m_t : T & \to & T \\ s & \to & ts \end{array}$. We will most often be interested in **right** actions of a **group** T.

Henceforth we will always assume unless explicitly indicated otherwise that T is a group, and that βT is provided with the semigroup structure of **1.2**. In the upcoming sections we will make extensive use of this semigroup structure and in particular the fact that it makes $(\beta T, T)$ into a flow. It is important to note that the assumption that T is a group, so that every element of T has an inverse does **not** guarantee that the elements of βT have inverses. In fact βT is a group only if T is finite and $\beta T = T$. In general, the only elements of βT which have inverses are the elements of T. This follows immediately from the fact that $p, q \in \beta T$ with $pq \in T$ implies that $p, q \in T$. Indeed if $pq = t \in T$, then $t \in L_p(\beta T) = L_p(\overline{T}) = \overline{L_p(T)}$ since L_p is continuous. On the other hand $T \subset \beta T$ has the discrete topology so $\{t\}$ is an open subset of βT. It follows that $t \in L_p(T)$ and there exists $s \in T$ with $ps = t$. But this implies that $p = ts^{-1} \in T$ and $q = s \in T$.

We end this section with an elementary proposition which speaks to the naturality of the construction of βT.

Proposition 1.4 Let:

(i) T be a semigroup,
(ii) $\emptyset \neq H \subset T$, and
(iii) $j \colon \beta H \to \beta T$ be the continuous extension to βH of the inclusion $H \to \beta T$.

Then:

(a) j is injective,
(b) $\operatorname{im} j = \overline{H}$, and

(c) if H is a subsemigroup of T, then $j(pq) = j(p)j(q)$ for all $p, q \in \beta H$. (Thus we will identify βH with $\overline{H} \subset \beta T$.)

PROOF: (a) 1. Let $h_0 \in H$.

2. Let $\varphi\colon T \to \beta H$ be defined by $\varphi(t) = \begin{cases} t & \text{if } t \in H \\ h_0 & \text{if } t \notin H \end{cases}$.

3. Let $\hat{\varphi}\colon \beta T \to \beta H$ be the continuous extension of φ to βT.

4. Let $\psi = \hat{\varphi} \circ j\colon \beta H \to \beta H$.

5. $\psi(h) = h$ for all $h \in H$. (by 2, 3, 4)

6. $\psi(p) = p$ for all $p \in \beta H$. (by 3, (iii), **1.1**(iii))

7. j is injective. (by 4, 6)

(b) and (c) We leave these to the reader.

APPENDIX TO SECTION 1: ULTRAFILTERS AND THE CONSTRUCTION OF βT

Our goal here is the construction of the compact Hausdorff space βT, which is characterized up to homeomorphism by **1.1**. Those readers already familiar with ultrafilters will recall that a topological space X is compact if and only if every ultrafilter on X converges to a point in X (see **1.A.16** and **Ex. 1.5**). This motivates the approach we will take; in analogy with the construction of the real numbers as Cauchy sequences of rational numbers, βT will be identified as the collection of ultrafilters on T. We have attempted to make this presentation self-contained, so that filters and ultrafilters are defined, and the elementary properties necessary for the construction explicitly introduced. We make use of one of these properties, namely **1.A.8**, in the appendix to section 5. All the other sections of the book, while occasionally using the terminology of this appendix, rely only on the results of section 1 itself. In the interest of brevity, proofs of some of the results in this appendix are left as exercises for the reader. We begin with some background material on filters and ultrafilters.

Definition 1.A.1 Let T be a nonempty set and $\mathcal{F}$ a collection of nonempty subsets of T. We make the following definitions:

(a) $\mathcal{F}$ is a *filter base on* T if

$$F_1, \ldots, F_n \in \mathcal{F} \implies \text{there exists } F \in \mathcal{F} \text{ with } F \subset F_1 \cap \cdots \cap F_n.$$

(b) $\mathcal{F}^c = \{A \mid A \subset T \text{ and there exists } F \in \mathcal{F} \text{ with } F \subset A\}$.

(c) $\mathcal{F}$ is a *filter on* T if $\mathcal{F}$ is a filter base on T and $\mathcal{F}^c = \mathcal{F}$. Thus if $\mathcal{F}$ is a filter, then it has the *finite intersection property* (F.I.P.), meaning

$$F_1, \ldots, F_n \in \mathcal{F} \implies F_1 \cap \cdots \cap F_n \in \mathcal{F}.$$

(d) $\mathcal{U}$ is an *ultrafilter on T* if $\mathcal{U}$ is a filter on T such that

$$\mathcal{F} \text{ a filter on } T \text{ with } \mathcal{U} \subset \mathcal{F} \Longrightarrow \mathcal{U} = \mathcal{F}$$

(so that $\mathcal{U}$ is a maximal filter on T).

The neighborhoods of a point x in a topological space provide an important motivating example; we leave it as an exercise for the reader to verify this.

Example 1.A.2 Let X be a topological space and $x \in X$. Then the collection

$$\mathcal{N}_x = \{A \mid \text{there exists } U \text{ open in } X \text{ with } x \in U \subset A\}$$

is a filter on X. We refer to $\mathcal{N}_x$ as the *neighborhood filter at x*.

Another elementary example which plays a fundamental role in the construction of βT is the following:

Example 1.A.3 Let $t \in T$. Then the collection

$$h(t) = \{A \mid t \in A \subset T\}$$

is an ultrafilter on T. Moreover $h(t)$ is the only ultrafilter on T which contains the singleton set $\{t\}$. We refer to $h(t)$ as the *principal ultrafilter generated by t*.

PROOF: We leave the proof as an exercise for the reader.

According to **1.A.3**, every $t \in T$ generates an ultrafilter on T; we now observe that any filter is contained in some ultrafilter. Suppose that $\{\mathcal{F}_i \mid i \in I\}$ is a collection of filters on T, where I is a totally ordered set. Assume further that if $i < j \in I$, then $\mathcal{F}_i \subset \mathcal{F}_j$. (These assumptions amount to saying that this collection is an increasing chain of filters on T.) Then it is straightforward to check that the union $\bigcup_{i \in I} \mathcal{F}_i$ is a filter on T. This shows that every increasing chain of filters has a maximal element; hence as an immediate consequence of Zorn's lemma (see also **3.3** for a statement) every filter is contained in some maximal filter (i.e an ultrafilter). We state this result as a lemma for future reference:

Lemma 1.A.4 Let $\mathcal{F}$ be a filter (or filter base) on T. Then there exists an ultrafilter $\mathcal{U}$ on T such that $\mathcal{F} \subset \mathcal{U}$.

The next few results examine the structure of ultrafilters on T. In particular they allow us to characterize those filters which are ultrafilters. In fact a filter $\mathcal{F}$ is an ultrafilter if and only if for every $\emptyset \neq A \subset T$, either A or its complement lie in $\mathcal{F}$. (This is the content of **1.A.6** and **1.A.7**.)

Proposition 1.A.5 Let:

(i) $\mathcal{U}$ be an ultrafilter on T, and
(ii) $A \subset T$.

Then $A \in \mathcal{U}$ if and only if $A \cap U \neq \emptyset$ for all $U \in \mathcal{U}$.

PROOF: 1. Since $\mathcal{U}$ is a filter it is clear that $A \in \mathcal{U} \Rightarrow A \cap U \neq \emptyset$ for all $U \in \mathcal{U}$.
2. Assume that $A \cap U \neq \emptyset$ for all $U \in \mathcal{U}$.
3. Let $\mathcal{G} = \{G \subset T \mid A \cap U \subset G \text{ for some } U \in \mathcal{U}\}$.
4. Let $G_1, \ldots, G_n \in \mathcal{G}$.
5. There exist $U_i \in \mathcal{U}$ such that $A \cap U_i \subset G_i$ for $1 \leq i \leq n$. (by 3, 4)
6. $U = U_1 \cap \cdots \cap U_n \in \mathcal{U}$. (by 5, (i))
7. $A \cap U \subset G_1 \cap \cdots \cap G_n$. (by 5, 6)
8. $G_1 \cap \cdots \cap G_n \in \mathcal{G}$. (3, 7)
9. $\mathcal{G}$ is a filter on T. (by 4, 8)
10. $\mathcal{U} \subset \mathcal{G}$. (by 3)
11. $\mathcal{U} = \mathcal{G}$. (by 10, (i))
12. $A \in \mathcal{U}$. (by 3, 11)

Corollary 1.A.6 Let:

(i) $\mathcal{U}$ be an ultrafilter on T, and
(ii) $A \subset T$.

Then either $A \in \mathcal{U}$ or $T \setminus A \in \mathcal{U}$.

PROOF: 1. Assume that $A \notin \mathcal{U}$.
2. There exists $U \in \mathcal{U}$ such that $A \cap U = \emptyset$. (by 1, **1.A.5**)
3. $U \subset T \setminus A$. (by 2)
4. $T \setminus A \in \mathcal{U}$. (by 3, (i))

Proposition 1.A.7 Let:

(i) $\mathcal{F}$ be a filter on T, and
(ii) $A \in \mathcal{F}$ or $T \setminus A \in \mathcal{F}$ for all $A \subset T$.

Then $\mathcal{F}$ is an ultrafiter on T.

PROOF: 1. Let $\mathcal{G}$ be a filter on T with $\mathcal{F} \subset \mathcal{G}$.
2. Let $G \in \mathcal{G}$.
3. $T \setminus G \notin \mathcal{G}$. (by 1, 2)
4. $T \setminus G \notin \mathcal{F}$. (by 1, 3)
5. $G \in \mathcal{F}$. (by 4, (ii))

6. $\mathcal{G} \subset \mathcal{F}$. (by 2, 5)
7. $\mathcal{F}$ is an ultrafilter on T. (by (i), 1, 5)

The following natural generalization of **1.A.6** will be useful here and is used in proposition **5.A.3** of the appendix to section 5.

Corollary 1.A.8 Let:

(i) $\mathcal{U}$ be an ultrafilter on T,
(ii) $A_1, \ldots, A_n$ be subsets of T, and
(iii) $A_1 \cup \cdots \cup A_n \in \mathcal{U}$.

Then there exists j with $A_j \in \mathcal{U}$.

PROOF: 1. Assume that $A_i \notin \mathcal{U}$ for all $i \neq j$.
2. $T \setminus A_i \in \mathcal{U}$ for all $i \neq j$. (by 1, (i), **1.A.6**)
3. $\bigcap_{i \neq j} (T \setminus A_i) \in \mathcal{U}$. (by 2, (i))
4. $A_j \cap \bigcap_{i \neq j} (T \setminus A_i) = (A_1 \cup \cdots \cup A_n) \cap \bigcap_{i \neq j} (T \setminus A_i) \in \mathcal{U}$. (by 3, (i), (iii))
5. $A_j \in \mathcal{U}$. (by 4, (i))
6. There exists j with $A_j \in \mathcal{U}$. (by 1, 5)

Having discussed a few of the elementary properties of ultrafilters, we are ready to define the Stone-Cech compactification βT of T. As a set βT simply consists of all the ultrafilters on T; the next step is to define a topology on βT. Describing this topology requires some notation.

Definition 1.A.9 Let T be a nonempty set. We define βT by

$$\beta T = \{\mathcal{U} \mid \mathcal{U} \text{ is an ultrafilter on } \mathcal{T}\}.$$

Definition and Notation 1.A.10 Let $\emptyset \neq A \subset T$. We define the *hull of A* by $h(A) = \{u \in \beta T \mid A \in u\}$.

Note that for $t \in T$ we have used the notation $h(t)$ for the **single element**

$$h(t) = \{A \mid t \in A \subset T\} \in \beta T,$$

whereas the hull $h(\{t\})$ as defined above is a **subset** of βT. This notation is justified by the fact that $h(\{t\}) = \{h(t)\}$ since $h(t)$ is the only ultrafilter which contains $\{t\}$. We will identify T with the subset $\{h(t) \mid t \in T\} \subset \beta T$ and thus write t for the element $h(t) \in \beta T$.

Note that if an ultrafilter $u \in h(A) \cap h(B)$, then $A \in u$ and $B \in u$. This implies that $A \cap B \in u$ and hence $u \in h(A \cap B)$. It follows that the collection

$$\{h(A) \mid A \subset T\}$$

is a base for a topology on βT. This gives us the following proposition.

Proposition 1.A.11 Let

$\mathcal{T} = \{\Gamma \subset \beta T \mid \text{for every } u \in \Gamma \text{ there exists } A \in u \text{ with } h(A) \subset \Gamma\}.$

Then $\mathcal{T}$ is a topology on βT.

Henceforth we will assume that βT is provided with the topology $\mathcal{T}$. In this topology if $A \subset T \subset \beta T$, then $h(A)$ is the closure of A in βT. We leave the proof as an exercise for the reader; one immediate consequence is the fact that $\overline{A}$ is both open and closed in βT, as we saw in **1.2**. On the other hand every ultrafilter on T contains the set T, so $\beta T = h(T) = \overline{T}$, in other words T is dense in βT. We restate this for emphasis.

Proposition 1.A.12 T is dense in βT.

Having constructed the space βT satisfying condition (i) of **1.1**, we now wish to show that βT is a compact Hausdorff space (condition (ii) of **1.1**).

Lemma 1.A.13 Let:

(i) $\{A_i \mid i \in I\}$ be nonempty subsets of T, and
(ii) $\bigcup_{i \in I} h(A_i) = \beta T$.

Then there exists a finite subset $F \subset I$ with $\bigcup_{i \in F} A_i = T$.

PROOF: 1. Assume that $B_F = T \setminus \left(\bigcup_{i \in F} A_i \right) \neq \emptyset$ for all finite sets $F \subset I$.
2. Let $\mathcal{B} = \{B_F \mid F \subset I \text{ finite}\}$.
3. $B_{F_1 \cup \cdots \cup F_n} = T \setminus \left(\bigcup_{i \in F_1 \cup \cdots \cup F_n} A_i \right) = \bigcap_{j=1}^{n} \left(T \setminus \left(\bigcup_{i \in F_j} A_i \right) \right) =$
$B_{F_1} \cap \cdots \cap B_{F_n}$. (by 1)
4. $\mathcal{B}$ is a filter base on T. (by 2, 3)
5. There exists an ultrafilter $\mathcal{U}$ on T with $\mathcal{B} \subset \mathcal{U}$. (by 4, **1.A.4**)
6. There exists $k \in I$ with $\mathcal{U} \in h(A_k)$, and hence $A_k \in \mathcal{U}$. (by 5, (ii))
7. $T \setminus A_k = B_{\{k\}} \in \mathcal{B} \subset \mathcal{U}$. (by 1, 2, 4)
8. $B_F = \emptyset$ and hence $\bigcup_{i \in F} A_i = T$ for some finite set $F \subset T$.
(6, 7 contradict 1)

Theorem 1.A.14 $(\beta T, \mathcal{T})$ is a compact Hausdorff topological space.

PROOF: 1. Let $\{\Gamma_i \mid i \in I\}$ be a family of open subsets of βT with $\bigcup_{i \in I} \Gamma_i = \beta T$.
2. For every $u \in \beta T$, there exists $A_u \in u$ and $i_u \in I$ with $h(A_u) \subset \Gamma_{i_u}$. (by 1)
3. $\bigcup_{u \in \beta T} h(A_u) = \beta T$. (by 2)
4. There exists a finite subset $F \subset \beta T$. such that $\bigcup_{u \in F} A_u = T$. (by 3, **1.A.13**)

5. Let $v \in \beta T$.
6. $A_u \in v$ for some $u \in F$. (by 4, 5, **1.A.8**)
7. $v \in h(A_u) \subset \Gamma_{i_u}$. (by 2, 6)
8. $\bigcup_{u \in F} \Gamma_{i_u} = \beta T$. (by 5, 7)
9. $(\beta T, \mathcal{T})$ is compact. (by 1, 8)
10. Let $u_1 \neq u_2 \in \beta T$.
11. There exists a subset $A \subset T$ such that $A \in u_1$ and $A \notin u_2$. (by 1)
12. $T \setminus A \in u_2$. (by 11, **1.A.7**)
13. $u_1 \in h(A)$ and $u_2 \in h(T \setminus A)$. (by 11, 12)
14. $h(A)$ and $h(T \setminus A)$ are disjoint open subsets of βT.
15. $(\beta T, \mathcal{T})$ is Hausdorff. (by 10, 13, 14)

In order to show that βT satisfies the final condition of **1.1**, we make use of the fact that in a compact Hausdorff space X, every ultrafilter converges to a unique point in X. This is the content of **1.A.16** whose proof we include in the interest of completeness. The converse also holds, but since we will not make explicit use of it, we have left its proof as an exercise.

Definition 1.A.15 Let X be a topological space, $x \in X$ and $\mathcal{F}$ be a filter on X. We say that $\mathcal{F}$ *converges to* x and write $\mathcal{F} \to x$, if the neighborhood filter at x, $\mathcal{N}_x \subset \mathcal{F}$.

Lemma 1.A.16 Let:

(i) X be a compact Hausdorff topological space,
(ii) $\emptyset \neq Y \subset X$,
(iii) $\mathcal{U}$ be an ultrafilter on X, and
(iv) $Y \in \mathcal{U}$.

Then there exists a unique $x \in \overline{Y}$ such that $\mathcal{U} \to x$.

PROOF: 1. Assume that $\mathcal{U} \not\to x$ for all $x \in \overline{Y}$.
2. For every $x \in \overline{Y}$ there exists $V_x \in \mathcal{N}_x$ with $V_x \notin \mathcal{U}$. (by 1)
3. There exists a finite subset $F \subset \overline{Y}$ such that $\overline{Y} \subset \bigcup_{x \in F} V_x$.
(by 2, ($\overline{Y}$ is compact by (i)))
4. $Y = \bigcup_{x \in F} (V_x \cap Y) \in \mathcal{U}$. (by 3, (iv))
5. There exists $x \in F$ such that $V_x \cap Y \in \mathcal{U}$. (by 4, (iii), **1.A.8**)
6. There exists $x \in \overline{Y}$ such that $\mathcal{U} \to x$. (5 contradicts 2)
7. Uniqueness follows from the fact that X is Hausdorff.

We now wish to prove that every mapping from T to a compact Hausdorff space Y can be extended to a continuous map $\beta T \to Y$. The proof uses the following elementary lemma, whose proof we leave as an exercise for the reader.

Lemma 1.A.17 Let:

(i) $f : T \to Y$,
(ii) $\mathcal{U}$ be an ultrafilter on T, and
(iii) $\bar{f}(\mathcal{U}) = \{B \subset Y \mid \text{there exists } A \in \mathcal{U} \text{ with } f(A) \subset B\}$.

Then $\bar{f}(\mathcal{U})$ is an ultrafilter on Y.

Theorem 1.A.18 (Compare with **1.1.**) Let:

(i) X be a compact Hausdorff topological space, and
(ii) $f : T \to X$.

Then there exists a unique continuous map $\hat{f} : \beta T \to X$ with $\hat{f}(t) = f(t)$ for all $t \in T$.

PROOF: 1. Let $u \in \beta T$.
2. $\bar{f}(u) = \{B \subset X \mid \text{there exists } A \in u \text{ with } f(A) \subset B\}$ is an ultrafilter on X. (by 1, **1.A.17**)
3. There exists a unique element $\hat{f}(u) \in X$ such that $\bar{f}(u) \to \hat{f}(u)$. (by 2, **1.A.16**)
4. Let $t \equiv h(t) \in T \subset \beta T$.
5. $\bar{f}(t) = \{B \subset X \mid f(t) \in B\} \supset \mathcal{N}_{f(t)}$. (by 2, 4)
6. $\hat{f}(t) = f(t)$. (by 3, 5)
7. Let $V \subset X$ be open and $u \in \beta T$ with $u \in \hat{f}^{-1}(V)$.
8. There exists $W \subset X$ open with $\hat{f}(u) \in W \subset \overline{W} \subset V$. (by 7, (i))
9. $\bar{f}(u) \to \hat{f}(u) \in W$. (by 3, 8, **1.A.16**)
10. $W \in \bar{f}(u)$. (by 8, 9)
11. There exists $B \in u$ with $f(B) \subset W$. (by 10)
12. Let $v \in h(B)$ (so that $B \in v$).
13. $W \in \bar{f}(v)$. (by 11, 12)
14. There exists $x \in \overline{W}$ such that $\bar{f}(v) \to x$. (by 13, **1.A.16**)
15. $\hat{f}(v) = x \in V$. (by 3, 8, 14)
16. $v \in \hat{f}^{-1}(V)$. (by 15)
17. $u \in h(B) \subset \hat{f}^{-1}(V)$. (by 11, 12, 16)
18. $\hat{f}^{-1}(V)$ is open and hence $\hat{f}$ is continuous. (by 7, 17)
19. $\hat{f}$ is unique because T is dense in βT (by **1.A.12**) and X is Hausdorff.

We turn now to the case where T is a semigroup. We saw in **1.3** that in this case there is a unique semigroup structure on βT which makes right multiplication by elements of T and left multiplication by all elements of βT continuous. Having constructed the topological space βT using the ultrafilters on T, we complete this appendix by showing that a semigroup structure on T allows us to define the product of two ultrafilters on T, giving βT a semigroup structure.

This definition is motivated by the fact that for an ultrafilter $u \in \beta T$, and an element $t \in T$, one expects their product to be given by:

$$ut = \{At \mid A \in u\} = \{A \mid At^{-1} \in u\}$$

which in fact makes sense when T is a **group** but may not be an ultrafilter when the right multiplication map R_t is not one-one. Instead we will use

$$ut = \{A \mid R_t^{-1}(A) \in u\},$$

which gives the same result when T is a group since $R_t^{-1} = R_{t^{-1}}$ in that case. Thus the ultrafilter $ut = uh(t)$ should be characterized by the fact that:

$$A \in ut \iff t \in \{s \mid R_s^{-1}(A) \in u\} \iff \{s \mid R_s^{-1}(A) \in u\} \in h(t).$$

This viewpoint generalizes in a natural way to the product uv of two ultrafilters:

$$A \in uv \iff \{s \mid R_s^{-1}(A) \in u\} \in v.$$

There are many details to check; for instance it needs to be shown that uv as defined above is an ultrafilter. We content ourselves with giving a precise outline of the notation and results involved while leaving the details of the proofs to the reader.

Definition and Notation 1.A.19 Let T be a semigroup, $t \in T$, $A \subset T$, and $u \in \beta T$. We use the notation:

$$\begin{array}{rcl} R_t : T & \to & T \\ s & \to & st \end{array} \quad \begin{array}{rcl} L_t : T & \to & T \\ s & \to & ts \end{array}$$

for the multiplication maps in T, and

$$At = R_t(A) = \{at \mid a \in A\}.$$

We define

$$A * u = \{s \in T \mid R_s^{-1}(A) \in u\}.$$

Note that

$$R_s^{-1}(A) \in h(t) \iff t \in R_s^{-1}(A) \iff ts \in A \iff s \in L_t^{-1}(A),$$

so that $A * h(t) = L_t^{-1}(A)$.

Proposition 1.A.20 Let:

(i) T be a semigroup,
(ii) $u, v \in \beta T$, and
(iii) $w = \{A \subset T \mid A * u \in v\}$.

Then w is an ultrafilter on T.

PROOF: We leave the proof as an exercise for the reader.

Definition 1.A.21 Let T be a semigroup, and $u, v \in \beta T$. We define

$$uv = \{A \subset T \mid A * u \in v\},$$

so that $uv \in \beta T$ by **1.A.20.**

For any two elements $s, t \in T$, applying **1.A.21** yields:

$h(s)h(t) = \{A \mid A * h(s) \in h(t)\} = \{A \mid t \in L_s^{-1}(A)\} = \{A \mid st \in A\} = h(st)$,
so this definition of a product on βT agrees with the semigroup structure on T. It follows immediately from the next lemma that this product is associative and hence gives βT a semigroup structure which extends the semigroup structure on T.

Lemma 1.A.22 Let:

(i) T be a semigroup,
(ii) $u, v \in \beta T$, and
(iii) $\emptyset \neq A \subset T$.

Then $A * (uv) = (A * u) * v$.

PROOF: We leave the proof as an exercise for the reader.

Proposition 1.A.23 Let:

(i) T be a semigroup, and
(ii) $u, v, w \in \beta T$.

Then $(uv)w = u(vw)$.

PROOF: $A \in (uv)w \iff (A * u) * v = A * (uv) \in w \iff A * u \in vw \iff A \in u(vw)$. (by **1.A.22**)

We complete our discussion of βT by indicating how to verify that the maps R_t and L_p are continuous for all $t \in T$ and $p \in \beta T$ respectively.

Lemma 1.A.24 Let:

(i) T be a semigroup,
(ii) $\emptyset \neq A \subset T$,

(iii) $t \in T$,
(iv) $u \in \beta T$,
(v) $R_t : \beta T \to \beta T$, and
$v \to vt$
(vi) $L_u : \beta T \to \beta T$.
$v \to uv$

Then
(a) $R_t^{-1}(h(A)) = h(R_t^{-1}(A))$,
(b) $L_u^{-1}(h(A)) = h(A * u)$.

PROOF: We leave the proof as an exercise for the reader.

Proposition 1.A.25 (Compare with **1.3**) Let:

(i) T be a semigroup,
(ii) $t \in T$,
(iii) $u \in \beta T$,
(iv) $R_t : \beta T \to \beta T$, and
$v \to vt$
(v) $L_u : \beta T \to \beta T$.
$v \to uv$

Then R_t and L_u are both continuous.

PROOF: This follows immediately from **1.A.24**.

EXERCISES FOR CHAPTER 1

Exercise 1.1 (See **1.4**) Let:

(i) T be a semigroup,
(ii) $\emptyset \neq H \subset T$ be a subsemigroup, and
(iii) $j : \beta H \to \beta T$ be the continuous extension to βH of the inclusion $H \to \beta T$.

Then $j(pq) = j(p)j(q)$ for all $p, q \in \beta H$ (so that $\beta H \equiv \overline{H}$ is a subsemigroup of βT).

Exercise 1.2 Let X be a topological space and $x \in X$. Prove that the collection

$$\mathcal{N}_x = \{A \mid \text{there exists } U \text{ open in } X \text{ with } x \in U \subset A\}$$

is a filter on T.

Exercise 1.3 Let:

(i) $\mathcal{F}$ be a filter on T, and
(ii) $\mathcal{U}(\mathcal{F}) = \{\mathcal{U} \mid \mathcal{U}$ is an ultrafilter on T with $\mathcal{F} \subset \mathcal{U}\}$.

Prove that $\mathcal{F} = \bigcap_{\mathcal{U} \in \mathcal{U}(\mathcal{F})} \mathcal{U}$.

Exercise 1.4 Let $t \in T$. Prove that the collection

$$h(t) = \{A \mid t \in A \subset T\}$$

is the unique ultrafilter on T containing $\{t\}$.

Exercise 1.5 Let X be a topological space. Show that X is compact if and only if every ultrafilter on X converges. (Compare with **1.A.16.**)

Exercise 1.6 Let:

(i) $f: T \to Y$,
(ii) $\mathcal{U}$ be an ultrafilter on T, and
(iii) $\bar{f}(\mathcal{U}) = \{B \subset Y \mid$ there exists $A \in \mathcal{U}$ with $f(A) \subset B\}$.

Show that $\bar{f}(\mathcal{U})$ is an ultrafilter on Y.

Exercise 1.7 Let

$$\mathcal{T} = \{\Gamma \subset \beta T \mid \text{for every } u \in \Gamma \text{ there exists } A \in u \text{ with } h(A) \subset \Gamma\}.$$

Show that $\mathcal{T}$ is a topology on βT; since $\emptyset, \beta T \in \mathcal{T}$ this amounts to proving:

(a) if $\{\Gamma_i \mid i \in I\} \subset \mathcal{T}$, then $\bigcup_{i \in I} \Gamma_i \in \mathcal{T}$, and
(b) if $\{\Gamma_1, \ldots, \Gamma_n\} \subset \mathcal{T}$, then $\Gamma_1 \cap \cdots \cap \Gamma_n \in \mathcal{T}$.

Exercise 1.8 Let:

(i) $\mathcal{T} = \{\Gamma \subset \beta T \mid \forall u \in \Gamma,\ \exists A \in u$ with $h(A) \subset \Gamma\}$, and
(ii) $A \subset T$.

Show that $\overline{A} = h(A)$; so that $\overline{A}$ is both open and closed with respect to the topology $\mathcal{T}$.

Exercise 1.9 Let:

(i) T be a semigroup,
(ii) $u, v \in \beta T$, and
(iii) $uv = \{A \subset T \mid A * u \in v\}$.

Show that

(a) uv is an ultrafilter on T, and
(b) $A * (uv) = (A * u) * v$ for all $\emptyset \neq A \subset T$.

Exercise 1.10 Let:

(i) T be a semigroup,
(ii) $\emptyset \neq A \subset T$,
(iii) $t \in T$,
(iv) $u \in \beta T$,
(v) $R_t : \beta T \rightarrow \beta T$, $v \rightarrow vt$ and
(vi) $L_u : \beta T \rightarrow \beta T$, $v \rightarrow uv$

Show that

(a) $R_t^{-1}(h(A)) = h(R_t^{-1}(A))$, and
(b) $L_u^{-1}(h(A)) = h(A * u)$.

2

Flows and their enveloping semigroups

For us a flow will be a compact Hausdorff space X provided with a continuous (right) action of a group T on X. In Topological Dynamics we are concerned with the so-called asymptotic behavior of this action. This motivates the consideration of not just the collection T thought of as a subset of the set X^X of self-mappings of X, but all of the limit points of T in X^X. This gives rise to the notion of the *enveloping semigroup* $E(X, T)$ of the flow. The composition of functions gives a natural semigroup structure on $E(X, T)$ which as we will see in subsequent sections, can be exploited to study the asymptotics of the original flow (X, T). The semigroup structure on βT discussed in the previous section and its appendix serves as a prototype example. Indeed since $T \subset \beta T$, this semigroup structure makes $(\beta T, T)$ itself a flow, and we will see in proposition **2.9** that $E(\beta T, T) \cong \beta T$ in a natural way.

In this section we introduce the appropriate notation, the details of the construction, and give an exposition of some of the elementary properties of $E(X, T)$. Many of these properties will be used directly, and serve as motivation in what follows. Several accounts of this material appear in the textbook-literature (see for example [Auslander, J., (1988)] and [Ellis, R., (1969)]. However, the point of view we have adopted here is slightly different. In order to emphasize the connection between the two, we have involved βT in the definition of $E(X, T)$ (see **2.8**). In particular the fact that $E(X, T)$ is a homomorphic image of βT, both as a flow and as a semigroup has important consequences. We begin with some basic notation and definitions.

Definition 2.1 Let X be a set and S a semigroup. Then an *action of S on X* is a function

$$\pi : X \times S \to X$$
$$(x, s) \to xs \qquad \text{such that} \qquad x(st) = (xs)t$$

for all $x \in X$ and $s, t \in S$. If S has an identity e we require that $xe = x$ for all $x \in X$. If π is an action of S on X, we say that S acts on X via π or simply that S acts on X.

Definition 2.2 A *flow* is a triple (X, T, π) where X is a compact Hausdorff space, T is a topological group, and π is an action of T on X such that the map π is continuous. Let $A \subset X$. We say that A is *invariant* if $AT \equiv \{at \mid a \in A, t \in T\} \subset A$. If A is invariant, then the restriction of π to $A \times T$ defines an action of T on A. If A is also closed, the resulting flow (A, T, π) is called a *subflow* of the flow (X, T, π). Most of the time the symbol π is suppressed, i.e. the flow (X, T, π) is denoted (X, T) or simply, X.

For all of the flows (X, T, π) considered here we will assume that T is provided with the discrete topology. In this case for π to be continuous it suffices that the maps $\begin{array}{rcl} \pi^t : X & \to & X \\ x & \to & xt \end{array}$ be continuous for all $t \in T$.

As we mentioned above, it follows from **1.3** that the map $\pi : \beta T \times T \to \beta T$ defined by $\pi(p, t) = pt$ makes $(\beta T, T)$ a flow.

We now make explicit the definition of a homomorphism of flows mentioned earlier.

Definition 2.3 Let (X, T), (Y, T) be flows and $f : X \to Y$. Then f is a *homomorphism* if f is continuous and $f(xt) = f(x)t$ for all $x \in X$, and $t \in T$. The set of homomorphisms from X to Y will be denoted *Hom*(X, Y). The set of *automorphisms* (bijective elements of $Hom(X, X)$) of X, will be denoted *Aut*(X).

One example of an asymptotic property of a flow is point transitivity as defined below.

Definition 2.4 Let (X, T) be a flow. We say that (X, T) is *point transitive* if there exists $x_0 \in X$ with $\overline{x_0 T} = X$.

Clearly the flow $(\beta T, T)$ is point transitive since $\overline{eT} = \overline{T} = \beta T$. Note also that if $f : (X, T) \to (Y, T)$ is an epimorphism (surjective homomorphism), and (X, T) is point transitive, then (Y, T) is point transitive. Thus any homomorphic image of $(\beta T, T)$ is point transitive. As we show in **2.5**, the converse of this statement also holds. For this reason the flow $(\beta T, T)$ is often called the *universal point transitive flow*.

Proposition 2.5 Let (X, T) be a point transitive flow. Then there exists an epimorphism $f : (\beta T, T) \to (X, T)$.

PROOF: 1. There exists $x_0 \in X$ with $\overline{x_0 T} = X$.

2. The map $\begin{array}{rcl} T & \to & X \\ t & \to & x_0 t \end{array}$ has a unique continuous extension $f : \begin{array}{rcl} \beta T & \to & X \\ p & \to & x_0 p \end{array}$.
(by **1.1**)

3. Let $t \in T$.

4. The maps

$$\begin{array}{rcl} \beta T & \to & X \\ p & \to & x_0(pt) \end{array} \quad \text{and} \quad \begin{array}{rcl} \beta T & \to & X \\ p & \to & (x_0 p)t \end{array}$$

are continuous extensions of the map

$$\begin{array}{rcl} T & \to & X \\ t' & \to & (x_0 t')t = x_0(t't). \end{array}$$

5. $x_0(pt) = (x_0 p)t$ and hence f is a homomorphism. (by 2, 3, 4, **1.1**)

6. $x_0 T = f(T) \subset f(\beta T)$. (by 1)

7. $X = \overline{x_0 T} \subset f(\beta T)$.
(by 2, 6, since βT and X are compact Hausdorff spaces)

In order to describe the so-called enveloping semigroup of a flow (X, T) we introduce the space of self-maps of X, along with some useful notation. Since our actions are on the right, it is natural write $x\pi^t = xt$ for the value at x of the element π^t of X^X associated with $t \in T$. For this reason we will write all of the functions in X^X on the right.

Notation 2.6 Let X be a compact Hausdorff space. Then X^X will denote the set of maps of X to X provided with the *topology of pointwise convergence*. Let $f, g \in X^X$ and $x \in X$. Then xf will denote the image of x under f, and fg the composite map first f then g. Thus $x(fg) = (xf)g$. Finally ρ: $X^X \times X^X \to X^X$ will denote the map defined by $\rho(f, g) = fg$ for all $f, g \in X^X$.

We will make use of the following elementary properties of X^X.

Proposition 2.7 Let X be compact Hausdorff. Then:

(a) X^X is compact Hausdorff,

(b) ρ provides X^X with a semigroup structure,

(c) the maps $\rho^f : \begin{array}{rcl} X^X & \to & X^X \\ g & \to & gf \end{array}$ are continuous for all continuous $f \in X^X$,

(d) the maps $\rho_f : \begin{array}{rcl} X^X & \to & X^X \\ g & \to & fg \end{array}$ are continuous for all $f \in X^X$, and

(e) ρ defines an action of the semigroup X^X on the set X^X.

PROOF: (a) and (b) are standard.

(c) 1. Let $g_\alpha \longrightarrow g$ be a convergent net, and f be continuous.

2. $xg_\alpha \longrightarrow xg$ for all $x \in X$.

3. $x(g_\alpha f) = (xg_\alpha)f \underset{1,2}{\longrightarrow} (xg)f = x(gf)$ for all $x \in X$.

4. $g_\alpha f \longrightarrow gf$. (by 3)

(d) 1. Let $g_\alpha \longrightarrow g$.

2. $yg_\alpha \longrightarrow yg$ for all $y \in X$. (by 1)

3. $x(fg_\alpha) = (xf)g_\alpha \underset{2}{\longrightarrow} (xf)g = x(fg)$ for all $x \in X$.

4. $fg_\alpha \longrightarrow fg$. (by 3)

(e) This follows immediately from parts (b) and (c).

It is clear that the map

$$\begin{array}{ccc} X \times X^X & \to & X \\ (x, f) & \to & xf \end{array}$$

defines an action of the semigroup X^X on X. Thus if T is any subgroup of X^X which consists entirely of continuous maps, then (X, T) is a flow. Conversely given any flow (X, T), the set $\{\pi^t \mid t \in T\}$ is a subgroup of X^X consisting of continuous maps. In this case we obtain a group homomorphism of T into a subgroup of X^X, allowing the following definition of the *enveloping semigroup* of the flow (X, T).

Definition 2.8 Let (X, T) be a flow. By **1.1** the map $\begin{array}{ccc} T & \to & X^X \\ t & \to & \pi^t \end{array}$ has a continuous extension $\Phi_X : \beta T \to X^X$. The image of Φ_X,

$$\Phi_X(\beta T) = \overline{\{\pi^t \mid t \in T\}} \equiv E(X, T)$$

is clearly a subsemigroup of X^X, which we call the *enveloping semigroup of the flow* (X, T) and denote by $E(X, T)$ or simply $E(X)$. The map Φ_X will be referred to as the *canonical map of βT onto* $E(X)$. Since we write xt for $x\pi^t = x\Phi_X(t)$ for all $x \in X$ and $t \in T$, it will be convenient to write xp for $x\Phi_X(p)$ for all $x \in X$ and $p \in \beta T$.

Some of the properties of $E(X)$ which follow directly from the definition are collected and detailed below in order that they may be referred to as needed. We leave it as an exercise for the reader to provide detailed proofs.

Proposition 2.9 Let (X, T) be a flow. Then:

(a) The action

$$\begin{array}{ccc} E(X) \times T & \to & E(X) \\ (p,t) & \to & p\pi^t \end{array}$$

where

$$\begin{array}{rcl} \pi^t = R_t \colon E(X) & \to & E(X) \\ q & \to & qt \end{array}$$

makes $E(X,T)$ a point transitive flow.

(b) The canonical map $\Phi_X \colon \beta T \to E(X)$ is both a flow and a semigroup homomorphism.

(c) The map $\begin{array}{ccc} E(X) & \to & X \\ p & \to & xp \end{array}$ is a flow homomorphism for all $x \in X$.

(d) The map $\Phi_{\beta T} \colon \beta T \to E(\beta T)$ is an isomorphism.

(e) Let $f \colon (X,T) \to (Y,T)$ be a homomorphism of flows. Then $f(xp) = f(x)p$ for all $x \in X$ and $p \in \beta T$.

The association of $E(X,T)$ to a flow (X,T) is natural in the sense outlined in the following proposition.

Proposition 2.10 Let $f \colon (X,T) \to (Y,T)$ be a surjective flow homomorphism. Then there exists a map $\theta \colon E(X) \to E(Y)$ such that:

(a) the following diagram is commutative

$$\begin{array}{cccccc} & \beta T & = & \beta T & \\ & \Phi_X \downarrow & & \downarrow \Phi_Y & \\ a & E(X) & \xrightarrow{\theta} & E(Y) & b \\ \downarrow & \downarrow & & \downarrow & \downarrow \\ xa & X & \xrightarrow{g} & Y & g(x)b \end{array}$$

for all homomorphisms $g \colon X \to Y$ and $x \in X$,

(b) θ is surjective and continuous,

(c) $\theta(pq) = \theta(p)\theta(q)$ for all $p, q \in E(X)$ so that θ is both a flow and a semigroup homomorphism, and

(d) if $\psi \colon E(X) \to E(Y)$ is a homomorphism with $\psi \circ \Phi_X = \Phi_Y$, then $\theta = \psi$.

PROOF: (a) 1. Let $a \in E(X)$ and define $\theta(a) = \Phi_Y(p)$ where $p \in \beta T$ with $\Phi_X(p) = a$.

2. To see that θ is well-defined, suppose $p, q \in \beta T$ with $\Phi_X(p) = \Phi_X(q)$ and let $y \in Y$.

3. Since f is onto, there exists $x \in X$ with $f(x) = y$.

4. $y\Phi_Y(p) \underset{3}{=} f(x)\Phi_Y(p) \underset{\mathbf{2.9}(e)}{=} f(x\Phi_X(p)) \underset{2}{=} f(x\Phi_X(q)) \underset{\mathbf{2.9}(e)}{=} f(x)\Phi_Y(q)$
$= y\Phi_Y(q)$.
5. Thus θ is well-defined and $\theta \circ \Phi_X = \Phi_Y$, so the top half of the diagram is commutative.
6. Now let $x \in X$, $a \in E(X)$, and $p \in \beta T$ with $\Phi_X(p) = a$.
7. Then $g(xa) = g(x\Phi_X(p)) \underset{\mathbf{2.9}(e)}{=} g(x)\Phi_Y(p) \underset{5}{=} g(x)\theta(\Phi_X(p)) = g(x)\theta(a)$ which shows that the bottom half of the diagram is commutative.
(b) 1. θ is surjective because Φ_Y is.
2. Let $C \subset E(Y)$ be closed.
3. $\theta^{-1}(C) = \Phi_X(\Phi_Y^{-1}(C))$ is closed.
(Φ_X, Φ_Y are continuous, βT is compact)
4. θ is continuous. (by 2, 3)
(c) 1. Let $a, b \in E(X)$, and $p, q \in \beta T$ with $\Phi_X(p) = a$ and $\Phi_X(q) = b$.
2. $\theta(ab) = \theta(\Phi_X(p)\Phi_X(q)) \underset{\mathbf{2.9}(b)}{=} \theta(\Phi_X(pq))$

$$= \Phi_Y(pq) \underset{\mathbf{2.9}(b)}{=} \Phi_Y(p)\Phi_Y(q) = \theta(\Phi_X(p))\theta(\Phi_X(q)) = \theta(a)\theta(b).$$

(d) 1. Assume that $\psi : E(X) \to E(Y)$ satisfies $\psi \circ \Phi_X = \Phi_Y$.
2. For $a = \Phi_X(p)$, $f(xa) = f(x\Phi_X(p)) \underset{\mathbf{2.9}(e)}{=} f(x)\Phi_Y(p) \underset{1}{=} f(x)\psi(\Phi_X(p))$
$= f(x)\psi(a)$.
3. $f(x)\psi(a) = f(x)\theta(a)$ for all $x \in X$ and $a \in E(X)$. (by 2 and part (a))
4. $\psi(a) = \theta(a)$ for all $a \in E(X)$. (by 2, 3, because f is onto)

We end this section with a few examples of how **2.10** is used to identify certain enveloping semigroups. Note that if (X, T) is a flow, then T acts diagonally on the Cartesian product $X \times X$, so that $(x, y)t = (xt, yt)$, making $(X \times X, T)$ a flow.

Corollary 2.11 Let (X, T) be a flow. Then $E(X, T) \cong E(X \times X, T)$.

PROOF: 1. The maps $g_1 : \begin{matrix} X \times X & \to & X \\ (x, y) & \to & x \end{matrix}$ and $g_2 : \begin{matrix} X \times X & \to & X \\ (x, y) & \to & y \end{matrix}$ are surjective homomorphisms of flows.
2. There exists a surjective flow and semigroup homomorphism θ: $E(X \times X) \to E(X)$. (by 1, **2.10**)
3. Let $p, q \in \beta T$ with $\theta(\Phi_{X\times X}(p)) = \theta(\Phi_{X\times X}(q))$.
4. $\Phi_X(p) = \Phi_X(q)$. (by 2, 3, **2.10**)
5. Let $(x, y) \in X \times X$.
6. $g_1((x, y)\Phi_{X\times X}(p)) \underset{\mathbf{2.10}}{=} g_1(x, y)\Phi_X(p) = x\Phi_X(p) \underset{4}{=} x\Phi_X(q)$
$\underset{\mathbf{2.10}}{=} g_1((x, y)\Phi_{X\times X}(q))$.

7. $g_2((x, y)\Phi_{X\times X}(p)) \underset{2.10}{=} g_2(x, y)\Phi_X(p) = y\Phi_X(p) \underset{4}{=} y\Phi_X(q) \underset{2.10}{=} g_2((x, y)$ $\Phi_{X\times X}(q))$.

8. $$\begin{aligned}(x, y)\Phi_{X\times X}(p) &= (g_1((x, y)\Phi_{X\times X}(p)), g_2((x, y)\Phi_{X\times X}(p)))\\ &\underset{7}{=} (g_1((x, y)\Phi_{X\times X}(q)), g_2((x, y)\Phi_{X\times X}(q)))\\ &= (x, y)\Phi_{X\times X}(q).\end{aligned}$$

9. $\Phi_{X\times X}(p) = \Phi_{X\times X}(q)$. (by 5, 8)

10. θ is an isomorphism. (by 2, 3, 9)

Proposition 2.12 Let:

(i) (X, T) be a flow,
(ii) $x_0 \in X$ with $\overline{x_0T} = X$ (so that (X, T) is point transitive), and
(iii) $f : E(X) \to X$ be defined by $f(p) = x_0p$ for all $p \in E(X)$.

Then f induces an isomorphism, $\theta : E(E(X)) \cong E(X)$.

PROOF: We leave the proof as an exercise for the reader.

Proposition 2.13 Let (X, T) be a flow. Then $E(E(E(X))) \cong E(E(X))$.

PROOF: This follows from **2.12** and the fact that $(E(X, T), T)$ is point transitive.

EXERCISES FOR CHAPTER 2

Exercise 2.1 (see **2.9**) Let:

(i) (X, T) and (Y, T) be flows, and
(ii) $f : (X, T) \to (Y, T)$ be a homomorphism.

Show that

(a) The canonical map $\Phi_X : \beta T \to E(X)$ is both a flow and a semigroup homomorphism.
(b) The map $\begin{array}{ccc} E(X) & \to & X \\ p & \to & xp \end{array}$ is a flow homomorphism for all $x \in X$.
(c) The map $\Phi_{\beta T} : \beta T \to E(\beta T)$ is an isomorphism.
(d) $f(xp) = f(x)p$ for all $x \in X$ and $p \in \beta T$.

Exercise 2.2 (see **2.12**) Let:

(i) (X, T) be a flow,
(ii) $x_0 \in X$ with $\overline{x_0 T} = X$ (so that (X, T) is point transitive), and
(iii) $f\colon E(X) \to X$ be defined by $f(p) = x_0 p$ for all $p \in E(X)$.

Then f induces an isomorphism, $\theta\colon E(E(X)) \cong E(X)$.

Exercise 2.3 Let:

(i) E be a compact Hausdorff space provided with a semigroup structure,
(ii) $\begin{array}{rcl} L_p : E & \to & E \\ m & \to & pm \end{array}$ for all $p \in E$,
(iii) $\begin{array}{rcl} R_p : E & \to & E \\ m & \to & mp \end{array}$ for all $p \in E$, and
(iv) $\varphi\colon E \to E^E$ be defined by $\varphi(p) = R_p$ for all $p \in E(X)$.

Then

(a) φ is a semigroup homomorphism, and
(b) φ is continuous if and only if L_p is continuous for all $p \in E$. In this case φ identifies E with a subsemigroup of E^E, and E is referred to as an *E-semigroup* (see also [Akin, E. (1997)]). It is immediate from **2.7**, that for any flow (X, T), its enveloping semigroup $E(X, T)$ is an E-semigroup.

3

Minimal sets and minimal right ideals

A subset $M \subset X$ of a flow (X, T) is *minimal* (see **3.1**) if it is a closed non-empty invariant set which is minimal with respect to those properties. One illustration of the interplay between the algebraic and topological properties of the enveloping semigroup is the fact that the minimal subsets of $E(X, T)$ are exactly the minimal right ideals in $E(X, T)$ with respect to its semigroup structure. This motivates a study of the algebraic structure of $E(X, T)$ and its minimal ideals in particular.

The key to understanding the structure of a minimal ideal $I \subset E(X)$, is an investigation of the idempotents in I (those $u \in I$ with $u^2 = u$). In a group, of course, the identity is the only idempotent. We will see that in $E(X)$, any closed sub-semigroup contains an idempotent, so that in general $E(X)$ contains many idempotents in addition to the identity. In particular any right ideal contains an idempotent. In **3.12** we show that if I is a minimal right ideal, then I is a disjoint union $\biguplus\{Iv \mid v \in J\}$, where $J = \{v \in I \mid v^2 = v\}$ is the set of idempotents in I. In fact all of the sets Iv, with $v \in J$ are **subgroups** of I which are isomorphic to one another.

We saw in **2.9** that $E(\beta T, T) \cong \beta T$. Thus the preceding discussion applies to any minimal right ideal M in βT. In this case the flow (M, T) is a *universal minimal flow* meaning that every minimal flow, is the image of M under some epimorphism. Thus every minimal flow can be identified with the quotient flow $(M/R, T)$ for some closed, invariant equivalence relation (*icer*) R on M. This is the basis for our approach to the algebraic theory of minimal flows, in which the structure of M plays a crucial role.

This section begins with some background material on minimal sets. We then make explicit the relationship between the minimal sets, and minimal ideals in the enveloping semigroup. We go on to describe the structure of the minimal ideals in $E(X, T)$, and the section closes with some brief remarks on

M which motivate the notation, terminology and approach which will be used in the later sections.

Definition 3.1 A subset, M of the flow X is *minimal* if:

(i) $\emptyset \neq M$,
(ii) M is closed,
(iii) M is *invariant*, meaning $Mt \subset M$ for all $t \in T$, and
(iv) if $N \subset M$ satisfies (i), (ii), (iii), then $N = M$. That is M is minimal with respect to (i),(ii), and (iii).

The flow X is *minimal* (or a *minimal set*) if X is minimal. Notice that if M is a minimal subset of X, then the flow (M, T) is a minimal set.

We begin with an elementary characterization of minimal sets. This shows that the minimal subsets of a flow reflect its asymptotic properties in the sense that minimality can be characterized in terms of orbit closures.

Proposition 3.2 Let:

(i) (X, T) be a flow, and
(ii) $\emptyset \neq M \subset X$ be closed and invariant.

Then the following are equivalent:

(a) M is minimal,
(b) $\overline{xT} = M$ for all $x \in M$, and
(c) if $U \subset X$ is open with $M \cap U \neq \emptyset$, then $M = (U \cap M)T$.

PROOF:

(a) $\Longrightarrow$ (b)

1. Assume that M is minimal and let $x \in M$.
2. $\overline{xT} \subset M$ satisfies (i), (ii), and (iii) of **3.1**.
3. $\overline{xT} = M$. (by 1, 2)

(b) $\Longrightarrow$ (c)

1. Assume that $\overline{xT} = M$ for all $x \in M$ and let $U \subset X$ be open with $U \cap M \neq \emptyset$.
2. $xT \cap U \neq \emptyset$ for all $x \in M$. (by 1)
3. $x \in UT$ for all $x \in M$. (by 2)

(c) $\Longrightarrow$ (a)

1. Assume that (c) holds and $N \subset M$ is a nonempty closed, invariant subset of M.
2. Let $U = X \setminus N$.
3. U is open. (by 1, 2)
4. $N \cap (U \cap M)T \subset NT \cap U = N \cap U = \emptyset$. (by 1, 2)
5. $(U \cap M)T \neq M$. (by 1, 4)

6. $U \cap M = \emptyset$. (by 1, 5)
7. $N = M$. (by 1, 2, 6)

We use the axiom of choice in the form of Zorn's lemma both in the following proposition, to show that minimal sets exist, and later to show that certain semigroups contain idempotents. Zorn's lemma was also used in the appendix to section 1; for the sake of completeness we give a statement here.

Zorn's Lemma 3.3 If $(S, \leq)$ is a partially ordered set such that any increasing chain $s_1 \leq \cdots \leq s_i \leq \cdots$ has a supremum in S, then S itself has a maximal element.

Proposition 3.4 Let (X, T) be a flow. Then there exists a minimal subset in X.

PROOF: 1. Let $\mathcal{C} = \{\emptyset \neq C \subset X \mid C \text{ is closed and invariant}\}$.
2. $\mathcal{C}$ is partially ordered by the relation $C_1 \supset C_2$.
3. Let $\Gamma = \{C_1 \supset C_2 \supset \cdots C_i \supset \cdots\}$ be an increasing chain of elements of $\mathcal{C}$.
4. $C = \bigcap C_i$ is nonempty, closed and invariant, hence $C \in \mathcal{C}$ is a supremum for Γ. (X is compact)
5. $\mathcal{C}$ has a maximal element $M \subset X$. (by Zorn's lemma)
6. M is a minimal subset of X. (by 1, 2, 5)

It is an elementary but important fact, detailed in the next proposition, that minimality is preserved by homomorphisms.

Proposition 3.5 Let $\varphi \colon X \to Y$ be a flow homomorphism.

(a) If M is a minimal subset of X, then $\varphi(M)$ is a minimal subset of Y.
(b) If N is a minimal subset of $\varphi(X)$, then there exists a minimal subset M of X with $\varphi(M) = N$.

PROOF: (a) 1. Let K be a non-empty closed invariant subset of $\varphi(M)$.
2. $\varphi^{-1}(K) \cap M$ is a closed non-empty invariant subset of the minimal set M.
3. $\varphi^{-1}(K) \cap M = M$.
4. $K = \varphi(M)$. (by 3)
(b) 1. $\varphi^{-1}(N)$ is a non-empty closed invariant subset of X.
2. There exists a minimal subset M of $\varphi^{-1}(N)$. (by **3.4**)
3. $\varphi(M)$ is a minimal subset of the minimal set N, so $\varphi(M) = N$. (by (a))

Now we turn to a discussion of the minimal ideals in the enveloping semigroup $E(X, T)$ of a flow (X, T). We first show that they coincide with the minimal sets of the flow $(E(X, T), T)$; then we will describe the structure of these minimal ideals.

Definition 3.6 Let X be a flow, and E its enveloping semigroup. Then a nonempty subset I of E is a (right) *ideal* if $IE \subset I$. The ideal is *minimal* if it contains no ideals as proper subsets.

Note that if $I \subset E(X, T)$ is an ideal, then $IT \subset I$, whence if I is closed, (I, T) is a flow. In fact as the following proposition shows, (I, T) is a minimal flow.

Proposition 3.7 Let:

(i) (X, T) be a flow,
(ii) $E = E(X, T)$, and
(iii) I an ideal in E.

Then I is a minimal ideal if and only if I is closed and the flow (I, T) is minimal.

PROOF: $\Longrightarrow$ 1. Assume that I is minimal and let $p \in I$.

2. $\overline{pT} = pE \subset I$. ($I$ is an ideal)
3. $L_p(E) = pE = I$. (pE is an ideal and I is minimal)
4. I is closed. (by 3, E is compact and L_p is continuous by **2.7**)
5. (I, T) is minimal. (by 2, 3, **3.2**)

$\Longleftarrow$ 1. Assume that (I, T) is a minimal flow, let $J \subset I$ be an ideal, and $p \in J$.

2. $I = \overline{pT} = pE \subset J$.
3. $I = J$. (by 1, 2)

Corollary 3.8 Let:

(i) (X, T) be a flow,
(ii) $E = E(X, T)$, and
(iii) I an ideal in E.

Then I contains a minimal ideal.

PROOF: This follows from **3.7** and **3.4**; we leave the details to the reader.

The description of the structure of a minimal ideal $I \subset E(X, T)$ relies on the existence of idempotents $u^2 = u \in I$. Thus the importance of the following theorem, which is an interesting example of the interplay between the topological and algebraic structure in a topological space which is also a semigroup.

Theorem 3.9 Let X be a compact T_1 (single points are closed) semigroup such that the left-multiplication maps

$$\begin{array}{rcl} L_x \colon X & \to & X \\ y & \to & xy \end{array}$$

are continuous and closed for all $x \in X$. Then there exists an *idempotent* u in X. (u is an idempotent if $u^2 = u$.)

PROOF: 1. Let $\Sigma = \{S \subset X \mid \emptyset \neq S = \overline{S} \text{ and } S^2 \subset S\}$.
2. Then $\Sigma \neq \emptyset$ and by Zorn's lemma there exists a minimal element, S of Σ when the latter is ordered by inclusion.
3. Let $s \in S$.
4. $sS = L_s(S)$ is a closed subset of S. (by 1, 2, 3, L_s is closed)
5. $sSsS \subset sSS \subset sS \subset S$, whence $sS = S$ by the minimality of S.
6. Let $R = \{t \in S \mid st = s\}$.
7. $\emptyset \neq R$. (by 3, 5)
8. $R^2 \subset R = L_s^{-1}\{s\} = \overline{R}$. ($L_s$ is continuous, X is T_1)
9. $R = S$. (by 2, 6, 7, 8)
10. $s^2 = s$ (by 3, 6, 9)

Since every continuous map from a compact space to a Hausdorff space is a closed map, the following is an immediate consequence of **3.9**.

Corollary 3.10 Let X be a compact Hausdorff semigroup such that the maps

$$L_x \colon X \to X$$

are continuous for all $x \in X$. (We often refer to such a semigroup as an *E-semigroup*, see **2.E.3**.) Then there exists an idempotent u in X.

Corollary 3.11 Let X be a compact T_1 group such that left multiplication is continuous, and let S be a closed sub-semigroup of X. Then S is a subgroup of X.

PROOF: 1. Let $x \in S$.
2. $xS = L_{x^{-1}}^{-1}(S)$ is a closed subsemigroup of S. ($L_{x^{-1}}$ is continuous)
3. There exists an idempotent $u \in xS$. (by 1, 2, **3.9**)
4. $u = id$. (X is a group)
5. $x^{-1} \in S$. (by 3, 4)

For any flow (X, T), the enveloping semigroup $E(X, T)$ is an E-semigroup in the sense of **3.10**, and hence any minimal ideal $I \subset E(X)$ is also an E-semigroup and therefore contains idempotents. It turns out that each of the idempotents in I acts as a left-identity on I. In fact the ideal I can be partitioned into a disjoint union of groups, each one containing exactly one idempotent (which serves as the identity for that group). We give the details in the following theorem.

Theorem 3.12 Let:

(i) (X, T) be a flow,
(ii) $E = E(X, T)$, and
(iii) $I \subset E$ be a minimal ideal in E.

Then:

(a) The set J of idempotents of I is non-empty,
(b) $vp = p$ for all $v \in J$ and $p \in I$,
(c) Iv is a group with identity v, for all $v \in J$,
(d) $\{Iv \mid v \in J\}$ is a partition of I, and
(e) if we set $G = Iu$ for some $u \in J$, then $I = \biguplus\{Gv \mid v \in J\}$ (disjoint union).

PROOF: (a) 1. This follows immediately from **3.10**.

(b) 1. Let $v \in J$, and $p \in I$.
2. $vI \subset I$. (*I* is an ideal)
3. $vI = I$. (by 2, *I* is minimal)
4. There exists $q \in I$ with $vq = p$. (by 3)
5. $vp \underset{4}{=} vvq = vq \underset{4}{=} p$.

(c) 1. Let $q \in Iv$.
2. There exists $p \in I$ with $q = pv$.
3. $qv = pvv = pv = q$.
4. v is both a left and right identity for Iv. (by (b), and 3)
5. qI is an ideal. (*I* is an ideal)
6. $qI = I$. (*I* is minimal)
7. There exists $q' \in I$ with $qq' = v$. (by 6)
8. $q(q'v) = (qq')v = vv = v$. (by 7)
9. $(q'q)(q'q) = q'(qq')q \underset{7}{=} q'(vq) \underset{(b)}{=} q'q$.
10. $(q'v)q \underset{(b)}{=} q'q \underset{3}{=} (q'q)v \underset{9,(b)}{=} v$.
11. $q'v$ is a left and right inverse of q in Iv. (by 8, 10)

(d) 1. Let $p \in I$.
2. As before $pI = I$. (*I* is a minimal ideal)
3. $K = \{q \in I \mid pq = p\} = L_p^{-1}(p)$ is a nonempty closed subsemigroup of I. (by 2, **2.9**)
4. There exists an idempotent $u \in J$ with $pu = p$. (by **3.9**)
5. $p \in Iu$. (by 4)

6. $I = \bigcup\{Iv \mid v \in J\}$. (by 1, 5)
7. Finally let $u, v \in J$ and $p \in Iv \cap Iu$.
8. $p = pu = pv$ and there exists $q \in Iv$ with $qp = v$. (by 8, (c))
9. $u \underset{(b)}{=} vu = (qp)u = q(pu) \underset{8}{=} qp \underset{8}{=} v$.

(e) This is just a restatement of (d).

Note that each of the groups Iv in proposition **3.12** is isomorphic to the group $G = Iu$; indeed the map $pu \to pv$ is an isomorphism since $(pv)(qv) = p(vq)v = (pq)v$. It is interesting to note that given an abstract group G and an index set J, one can define a semigroup structure on the disjoint union $I = \biguplus_{v \in J} G_v$ of copies of G, in which J can be identified with the set of idempotents. We simply define $p_v q_w = (pq)_w \in G_w$ for $p, q \in G$ and $v, w \in J$. Then $\{e_v \mid v \in J\}$ is the set of idempotents in I. Proposition **3.12** shows that every minimal ideal in $E(X, T)$ has this structure.

For any semigroup E one can define an equivalence relation on the set J of idempotents in E as follows:

$$u \sim v \iff uv = u \text{ and } vu = v.$$

This relation is clearly reflexive and symmetric; to check transitivity we observe that:

$$u \sim v \sim w \Rightarrow uw = (uv)w = u(vw) = uv = u$$
$$\text{and } wu = (wv)u = w(vu) = wv = w.$$

This motivates the following definition.

Definition 3.13 Let (X, T) be a flow. An idempotent $u^2 = u \in E(X, T)$ is said to be a *minimal idempotent* if u is contained in some minimal ideal $I \subset E(X, T)$. If $u, v \in E(X, T)$ are idempotents with $uv = u$ and $vu = v$, we write $u \sim v$ and refer to u and v as *equivalent idempotents.*

When u and v are in the same minimal ideal I, then $u \sim v \Rightarrow u = v$ since by **3.12** both u and v act as left-identities on I. Thus for any minimal idempotent, the equivalence class $[u]$ intersects each minimal ideal at most once; we leave it as an exercise for the reader to check that $[u]$ contains only **minimal** idempotents. The following proposition says that $[u]$ intersects every minimal ideal exactly once, so that $[u]$ consists of one idempotent from every minimal ideal in $E(X, T)$.

Proposition 3.14 Let:

(i) (X, T) be a flow,
(ii) $E = E(X, T)$,

(iii) $I, K \subset E$ be minimal ideals in E, and
(iv) $u^2 = u \in I$ be an idempotent.

Then there exists a unique idempotent $v \in K$ with $uv = u$ and $vu = v$.

PROOF: 1. Let $u^2 = u \in I$.
2. uK is a closed ideal in I, whence $uK = I$.
3. $N \equiv \{k \in K \mid uk = u\} \neq \emptyset$. (by 2)
4. $N = L_u^{-1}(u) \cap K$ is closed, and $N^2 \subset N$.
5. There exists $v^2 = v \in N$. (by 4, **3.10**)
6. $uv = u$. (by 3, 5)
7. Similarly there exists $w^2 = w \in I$ with $vw = v$. (applying 1-6 to $v \in K$)
8. $w \underset{1,\mathbf{3.12}}{=} uw \underset{6}{=} uvw \underset{7}{=} uv \underset{6}{=} u$.
9. $vu = v$. (by 7, 8)
10. Now suppose $\eta^2 = \eta \in K$ with $u\eta = u$ and $\eta u = \eta$.
11. $\eta \underset{10}{=} \eta u \underset{6}{=} \eta uv \underset{10}{=} \eta v$.
12. $\eta \in K\eta \cap Kv$. (by 10, 11)
13. $\eta = v$. (by **3.12**)

As an immediate consequence of **3.12** we see that any two minimal ideals in $E(X, T)$ are isomorphic as minimal flows in a natural way.

Proposition 3.15 Let:

(i) (X, T) be a flow,
(ii) $E = E(X, T)$,
(iii) $I, K \subset E$ be minimal ideals in E,
(iv) $u^2 = u \in I$ be an idempotent, and
(v) $v^2 = v \in K$ with $u \sim v$.

Then the map $\begin{array}{rcl} L_v : (I, T) & \to & (K, T) \\ p & \to & vp \end{array}$ is an isomorphism, its inverse being the map L_u.

PROOF: We leave the proof as an exercise for the reader.

The structure of the minimal ideals in the enveloping semigroup $E(X, T)$ of any flow (X, T) described above, and the minimal idempotents themselves play an important role (see in particular section 4) in the study of the dynamics of (X, T). The minimal ideals in the semigroup βT have exactly the same structure. This can be seen by noting that the proofs of **3.12**, **3.14**, and **3.15** rely only on properties of $E(X, T)$ which are shared by βT. On the other hand we saw in **2.9**, that $E(\beta T) \cong \beta T$ as semigroups. Thus **3.12**, **3.14**, and **3.15** can be applied directly to βT. In particular, we can speak of minimal idempotents in βT, and the minimal subsets (ideals) of βT are isomorphic to one another. The importance of the minimal ideals in βT stems from the fact that any such minimal ideal $M \subset \beta T$ is a *universal minimal set*, meaning that every minimal flow (X, T) is a homomorphic image of the flow (M, T).

Theorem 3.16 Let M be a minimal subset of βT, and (X, T) be a minimal flow. Then there exists an epimorphism $f: M \to X$.

PROOF: 1. Let $x \in X$.
2. The map $T \to X$ has a continuous extension $g: \beta T \to X$. (by **1.1**)
$t \to xt$
3. g is a homomorphism of flows.
4. The restriction f of g to M is an epimorphism. (X is minimal)

We end this section with an observation which emphasizes the importance of the preceding theorem. Let $f: M \to X$ be an epimorphism and

$$R_f = \{(p, q) \in M \times M \mid f(p) = f(q)\}.$$

Then R_f is an invariant closed equivalence relation (*icer* for short) on M, such that $M/R_f \cong X$. Thus theorem **3.16** shows that every minimal flow can be obtained as a quotient of M by an icer. Conversely for every icer R on M, the quotient M/R is a minimal flow. This follows from the purely topological result (included here as **6.2**) that the quotient of any compact Hausdorff space by a closed equivalence relation is itself a compact Hausdorff space. Minimal flows and their properties can therefore be studied by studying icers on the universal minimal set M. This is one of the major themes of this book. We begin the development of this theme in part II. In particular the structure of M given in **3.12** will be used in section 7 to facilitate the study of icers on M, and lay the foundation for our approach to the algebraic theory of minimal sets.

EXERCISES FOR CHAPTER 3

Exercise 3.1 (see **3.15**) Let:

(i) (X, T) be a flow,
(ii) $E = E(X, T)$,
(iii) $I, K \subset E$ be minimal ideals in E,
(iv) $u^2 = u \in I$ be an idempotent, and
(v) $v^2 = v \in K$ with $u \sim v$.

Show that $\begin{array}{rcl} L_v : (I, T) & \to & (K, T) \\ p & \to & vp \end{array}$ is an isomorphism, its inverse being the map L_u.

Exercise 3.2 Let:

(i) (X, T) be a flow,
(ii) $E = E(X, T)$,
(iii) u be a minimal idempotent in E, and
(iv) $v \in E$ be an idempotent with $u \sim v$.

Show that v is a minimal idempotent in E.

Exercise 3.3 Show that a compact Hausdorff semigroup X is an E-semigroup if and only if the map $\begin{array}{rcl} X & \to & X^X \\ x & \to & R_x \end{array}$ is continuous. (see **3.10**)

Exercise 3.4 Prove analogs of **3.12**, **3.14**, and **3.15** for E-semigroups. (see **3.10**)

Exercise 3.5 Show that any topological group S which is T_1 must be Hausdorff.

4
Fundamental notions

Certain important notions in topological dynamics serve as the language, foundation and motivation for the theory. These include pointwise almost periodicity, minimality, distality, proximality, weak-mixing, and topological transitivity for flows. This section is devoted to defining and discussing these concepts and some of their analogs and generalizations to homomorphisms of flows.

Our exposition emphasizes the role played by the semigroups βT, and $E(X, T)$, and the minimal ideals therein in understanding the properties of the flow (X, T). Many of the fundamental notions can be cast in terms of the algebraic structure of these semigroups, their minimal ideals and idempotents. One example of this is proposition **4.9** which shows that (X, T) is distal if and only if $E(X, T)$ is a group. This algebraic approach also leads to **4.12**, where it is shown that every distal flow is pointwise almost periodic.

The later part of this section is devoted to a discussion of topological transitivity and related questions. For metric flows, the notions of point transitivity and topological transitivity are quite easily seen to be equivalent (see **4.18**). This allows certain results (notably the Furstenberg structure theorem for distal flows) to be deduced for metric flows in a straightforward way. On the other hand topological transitivity and point transitivity are not equivalent in general for flows on compact Hausdorff spaces. Despite this, some deep results along these lines can be obtained by deducing the general case from metric considerations. As one example of this technique we show in **4.24** that every flow which is both topologically transitive and distal, must be minimal. This result has the general case of the Furstenberg theorem as an immediate consequence. This result can also be generalized to the case of homomorphisms of minimal flows. Here one can prove that a homomorphism of minimal flows which is both weak mixing (so that the corresponding equivalence relation is topologically transitive) and distal must be trivial. Our proof (given in section 9) of this more general result again involves reducing the argument to the metric case, but also requires the introduction of what we call the quasi-relative product of two icers.

We begin with the definition of an *almost periodic point* which is given here in terms of minimal sets.

Definition 4.1 Let (X, T) be a flow and $x \in X$. Then x is an *almost periodic point* if the *orbit closure* of x, denoted $\overline{xT}$, is minimal. We say that the flow (X, T) is *pointwise almost periodic* if every $x \in X$ is an almost periodic point.

It is clear that any minimal flow is pointwise almost periodic, and that any pointwise almost periodic flow is a disjoint union of minimal sets. For semigroups, which we do not assume contain the identity, the definition needs to be modified slightly. In this case $\overline{xT}$ need not contain x, so we require that an almost periodic point satisfy $x \in \overline{xT}$. Thus any pointwise almost periodic semigroup action on X partitions X into a disjoint union of minimal sets.

Note that if a point $x \in X$ is fixed by every element of T, or more generally if the orbit xT is finite (where we might think of x as a periodic point), then x is an almost periodic point of the flow (X, T). The terminology is further motivated by the case $T = \mathbf{Z}$, the group of integers. Here by abuse of notation we think of

$$T = \{\dots, T^{-n}, \dots, T^{-2}, T^{-1}, id, T, T^2, \dots, T^n, \dots\},$$

where $T\colon X \to X$ is a homeomorphism. For this special case a periodic point of period n has as set of *return times* the set

$$A(x) = \{T^j \,|\, xT^j = x\} = \{T^{nk} \,|\, k \in \mathbf{Z}\}.$$

This set is large in the sense that its product with the finite set $\{T^1, \dots, T^{n-1}\}$ gives the whole group T. This way of characterizing a periodic point when $T = \mathbf{Z}$ has a natural generalization which makes sense for any group T. For any $x \in X$ and neighborhood U of x we consider the set of *return times to* U:

$$A(U) = \{t \in T \mid xt \in U\}.$$

The next proposition proves that $A(U)$ is large in the sense mentioned above if and only if x is an almost periodic point, thus motivating the terminology.

Proposition 4.2 Let:

(i) (X, T) be a flow,
(ii) $x \in X$,
(iii) $\mathcal{N}_x = \{U \subset X \,|\, x \in U \text{ and } U \text{ is open}\}$, and
(iv) $A(U) = \{t \in T \,|\, xt \in U\}$.

Then x is an almost periodic point if and only if for every $U \in \mathcal{N}_x$, there exists a finite set $F \subset T$, such that $A(U)F = T$. In other words $A(U)$ is a *syndetic* subset of T.

PROOF: $\Longrightarrow$

1. Assume that x is an almost periodic point and let $U \in \mathcal{N}_x$.
2. $\overline{xT} \subset (U \cap \overline{xT})T$. $\quad$ $((\overline{xT}, T)$ is minimal by 1)
3. There exists a finite set $F \subset T$ such that $\overline{xT} \subset UF$. (by 2, $\overline{xT}$ is compact)
4. Let $t \in T$.
5. $xt \in Us$ for some $s \in F$. $\quad$ (by 3)
6. $ts^{-1} \in A(U)$. $\quad$ (by 5, (iv))
7. $t \in A(U)F$. $\quad$ (by 6)
8. $A(U)F = T$. $\quad$ (by 4, 7)

$\Longleftarrow$

1. Assume that $A(U)$ is syndetic for every $U \in \mathcal{N}_x$.
2. Let $W \subset X$ be open with $W \cap \overline{xT} \neq \emptyset$.
3. Let W_0 be open with $\overline{W_0} \subset W$ and $W_0 \cap \overline{xT} \neq \emptyset$.
4. There exists $t \in T$ such that $W_0 t \in \mathcal{N}_x$. $\quad$ (by 3, (iii))
5. There exists a finite subset $F \subset T$ such that $A(W_0 t)F = T$. $\quad$ (by 1, 4)
6. $\overline{xT} \underset{4}{\subset} \overline{xA(W_0 t)F} \underset{4}{=} \overline{xA(W_0 t)}F \subset (\overline{W_0 t} \cap \overline{xT})F \subset (\overline{W_0} \cap \overline{xT})tF \underset{3}{\subset} (W \cap \overline{xT})T$.
7. $\overline{xT}$ is minimal. $\quad$ (by 2, 6)

We now characterize the almost periodic points of (X, T) in terms of the minimal idempotents in the enveloping semigroup $E(X, T)$.

Proposition 4.3 Let:

(i) (X, T) be a flow,
(ii) $E = E(X, T)$,
(iii) $I \subset E$ be a minimal ideal in E, and
(iv) $x \in X$.

Then the following are equivalent:

(a) x is an almost periodic point of X,
(b) $\overline{xT} = xI \equiv \{xp \mid p \in I\}$, and
(c) there exists $u^2 = u \in I$ with $xu = x$.

PROOF: (a) $\Rightarrow$ (b) 1. Assume that x is an almost periodic point of (X, T).
2. $\begin{array}{rcl} \varphi_x : E & \to & X \\ p & \to & xp \end{array}$ is a homomorphism of E onto $\overline{xT}$. $\quad$ (by **2.9**)

3. $xI = \varphi_x(I)$ is a closed invariant subset of $\overline{xT}$. (by (iii), **3.7**)
4. $xI = \overline{xT}$. (by 1, 3)

(b) $\Rightarrow$ (c) 1. Assume that $\overline{xT} = xI$.
2. The set $S = \{p \in I \mid xp = x\}$ is a closed nonempty subsemigroup of I. (by 1)
3. There exists $u^2 = u \in S$. (by 2, **3.10**)

(c) $\Rightarrow$ (a) 1. Assume that there exists $u = u^2 \in I$ with $xu = x$.
2. $xT = (xu)T = x(uT) \subset xI$.
3. $\overline{xT} \subset xI$. (by 2 and **3.7**)
4. $\overline{xT} = xI$. (xI is minimal by **3.7**)
5. x is an almost periodic point. (by 4)

Continuing with our theme of characterizing dynamical notions in terms of the semigroups βT and $E(X, T)$, the following proposition considers notions which are equivalent to what we call proximality.

Proposition 4.4 Let (X, T) be a flow and $x, y \in X$. Then the following are equivalent:

(a) there exists a net $\{t_i\} \subset T$ with $\lim xt_i = \lim yt_i$,
(b) $\overline{(x, y)T} \cap \Delta \neq \emptyset$, (here $\Delta = \{(x, x) \mid x \in X\}$ is the diagonal),
(c) there exists $p \in \beta T$ with $xp = yp$,
(d) there exists $r \in E(X)$ with $xr = yr$,
(e) there exists a minimal right ideal $I \subset E(X)$ with $xr = yr$ for all $r \in I$, and
(f) there exists a minimal right ideal $K \subset \beta T$ with $xq = yq$ for all $q \in K$.

PROOF: (a) $\Longrightarrow$ (b)

1. Assume that $\{t_i \mid i \in L\} \subset T$ is a net with $\lim xt_i = z = \lim yt_i$.
2. Let $\mathcal{V} \subset X \times X$ be a neighborhood of (z, z).
3. There exists $N \subset X$ open with $(z, z) \in N \times N \subset \mathcal{V}$. (by 2)
4. There exist $i_1, i_2 \in L$ such that $i > i_1 \Rightarrow xt_i \in N$ and $i > i_2 \Rightarrow yt_i \in N$. (by 1, 3)
5. Let $i \in L$ with $i > i_1$ and $i > i_2$.
6. $(x, y)t_i = (xt_i, yt_i) \in N \times N \subset \mathcal{V}$. (by 3, 4, 5)
7. $((x, y)T) \cap \mathcal{V} \neq \emptyset$. (by 6)
8. $(z, z) \in \overline{(x, y)T} \cap \Delta$. (by 2, 7)

(b) $\Longrightarrow$ (c)

This follows from the fact that $\overline{(x, y)T} = (x, y)\beta T$.

(c) $\Longrightarrow$ (d)

1. Let $p \in \beta T$ with $xp = yp$.

2. Let $\Phi\colon \beta T \to E(X)$ be the canonical map. (see **2.8**)
3. Set $r = \Phi(p) \in E(X)$.
4. $xr = x\Phi(p) = xp = yp = yr$.

(d) $\Longrightarrow$ (e)

1. Assume that $xr = yr$ for some $r \in E(X)$.
2. $xrq = yrq$ for all $q \in E(X)$.
3. There exists a minimal ideal $I \subset rE(X)$. (by **3.8**)
4. $xm = ym$ for all $m \in I$. (by 2, 3)

(e) $\Longrightarrow$ (f)

1. Assume that $xq = yq$ for all $q \in I$, a minimal ideal in $E(X)$.
2. $\Phi^{-1}(I) \subset \beta T$ is a right ideal in βT.
3. There exists a minimal ideal $K \subset \Phi^{-1}(I)$. (by **2.9** and **3.8**)
4. $xq = x\Phi(q) = y\Phi(q) = yq$ for all $q \in K$.

(g) $\Longrightarrow$ (a)

1. Assume $xq = yq$ for all $q \in K$ with K a minimal right ideal in βT.
2. Let $k \in K$ and $\{t_i\} \subset T$ with $t_i \to k$.
3. $\lim xt_i = x \lim t_i = xk = yk = y \lim t_i = \lim yt_i$.

Definition 4.5 Let (X, T) be a flow and $x, y \in X$. Then the pair (x, y) is *proximal* if it satisfies any one of the equivalent conditions of **4.4**. The collection of proximal pairs in $X \times X$ will be denoted $P(X)$, so that

$$P(X) = \{(x, y) \in X \times X \mid \overline{(x, y)T} \cap \Delta \neq \emptyset\}.$$

We will also be interested in the *proximal cells*; anticipating the notation of **6.1**, for general relations, we write

$$xP(X) = \{z \in X \mid (x, z) \in P(X)\}$$

for the proximal cell containing x. We say that the flow (X, T) is a *proximal flow* if every pair $(x, y) \in X \times X$ is proximal; that is $P(X) = X \times X$. We say that the flow (X, T) is a *distal flow* if the only proximal pairs are of the form $(x, x) \in X \times X$; that is $P(X) = \Delta \subset X \times X$.

The relation $P(X)$ described above is invariant reflexive and symmetric, but is in general not transitive. In section 13 we will study conditions under which $P(X)$ is transitive, i.e. when $P(X)$ is an equivalence relation. In this section we focus on how the proximal relation $P(X)$, and consequently the notion of distality, can be studied via the idempotents in βT or $E(X, T)$.

Let $u^2 = u \in \beta T$ be an idempotent in βT, and $x \in X$ where (X, T) is a flow. Since T is dense in βT, there exists a net $\{t_i\} \subset T$, such that $t_i \to u$ in βT. Thus

$$\lim(x, xu)t_i = (\lim xt_i, \lim(xu)t_i) = (xu, xu^2) = (xu, xu),$$

and hence x is proximal to xu. This elementary observation makes a key connection between proximality and the idempotents in βT, and is the basis for many interesting results. We describe some of them below, beginning with a characterization of the proximal relation.

Lemma 4.6 Let (X, T) be a minimal flow. Then:

(a) $P(X) = \{(x, xw) \mid x \in X \text{ and } w \text{ is a minimal idempotent in } \beta T\} \equiv K$, and
(b) $P(X) = \{(x, xw) \mid x \in X \text{ and } w \text{ is a minimal idempotent in } E(X)\} \equiv L$.

PROOF: The proof of parts (a) and (b) are similar. We give a proof of part (b), leaving the proof of part (a) to the reader.

Proof that $L \subset P(X)$:

1. Let $x \in X$ and w be any idempotent in $E(X)$.
2. $(x, xw)w = (xw, xw)$.
3. $(x, xw) \in P(X)$. (by 2, **4.4**)

Proof that $P(X) \subset L$:

4. Let $(x, y) \in P(X)$.
5. There exists a minimal ideal $I \subset E(X)$ such that $xq = yq$ for all $q \in I$.
6. There exists $w = w^2 \in I$ such that $yw = y$. (by **4.3**, since X is minimal)
7. $xw \underset{5}{=} yw \underset{6}{=} y$.
8. $(x, y) = (x, xw) \in L$.

The characterization of $P(X)$ given in **4.6** allows us to prove that, for minimal flows, the proximal relation is preserved by homomorphisms.

Proposition 4.7 Let:

(i) X and Y be minimal flows, and
(ii) $\pi : X \to Y$ be a homomorphism.

Then:

(a) $\pi(P(X)) = P(Y)$, and
(b) $\pi(xP(X)) = \pi(x)P(Y)$ for all $x \in X$.

PROOF: (a) 1. It is clear that $\pi(P(X)) \subset P(Y)$,
2. Let $(y_1, y_2) \in P(Y)$.

3. There exists a minimal right ideal I in βT with $y_1 p = y_2 p$ for all $p \in I$.
4. There exists $u \in I$ with $u^2 = u$ and $y_2 u = y_2$. (Y is minimal)
5. Let $x \in X$ with $\pi(x) = y_1$.
6. $(x, xu) \in P(X)$. (by **4.4**)
7. $(y_1, y_2) \underset{4}{=} (y_1, y_2 u) \underset{3,4}{=} (y_1, y_1 u) \underset{5}{=} (\pi(x), \pi(x)u) = \pi(x, xu) \in \pi(P(X))$.

(b) 1. If $(x, y) \in P(X)$, then $(\pi(x), \pi(y)) \in P(Y)$. (by part (a))
2. $\pi(xP(X)) \subset \pi(x)P(Y)$. (by 1)
3. Let $y \in \pi(x)P(Y)$.
4. There exists $\eta^2 = \eta$ in some minimal ideal in βT with $\pi(x)\eta = y$. (by 3, **4.6**)
5. $(x, x\eta) \in P(X)$. (by 4, **4.6**)
6. $y = \pi(x)\eta = \pi(x\eta) \in \pi(xP(X))$. (by 4, 5)

We observed earlier that for any flow (X, T) and $x \in X$, the pair (x, xu) is proximal for any idempotent $u \in E(X, T)$. This of course means that in a distal flow $xu = x$, so that the identity is the only idempotent in $E(X)$. This basic idea (made explicit in **4.8**), is exploited in **4.9** to characterize a distal flow in terms of its enveloping semigroup.

Proposition 4.8 Let (X, T) be a flow. Then (X, T) is distal if and only if $xu = x$ for all $x \in X$ and idempotents $u = u^2 \in \beta T$.

PROOF: 1. Assume that (X, T) is distal and let $u^2 = u \in \beta T$.
2. $(x, xu)u = (xu, xu^2) = (xu, xu)$. (by 1)
3. $(x, xu) \in P(X)$. (by 2, **4.4**)
4. $x = xu$. (by 1, 3)
5. Assume that $xu = x$ for all $u^2 = u \in \beta T$, and $x \in X$.
6. Let $(x, y) \in P(X)$.
7. There exists a minimal right ideal $K \subset \beta T$ such that $(x, y)p \in \Delta$ for every $p \in K$. (by 6, **4.4**)
8. There exists an idempotent $u^2 = u \in K$. (by 7, **3.10**)
9. $x = xu = yu = y$ (by 5, 7, 8)

The proof given above shows somewhat more: if $xu = x$ for all $x \in X$ and all *minimal* idempotents, then (X, T) is distal.

Proposition 4.9 Let (X, T) be a flow and $\Phi\colon \beta T \to E(X)$ be the canonical map. Then the following are equivalent:

(a) X is distal,
(b) e is the only idempotent in $E(X)$,
(c) $E(X)$ is a group,

(d) $E(X) = \Phi(M)$, and
(e) $E(X)$ is minimal.

PROOF: The proof is similar to that of **4.8**; we leave the details as an exercise for the reader.

The fact that distality is preserved by homomorphisms will be used extensively in later sections. This could be deduced immediately from **4.7** in the case of minimal flows, or directly from the definition. As a means of emphasizing the algebraic approach we give a very short proof using the idempotents in βT.

Proposition 4.10 Let:

(i) $f: (X, T) \to (Y, T)$ be a surjective homomorphism of flows, and
(ii) (X, T) be distal.

Then (Y, T) is distal.

PROOF: 1. Let $u^2 = u \in \beta T$ and $y = f(x) \in Y$.
2. $yu = f(x)u = f(xu) = f(x) = y$. (by 1, **4.8**)
3. (Y, T) is distal. (by 1, 2, **4.8**)

The following result, though it is an immediate consequence of the elementary observation made earlier, is of independent interest. Historically, the result which we derive from it, that every distal flow is pointwise almost periodic, was proven first. The Auslander-Ellis theorem was in some sense an afterthought.

Theorem 4.11 **(Auslander-Ellis)**: Let (X, T) be a flow and $x \in X$. Then there exists an almost periodic point $y \in X$ with $(x, y) \in P(X)$.

PROOF: 1. There exist a minimal ideal $I \subset E(X, T)$ and an idempotent $u \in I$. (by **3.8** and **3.12**)
2. $y = xu$ is an almost periodic point of (X, T). (by 1, and **4.3**)
3. $(x, y) \in P(X)$. (by 2, and **4.6**)

Theorem 4.12 Let (X, T) be a distal flow. Then (X, T) is pointwise almost periodic.

PROOF: 1. Let $x \in X$.
2. There exists $y \in X$ an almost periodic point with $(x, y) \in P(X)$. (by **4.11**)
3. $x = y$ and hence (X, T) is pointwise almost periodic.
(by 1, 2, since $P(X) = \Delta$)

It is worth noting that **4.12** is not at all obvious from the definitions, yet the algebraic techniques allow a very elementary proof. Another interesting connection between the notions of distal and pointwise almost periodic flows is given below. The result is of independent interest, but as we will see it also serves as motivation for many other results involving homomorphisms of flows.

Theorem 4.13 Let (X, T) be a flow, then the following are equivalent:

(a) (X, T) is a distal flow,
(b) $(X \times X, T)$ is a distal flow, and
(c) $(X \times X, T)$ is pointwise almost periodic.

PROOF: (a) $\Longrightarrow$ (b)

1. Assume that (X, T) is distal.
2. Let $(x, y) \in X \times X$ and $u^2 = u \in \beta T$.
3. $(x, y)u = (xu, yu) = (x, y)$. (by 1, 2, and **4.8**)
4. $(X \times X, T)$ is distal. (by 2, 4, and **4.8**)

(b) $\Longrightarrow$ (c)

This follows from **4.12**.

(c) $\Longrightarrow$ (a)

1. Assume that $(X \times X, T)$ is pointwise almost periodic and let $(x, y) \in P(X)$.
2. $\overline{(x, y)T} \cap \Delta_X \neq \emptyset$. (by 1, **4.4**)
3. $\overline{(x, y)T}$ is a minimal subset of $(X \times X, T)$. (by 1)
4. $\overline{(x, y)T} \subset \Delta_X$. (by 2, 3, Δ_X is closed and T-invariant)
5. $x = y$. (by 4)
6. $P(X) = \Delta_X$ and (X, T) is distal. (by 1, 5)

There are natural generalizations of the notions of proximal and distal to homomorphisms of flows. These notions are defined in such a way that the trivial homomorphism $(X, T) \to (\{pt\}, T)$ is proximal (resp. distal) if and only if the flow (X, T) is proximal (resp. distal).

Definition 4.14 Let $f: (X, T) \to (Y, T)$ be a homomorphism of flows. We say that f is a *proximal homomorphism* if whenever $f(x) = f(y)$, the pair (x, y) is proximal. We say that f is a *distal homomorphism* if whenever $f(x) = f(y)$, and (x, y) is proximal, we have $x = y$. When $f: (X, T) \to (Y, T)$ is a **surjective** homomorphism of flows, we often refer to X as an *extension* of Y, and Y as a *factor* of X. If f is a proximal homomorphism, then we refer to (X, T) as a *proximal extension* of (Y, T). If f is a distal homomorphism, then we refer to (X, T) as a *distal extension* of (Y, T). Using this terminology, a flow is proximal (resp. distal) if and only if it is a proximal (resp. distal) extension of the one-point flow.

Let $f: X \to Y$ be a homomorphism of flows. Generalizing the notation introduced at the end of section 3, the icer (closed invariant equivalence relation) R_f is defined by:

$$R_f = \{(x, y) \mid f(x) = f(y)\} \subset X \times X.$$

Then f is a proximal homomorphism if and only if $R_f \subset P(X)$. Similarly f is distal if and only if $R_f \cap P(X) = \Delta$, the diagonal in $X \times X$. These are two elementary examples of how the dynamics of the homomorphism f is related to the structure of the icer R_f. A deeper illustration of this idea where the *dynamics* of the flow (R_f, T) plays a role is motivated by **4.13**. For the constant homomorphism c, we have $R_c = X \times X$, so **4.13** says that c is distal if and only if R_c is pointwise almost periodic. This is true for any homomorphism under the assumption that the flow (X, T) is itself pointwise almost periodic (this is the best we can hope for since $\Delta \subset R_f$, so (R_f, T) pointwise almost periodic implies (X, T) pointwise almost periodic). The following theorem gives the details.

Theorem 4.15 Let $f: (X, T) \to (Y, T)$ be a homomorphism of flows, then:

(a) if (R_f, T) is pointwise almost periodic, then f is distal, and
(b) if (X, T) is pointwise almost periodic, and f is distal, then (R_f, T) is pointwise almost periodic.

PROOF: (a) The proof is similar to that of **4.13**, we leave it as an exercise for the reader.

(b) 1. Assume that (X, T) is pointwise almost periodic, and f is distal.
2. Let $(x, y) \in R_f$.
3. There exist a minimal ideal $I \subset E(X, T)$ and an idempotent $u \in I$ with $xu = x$. (by 1, **4.3**)
4. $(x, yu) = (xu, yu) = (x, y)u \in R_f u \subset R_f$. (by 3, R_f is closed and invariant)
5. $(y, yu) \in R_f$. (by 2, 4, R_f is an equivalence relation)
6. $(yu, yu) = (y, yu)u \in \overline{(y, yu)T}$.
7. $(y, yu) \in R_f \cap P(X) = \Delta_X$. (by 1, 5, 6)
8. $(x, y) = (x, y)u$, so (x, y) is an almost periodic point. (by 3, 7, **4.3**)

The remainder of this section is devoted to a discussion of topological transitivity, the related notion of weak mixing, and their relationship to the ideas introduced so far. Recall that an action of T on X is transitive if and only if the orbit $xT = X$. We have introduced the notions of minimality (where $\overline{xT} = X$ for all $x \in X$) and point-transitivity (where $\overline{x_0 T} = X$ for some $x_0 \in X$). Both of these are examples of "topological weakenings" of the notion of transitivity.

Another approach is to require for every pair $x, y \in X$, not that there exist $t \in T$ with $xt = y$, but that for every neighborhood U of x and V of y, there exist $x_U \in U$, and $t \in T$ with $x_U t \in V$. This gives us topological transitivity; we give an equivalent formulation as the definition.

Definition 4.16 We say that (X, T) is *topologically transitive* if for any pair V, W of nonempty open subsets of X, the intersection $VT \cap W \neq \emptyset$. This is equivalent to saying that $\overline{VT} = X$ for any nonempty open set $V \subset X$. We say that (X, T) is *weak mixing* if the flow $(X \times X, T)$ is topologically transitive. As in **4.14**, we generalize this terminology to homomorphisms, referring to a surjective homomorphism $f\colon (X, T) \to (Y, T)$ as a *weak mixing extension* if the corresponding icer $R_f \subset X \times X$ is topologically transitive.

Let (X, T) be a flow. If (X, T) is minimal, then for every $x \in X$ and open set $V \subset X$, $xT \cap V \neq \emptyset$, because $\overline{xT} = X$. It follows that $X = VT$ and (X, T) is topologically transitive. More generally, suppose that (X, T) is point transitive, that is there exists $x_0 \in X$ with $\overline{x_0 T} = X$. Then for every open subset $V \subset X$, there exists $t \in T$ with $x_0 t \in V$. Thus $x_0 T \subset VT$ so $\overline{VT} = X$. In other words any flow which is point transitive is also topologically transitive. It is worth noting that the argument above fails for semigroup actions; indeed, in general, a point transitive semigroup action need not be topologically transitive. For metric flows (where T is a group), as the following lemma states, the converse also holds; for completeness sake we give an outline of a proof. We first give an example of a topologically transitive flow which is not point transitive.

Example 4.17 Let Y be a compact Hausdorff space which is not separable (so Y has no countable dense subset). Consider the flow $(X, \mathbf{Z})$ where $\mathbf{Z}$ is the group of integers under addition,

$$X \equiv Y^{\mathbf{Z}} = \{f \mid f : \mathbf{Z} \to Y\},$$

is the *symbol space* with symbols in Y, and the action of $\mathbf{Z}$ on X is given by the homeomorphism (*shift map*) $\sigma : X \to X$ defined by:

$$(n)(f\sigma) = (n+1)f \quad \text{for all } n \in \mathbf{Z}.$$

Note that since X is not separable, and $\mathbf{Z}$ is countable, the flow $(X, \mathbf{Z})$ is not point transitive. We leave it as an exercise for the reader to check that $(X, \mathbf{Z})$ is topologically transitive.

Lemma 4.18 Let:

(i) (X, T) be a topologically transitive flow, and
(ii) X be metrizable.

Then there exists $x \in X$ with $\overline{xT} = X$. (That is (X, T) is point transitive.)

PROOF: 1. Let $\mathcal{V}$ be a countable base for the topology on X.
2. VT is open and dense in X for every $V \in \mathcal{V}$. (by 1, (i))
3. There exists $x \in \bigcap\{VT | V \in \mathcal{V}\}$. (1, 2, X is a Baire space by (ii))
4. $\overline{xT} = X$. (by 3)

The following immediate corollaries of **4.18** are of independent interest, and also motivate deeper investigations in the case of non-metric flows.

Corollary 4.19 Let:

(i) (X, T) be a topologically transitive flow,
(ii) (X, T) be pointwise almost periodic, and
(iii) X be metrizable.

Then (X, T) is minimal.

PROOF: This follows immediately from **4.18**.

Corollary 4.20 Let:

(i) (X, T) be a topologically transitive flow,
(ii) (X, T) be distal, and
(iii) X be metrizable.

Then (X, T) is minimal.

PROOF: This follows immediately from **4.12** and **4.19**.

Proposition 4.21 Let:

(i) $\pi : (X, T) \to (Y, T)$ be an epimorphism of flows, and
(ii) (X, T) be topologically transitive.

Then (Y, T) is topologically transitive.

PROOF: We leave the proof of this proposition as an exercise for the reader.

We will see in later sections that **4.19** has important consequences for minimal distal flows and distal extensions of minimal flows. As a preview note that if R is an icer on a minimal metrizable flow such that (R, T) is both topologically transitive and pointwise almost periodic, then it follows from **4.19** that $R = \Delta$. This fact is equivalent to the so-called generalized (relative) Furstenberg Structure Theorem (see section 20). It turns out that the metrizability assumption can be dropped, so that among icers on **any** minimal flow, Δ is the only one which is both pointwise almost periodic and topologically transitive. In the language of **4.16**, this says that any extension of minimal flows which is both distal and weak mixing must be trivial. The proof in the general case is considerably more difficult than in the metric case. Our proof, given in **9.13**, is an application of the quasi-relative product introduced in section 9.

The remainder of this section will be devoted to exploring various generalizations and consequences of **4.19** and **4.20** for compact Hausdorff (not necessarily metric) spaces, which are accessible without the use of the quasi-relative product. The proofs are a bit technical and rely on some familiarity with pseudo-metrics (see **15.A.7**). On a first reading, or for a reader focusing on the metric case, the technical details of the two proofs which follow might well be skipped. We begin with a technical lemma which we use to deduce **4.19** for general compact Hausdorff topological spaces X under the assumption that the group T is countable. We then use the same lemma to deduce **4.20** in general.

Lemma 4.22 Let:

(i) (X, T) be a topologically transitive flow,
(ii) $H \subset T$ be a countable subgroup of T, and
(iii) $V \subset X$ be an open set.

Then there exist a countable subgroup $K \subset T$, a surjective homomorphism

$$\pi : (X, K) \to (Y, K), \quad \text{and an open subset} \quad B \subset V,$$

such that:

(a) $H \subset K$,
(b) (Y, K) is a topologically transitive flow,
(c) Y is metrizable, and
(d) $\pi^{-1}(\pi(B)) = B$.

PROOF: 1. Let $y \in V$.
2. There exists a continuous psuedo-metric d on X and an $\epsilon > 0$ such that

$$B \equiv \{z \in X \mid d(y, z) < \epsilon\} \subset V.$$

3. For any subgroup $F \subset T$ set

$$R[F] = \{(a, b) \in X \times X \mid d(at, bt) = 0 \text{ for all } t \in F\}.$$

4. $(X/R[F], F)$ is a compact metrizable flow when F is countable. (by 3)
5. Set $H_0 = H$, and let $\pi_0 : X \to X_0 = X/R[H_0]$ be the canonical map.
6. There exists a countable base $\mathcal{U}_0$ for the topology on X_0. (by 4, 5)
7. $V = \bigcap\{\pi_0^{-1}(U)T \mid U \in \mathcal{U}_0\}$ is a residual subset of X. (X_0 is a Baire space)
8. Let $x_0 \in V$.
9. $\pi_0(x_0T) \cap U \neq \emptyset$ for all $U \in \mathcal{U}_0$. (by 7, 8)

10. There exists a countable subgroup H_1 of T with $H_0 \subset H_1$ and

$$\overline{\pi_0(x_0 H_1)} = X_0. \qquad \text{(by 6, 9)}$$

11. Set $X_1 = X/R[H_1]$ and let $\pi_1 : X \to X_1$ be the canonical map.
12. There exist $x_1 \in X$ and a countable subgroup H_2 of T with $H_1 \subset H_2$ and

$$\overline{\pi_1(x_1 H_2)} = X_1. \qquad \text{(apply 6-10 to } X_1\text{)}$$

13. There exist a countable subgroups

$$H = H_0 \subset \cdots \subset H_n \subset \cdots \subset T$$

and points $(x_n) \subset X$ such that

$$\overline{\pi_n(x_n H_{n+1})} = X/R[H_n], \quad \text{for } n = 0, 1, \ldots. \qquad \text{(by induction)}$$

14. Let $K = \bigcup H_n$ and $V_1 \neq \emptyset \neq V_2$ be two open subsets of $X/R[K]$.
15. Let $\phi_i : X/R[K] \to X/R[H_i]$ be the canonical map.
16. $X/R[K] = \lim\limits_{\leftarrow} X/R[H_n]$. (by 14, 15)
17. There exist n and open subsets $U_1, U_2 \subset X/R[H_n]$ with

$$\phi_n^{-1}(U_i) \subset V_i \text{ for } i = 1, 2. \qquad \text{(by 14, 16)}$$

18. There exist $h_1, h_2 \in H_{n+1} \subset K$ with

$$\phi_n(\pi_K(x_n h_i)) = \pi_n(x_n h_i) \in U_i \text{ for } i = 1, 2. \qquad \text{(by 13, 17)}$$

19. $\pi_K(x_n)h_i = \pi_K(x_n h_i) \underset{18}{\in} \phi_n^{-1}(U_i) \underset{17}{\subset} V_i$ for $i = 1, 2$.
20. $V_1 K \cap V_2 \neq \emptyset$. (by 19)
21. $(Y, K) \equiv (X/R[K], K)$ is topologically transitive. (by 14, 20)
22. Let $q \in \pi_K^{-1}(\pi_K(B))$ so $\pi_K(q) = \pi_K(b)$ for some $b \in B$.
23. $d(y, q) \leq d(y, b) + d(b, q) = d(y, b) < \epsilon$. (by 3, 22)
24. $q \in B$. (by 2, 23)
25. $\pi_K^{-1}(\pi_K(B)) = B$. (by 22, 24)

Proposition 4.23 Let:

(i) (X, T) be a topologically transitive flow,
(ii) (X, T) be pointwise almost periodic, and
(iii) T be countable.

Then (X, T) is minimal.

PROOF: 1. Let V be open in X.
2. There exists a topologically transitive metrizable flow (Y, T), an open subset $B \subset V$, and a surjective homomorphism

$\pi : (X, T) \to (Y, T)$ with $\pi^{-1}(\pi(B)) = B$. (by **4.22** with $H = T$)

3. (Y, T) is pointwise almost periodic. (by 2, (ii))
4. (Y, T) is minimal. (by 2, 3, **4.19**)
5. $\pi(B)$ is open in Y. (by 2)
6. $X = \pi^{-1}(Y) = \pi^{-1}(\pi(B)T) = \pi^{-1}(\pi(B))T = BT \subset VT$. (by 2, 4, 5)
7. (X, T) is minimal. (by 1, 6)

Proposition 4.24 Let (X, T) be topologically transitive and distal. Then (X, T) is minimal.

PROOF: 1. Let V be an open subset of X.
2. There exists a countable subgroup $K \subset T$, a surjective homomorphism

$$\pi : (X, K) \to (Y, K), \text{ and an open subset } B \subset V,$$

such that

(a) (Y, K) is topologically transitive flow,
(b) Y is metrizable, and
(c) $\pi^{-1}(\pi(B)) = B$. (by **4.22** with $H = \{id\}$)

3. (X, K) is distal. (since (X, T) is distal)
4. (Y, K) is distal. (by 2, 3, **4.10**)
5. (Y, K) is minimal. (by 2ab, 4, **4.20**)
6. $X = \pi^{-1}(Y) = \pi^{-1}(\pi(B)K) = \pi^{-1}(\pi(B))K = BK \subset VT$. (by 2, 4, 5)
7. (X, T) is minimal. (by 1, 6)

Corollary 4.25 Let:

(i) (X, T) be a distal flow, and
(ii) (X, T) be weak mixing.

Then $X = \{pt\}$.

PROOF: 1. $(X \times X, T)$ is distal. (by (i), **4.13**)
2. $(X \times X, T)$ is minimal. (by 1, (ii), **4.24**)
3. $X = \{pt\}$. (by 2)

Corollary 4.26 Let:

(i) (X, T) be a distal flow,
(ii) L be an icer on X, and
(iii) (L, T) be topologically transitive.

Then $L = \Delta_X$.

PROOF: 1. $(X \times X, T)$ is distal. (by (i), **4.13**)
2. (L, T) is distal. (by 1)
3. (L, T) is minimal. (by 2, (iii), **4.24**)
4. $L = \Delta_X$. (by 3)

Note that **4.26** uses the fact that when L is an icer on a distal flow (X, T), the flow (L, T) is distal. Conversely if a flow (X, T) admits an icer L such that (L, T) is distal, then (X, T) is distal (because $(X, T) \cong (\Delta_X, T) \subset (L, T)$). Thus we may regard **4.26** as saying that the only icer L on a flow (X, T) for which (L, T) is distal and topologically transitive is $L = \Delta_X$. As remarked earlier, we will prove a natural generalization of this result: the only icer L on a minimal flow (X, T) for which (L, T) is pointwise almost periodic and topologically transitive is $L = \Delta_X$ (see **9.13**). It should be pointed out that the situation for pointwise almost periodic icers is different from that for distal icers. If L is an icer on X and (L, T) is pointwise almost periodic, then (X, T) is pointwise almost periodic (again because $\Delta_X \subset L$). On the other hand when (X, T) is pointwise almost periodic, an icer L on X need not be pointwise almost periodic. Indeed an icer L on X is pointwise almost periodic if and only if (X, T) is pointwise almost periodic and the extension $X \to X/L$ is a distal extension (see **4.15**). Here (L, T) is pointwise almost periodic but will not be distal unless (X, T) is distal.

Note that the analog of **4.25** holds for homomorphisms of minimal flows. Namely if $f: (X, T) \to (Y, T)$ is a distal homomorphism of minimal flows and (R_f, T) is topologically transitive, then f is trivial (that is $X = Y$). This follows immediately from **4.15**, and **9.13** mentioned above: the icer R_f is both topologically transitive and pointwise almost periodic, so $R_f = \Delta$ and hence $X = Y$.

EXERCISES FOR CHAPTER 4

Exercise 4.1 Give the details of the proof of **4.6** (a).

Exercise 4.2 (See **4.9**) Let (X, T) be a flow and $\Phi: \beta T \to E(X)$ be the canonical map. Then the following are equivalent:

(a) X is distal,
(b) e is the only idempotent in $E(X)$,
(c) $E(X)$ is a group,
(d) $E(X) = \Phi(M)$, and
(e) $E(X)$ is minimal.

Exercise 4.3 Let X be distal. Show that $E(X)$ is distal.

Exercise 4.4 Formulate and prove analogs of **4.2**, **4.3**, **4.4**, **4.8**, **4.9**, **4.11**, **4.12**, and **4.13** for semigroup actions.

Exercise 4.5 Let:

(i) T be a group and $S \subset T$ a subsemigroup,
(ii) $sS = Ss$ for all $s \in S$ (this holds in particular if S is abelian), and
(iii) (X, T) be a flow.

Prove that (X, S) is distal $\iff$ (X, S^{-1}) is distal $\iff$ $(X, S \cdot S^{-1})$ is distal.

Exercise 4.6 Let:

(i) $\varphi \colon (X, T) \to (Y, T)$ be a homomorphism of flows, and
(ii) $\psi \colon (X, T) \to (X, T)$ be an isomorphism of flows.

Show that φ is proximal if and only if $\varphi \circ \psi$ is proximal.

Exercise 4.7 (see **4.15**) Let $f \colon (X, T) \to (Y, T)$ be a homomorphism of flows, and assume that (R_f, T) is pointwise almost periodic. Show that f is distal.

Exercise 4.8 (see **4.21**) Let:

(i) $\pi \colon (X, T) \to (Y, T)$ be an epimorphism of flows, and
(ii) (X, T) be topologically transitive.

Show that (Y, T) is topologically transitive.

Exercise 4.9 Verify that the symbol space $(X, \mathbf{Z})$ of **4.17** is topologically transitive.

Exercise 4.10 Let:

(i) (X, T) be pointwise almost periodic,
(ii) $f \colon (X, T) \to (Y, T)$ be a distal epimorphism of flows,
(iii) (R_f, T) be topologically transitive, and
(iv) X be metrizable.

Show that $X = Y$.

5
Quasi-factors and the circle operator

For a minimal flow (X, T), the set 2^X of non-vacuous closed subsets of X is a compact Hausdorff space when provided with the Vietoris topology (see the following appendix). There is a natural action of T on this space, resulting in a flow $(2^X, T)$. The minimal subsets of this flow provide many interesting examples of minimal flows. These are the so-called *quasi-factors* of (X, T). Following the themes developed in the preceding sections quasi-factors and more generally the flow $(2^X, T)$ can be studied using the extension of the action of T on 2^X to an action of βT on 2^X. Some elementary properties of this extension, which we refer to as the *circle operator*, are outlined in this section. The circle operator will be used in section 12 to construct icers on M and in section 17 to study open and highly proximal extensions.

Notation, Definition, and Assumptions 5.1

(a) All the topological spaces that occur are assumed to be compact and Hausdorff.

(b) For any topological space X, 2^X will denote the collection of non-vacuous **closed** subsets of X.

(c) If $\emptyset \neq A = \overline{A} \subset X$, then $[A]$ will denote the element of 2^X determined by A. Thus the locution "let $[A] \in 2^X$" means that $\emptyset \neq A = \overline{A} \subset X$, and $[A]$ is the corresponding element of 2^X.

(d) Let $\pi\colon X \to Y$ be continuous. Then the map:

$$2^\pi : 2^X \to 2^Y \quad \text{is defined by} \quad 2^\pi([A]) = [\pi(A)].$$

If in addition π is onto, it induces a map:

$$\pi^* : 2^Y \to 2^X \quad \text{given by} \quad \pi^*([B]) = [\pi^{-1}(B)].$$

By "abuse of notation" the map 2^π will often be designated simply by π.

We will need an understanding of the topology on 2^X which is provided by the following proposition; we relegate its proof to the appendix to this section.

Proposition 5.2 Let:

(i) X be a compact Hausdorff space,
(ii) $< V_1, \dots, V_k > = \{[A] \in 2^X \mid A \subset \bigcup_{i=1}^k V_i$ and $A \cap V_i \neq \emptyset$ for $1 \leq i \leq k\}$ for any finite collection $\{V_1, \dots, V_k\}$ of subsets of X, and
(iii) $\mathcal{C} = \{< V_1, \dots, V_k > \mid k = 1, 2, \dots$ and V_i is an open subset of X for $1 \leq i \leq k\}$.

Then $\mathcal{C}$ is a base for a compact Hausdorff topology, $\mathcal{V}$, on 2^X called the *Vietoris topology* on 2^X.

Using **5.2**, we deduce some elementary properties of the Vietoris topology.

Proposition 5.3 Let:

(i) X be a compact Hausdorff space,
(ii) $[A_i] \to [A]$ in $(2^X, \mathcal{V})$,
(iii) $[B_i] \to [B]$ in $(2^X, \mathcal{V})$, and
(iv) $A_i \subset B_i$ for all i.

Then $A \subset B$.

PROOF: 1. Let $a \in A$ and $U_a \subset X$ be open with $a \in U_a$.
2. There exists an open set $U \subset X$ with $a \in U \subset \overline{U} \subset U_a$. (by 1, (i))
3. Assume that $B \cap \overline{U} = \emptyset$.
4. $X \setminus \overline{U}$ is open with $B \subset X \setminus \overline{U}$. (by 3)
5. There exists i with $[A_i] \in < U >$ and $[B_i] \in < X \setminus \overline{U} >$. (by 2, 4, (ii), (iii))
6. $\emptyset \neq A_i = U \cap A_i \subset U \cap B_i \subset U \cap (X \setminus \overline{U}) = \emptyset$ (a contradiction). (by 5, (iv))
7. $\emptyset \neq \overline{U} \cap B \subset U_a \cap B$. (since 3 $\Rightarrow$ 6 which is false)
8. $a \in \overline{B} = B$. (by 1, 7)

Proposition 5.4 Let:

(i) X, Y be compact Hausdorff spaces, and
(ii) $\pi \colon X \to Y$ be continuous.

Then $2^\pi : 2^X \to 2^Y$ is continuous.

PROOF: For any open subsets $V_1, \dots, V_k$ of X,

$$(2^\pi)^{-1}(< V_1 \dots, V_k >) =< \pi^{-1}(V_1), \dots, \pi^{-1}(V_k) >.$$

Corollary 5.5 Let:

(i) (X, T), (Y, T) be flows, and
(ii) $\pi : X \to Y$ be a homomorphism of flows.

Then:

(a) The maps $\begin{array}{ccc} 2^X & \to & 2^X \\ [A] & \to & [At] \end{array}$ define an action of T on 2^X, and

(b) 2^π is a homomorphism of flows.

PROOF: (a.) 1. Each of the maps $[A] \to [A]t \equiv [At]$ is continuous. (by **5.2**)

2. $[A](ts) = [A(ts)] = [(At)s] = [At]s = ([A]t)s$, for all $[A] \in 2^X$, and $s, t \in T$.

3. $[A]e = [Ae] = [A]$ for all $[A] \in 2^X$.

(b) 1. 2^π is continuous. (by **5.4**)

2. $2^\pi([A]t) = 2^\pi([At]) = [\pi(At)] \underset{\text{(ii)}}{=} [\pi(A)t] = [\pi(A)]t = 2^\pi([A])t$ for all $t \in T$.

Lemma 5.6 Let:

(i) X be a compact Hausdorff space, and

(ii) $\sigma : \begin{array}{ccc} X & \to & 2^X \\ x & \to & [\{x\}] \end{array}$.

Then σ is continuous.

PROOF: We leave the proof as an exercise for the reader.

We have seen that for a continuous map $\pi : X \to Y$ the map $2^\pi : 2^X \to 2^Y$ is continuous. This is not true in general for the map $\pi^* : 2^Y \to 2^X$ which is defined when π is onto; in fact π^* is continuous if and only if the map π is open. The proof of this result given below requires careful use of the basis for the Vietoris topology described in **5.2**.

Proposition 5.7 Let:

(i) X, Y be compact Hausdorff spaces,

(ii) $\pi : X \to Y$ be continuous and onto, and

(iii) $\varphi : Y \to 2^X$ be defined by $\varphi(y) = [\pi^{-1}(y)]$ for all $y \in Y$.

Then the following are equivalent:

(a) π^* is continuous.

(b) φ is continuous.

(c) π is open.

PROOF: (a) $\Rightarrow$ (b)

1. Assume that π^* is continuous.
2. Let $\sigma\colon Y \to 2^Y$ be defined by $\sigma(y) = [\{y\}]$ for all $y \in Y$.
3. σ is continuous. (by **5.6**)
4. $\varphi = \pi^* \circ \sigma$ is continuous. (by 1, 2, 3)

(b) $\Rightarrow$ (c)

1. Assume that φ is continuous.
2. Let $U \subset X$ be open, $x \in U$ and $y = \pi(x)$.
3. Case 1: Assume that $\pi^{-1}(y) \subset U$.
   3.1 $< U >$ is a neighborhood of $\varphi(y) = [\pi^{-1}(y)]$. (by 3)
   3.2. There exists an open neighborhood V of y with $\varphi(V) \subset < U >$. (by 1, 3.1)
   3.3. $\varphi(z) = [\pi^{-1}(z)] \in < U >$ for all $z \in V$. (by 3.2)
   3.4. $\pi^{-1}(V) \subset U$. (by 3.3)
   3.5 $y \in V \subset \pi(U)$. (by 3.2, 3.4)
4. Case 2: Assume that $\pi^{-1}(y) \not\subset U$.
   4.1. $< U, X \setminus \{x\} >$ is an open neighborhood of $\varphi(y) = [\pi^{-1}(y)]$. (by 2, 4)
   4.2. There exists an open neighborhood V of y with $\varphi(V) \subset < U, X \setminus \{x\} >$. (by 1, 4.1)
   4.3. $\varphi(z) = [\pi^{-1}(z)] \in < U, X \setminus \{x\} >$ for all $z \in V$. (by 4.2)
   4.4. $\emptyset \neq \pi^{-1}(z) \cap U$ for all $z \in V$. (by 4.3)
   4.5. $y \in V \subset \pi(U)$. (by 3.2, 4.4)
5. π is open. (by 2, 4, 5)

(c) $\Rightarrow$ (a)

1. Assume that π is open.
2. Let $[B] \in 2^Y$ with $\pi^*([B]) = [\pi^{-1}(B)] \in < U_1, \ldots, U_k >$.
3. $\pi^{-1}(B) \subset \bigcup U_i \equiv U$. (by 2)
4. There exists an open set $W \subset Y$ with $B \subset W$ and $\pi^{-1}(\overline{W}) \subset U$. (by 1, 3)
5. Let $V_i = \pi(U_i) \cap W$ for $1 \leq i \leq k$.
6. V_i is open for $1 \leq i \leq k$. (by 1, 4)
7. $B \underset{\text{(ii)}}{=} \pi(\pi^{-1}(B)) \underset{3,4}{\subset} W \cap \pi(U) = W \cap \bigcup \pi(U_i) = \bigcup (W \cap \pi(U_i)) \underset{5}{=} \bigcup V_i$.
8. $B \cap V_i = B \cap \pi(U_i) \cap W$
   $\underset{4}{=} \pi(\pi^{-1}(B)) \cap \pi(U_i)$
   $\supset \pi(\pi^{-1}(B) \cap U_i) \underset{2}{\neq} \emptyset$ for $1 \leq i \leq k$.
9. $< V_1, \ldots, V_k >$ is an open neighborhood of $[B]$. (by 6, 7, 8)

10. Let $[C] \in < V_1, \ldots, V_k >$.
11. $\pi^{-1}(C) \underset{10}{\subset} \pi^{-1}\left(\bigcup V_i\right) = \bigcup \pi^{-1}(V_i) \underset{5}{\subset} \pi^{-1}(W) \underset{4}{\subset} U \underset{3}{=} \bigcup U_i$.
12. $\emptyset \neq C \cap V_i \subset C \cap \pi(U_i)$ for all $1 \leq i \leq k$. (by 5, 10)
13. $\pi^{-1}(C) \cap U_i \neq \emptyset$ for all $1 \leq i \leq k$. (by 12)
14. $\pi^*([C]) \in < U_1, \ldots, U_k >$. (by 11, 13)
15. $[B] \in < V_1, \ldots, V_k > \subset (\pi^*)^{-1}(< U_1, \ldots, U_k >)$. (by 9, 14)

Even though it follows from **5.7** that the map φ is not continuous in general, the following result will be useful.

Lemma 5.8 Let:

(i) X and Y be compact Hausdorff,
(ii) $\pi : X \to Y$ be continuous and onto,
(iii) $U \subset X$ be an open set, and
(iv) $V = \{y \mid \pi^{-1}(y) \subset U\}$.

Then V is an open subset of Y.

PROOF: 1. $Y \setminus V = \pi(X \setminus U) = \overline{\pi(X \setminus U)}$. (by (i), (ii), (iii), (iv))
2. V is open. (by 1)

Let (X, T) be a flow. Then according to **5.5**, $(2^X, T)$ is also a flow. Thus the action of T on 2^X may be extended to an action of βT on 2^X. We would like, given a closed non-empty subset $A \subset X$, which represents an element $[A] \in 2^X$, to describe explicitly the closed subset of X which represents the element $[A]p \in 2^X$ where $p \in \beta T$. For this we need the so-called *circle operator* defined below.

Definition 5.9 Let (X, T) be a flow, $\emptyset \neq A = \overline{A} \subset X$, and $p \in \beta T$. We define the *circle operation* of βT on X by

$$A \circ p = \{x \in X \mid \text{there exist nets } (a_i) \subset A, \text{ and } (t_i) \subset T \text{ with } t_i \to p \text{ and } a_i t_i \to x\}.$$

The following proposition shows that the circle operator does indeed characterize the action of βT on 2^X. This proposition and its corollary also detail some of the properties of the circle operator that will be used later.

Proposition 5.10 Let:

(i) (X, T) be a flow,
(ii) $\emptyset \neq A \subset X$, and
(iii) $t \in T$ and $p, q \in \beta T$.

Then:

(a) $A \circ t = \overline{A}t$,
(b) $A \circ p = \bigcap\{\overline{A(N \cap T)} \mid N$ is a neighborhood of p in $\beta T\}$,
(c) $Ap \subset A \circ p = \overline{A \circ p}$,
(d) $\overline{A} \circ p = A \circ p$,
(e) $[A \circ p] = [\overline{A}]p$, and
(f) $(A \circ p) \circ q = A \circ (pq)$.

PROOF: (a), (b) We leave these to the reader.

(c) 1. $Ap = \{ap \mid a \in A\} \subset A \circ p$. (this is clear from the definition)
2. Let $x \in \overline{A \circ p}$, and $U \subset X$ be open with $x \in U$.
3. $U \cap (A \circ p) \neq \emptyset$. (by 2)
4. Let $N \subset \beta T$ be open with $p \in N$.
5. There exist $a(U, N) \in A$ and $t(U, N) \in N \cap T$ with $a(U, N)t(U, N) \in U$. (by 3, **5.9**)
6. There exist $\{a(U, N)\} \subset A$, $\{t(U, N)\} \subset T$ with $t(U, N) \to p$ and $a(U, N)t(U, N) \to x$. (by 2, 4, 5)
7. $x \in A \circ p$. (by 6)

(d) We leave this to the reader.

(e) 1. Let $V_1, \ldots, V_k$ be open subsets of X with $[\overline{A}]p \in < V_1, \ldots, V_k >$.
2. There exist open sets $W_1, \ldots, W_k$ with $W_i \subset \overline{W_i} \subset V_i$ and $[\overline{A}]p \in < W_1, \ldots, W_k >$.
3. $\lim_{t \to p}[\overline{At}] = \lim_{t \to p}[\overline{A}]t = [\overline{A}]p$.
4. There exists $N \subset \beta T$ open with $p \in N$ such that $[\overline{At}] \in < W_1, \ldots, W_k >$ for all $t \in N \cap T$. (by 3)
5. $\overline{A} \circ p \subset \overline{\bigcup W_i} = \bigcup \overline{W}_i \subset \bigcup V_i$. (by 4)
6. $(\overline{A} \circ p) \cap V_i \supset (\overline{A} \circ p) \cap \overline{W}_i \neq \emptyset$. (by 4)
7. $[\overline{A} \circ p] \in < V_1, \ldots, V_k >$. (by 5, 6)
8. $[A \circ p] \underset{(d)}{=} [\overline{A} \circ p] \underset{(1,7)}{=} [\overline{A}]p$.

(f) $[(A \circ p) \circ q] \underset{(e)}{=} [A \circ p]q = ([\overline{A}]p)q = [\overline{A}](pq) = [A \circ (pq)]$.

Corollary 5.11 Let:

(i) (X, T), (Y, T) be flows, and
(ii) $\pi : X \to Y$ be a homomorphism of flows.

Then:

(a) $\pi(A \circ p) = \pi(A) \circ p$ for all $\emptyset \neq A \subset X$ and all $p \in \beta T$, and
(b) $\pi^{-1}(y) \circ p \subset \pi^{-1}(yp)$ for all $y \in Y$ and $p \in \beta T$.

PROOF: 1. $\pi(A \circ p) = \pi\left(\bigcap\{\overline{A(N \cap T)} \mid N \text{ is a neighborhood of } p \text{ in } \beta T\}\right)$
$\subset \bigcap\{\overline{\pi(A)(N \cap T)} \mid N \text{ is a neighborhood of } p \text{ in } \beta T\}$
$= \pi(A) \circ p.$ (by **5.10**)

2. Let $y \in \pi(A) \circ p = \bigcap\{\pi\left(\overline{A(N \cap T)}\right) \mid N \text{ is a neighborhood of } p \text{ in } \beta T\}$.
3. $\pi^{-1}(y) \cap \overline{A(N \cap T)} \neq \emptyset$ for all neighborhoods N of p. (by 2)
4. $\pi^{-1}(y) \cap \bigcap\{\overline{A(N \cap T)} \mid N \text{ is a neighborhood of } p \text{ in } \beta T\} \neq \emptyset$.
(by 3, compactness)
5. $y \in \pi\left(\bigcap\{\overline{A(N \cap T)} \mid N \text{ is a neighborhood of } p \text{ in } \beta T\}\right) = \pi(A \circ p)$.
(by **5.10**)

(b) 1. $\pi(\pi^{-1}(y) \circ p) = \pi(\pi^{-1}(y)) \circ p = \{y\} \circ p = \{yp\}$. (by part (a))
2. $\pi^{-1}(y) \circ p \subset \pi^{-1}(yp)$. (by 1)

APPENDIX TO SECTION 5: THE VIETORIS TOPOLOGY ON 2^X

In this appendix we show that when X is a compact Hausdorff space, the space 2^X provided with the Vietoris topology is also a compact Hausdorff space.

Notation 5.A.1 Let X be a set provided with a compact Hausdorff topology $\mathcal{T}$. We set

$$\mathcal{C} = \{< V_i, \ldots, V_k > \mid V_i \in \mathcal{T} \text{ for } 1 \leq i \leq k\}.$$

Recall that

$$< V_i, \ldots, V_k >= \{[A] \in 2^X \mid A \subset \bigcup_{i=1}^k V_i \text{ and } A \cap V_i \neq \emptyset \text{ for all } 1 \leq i \leq k\}.$$

Lemma 5.A.2 Let X be a compact Hausdorff space.
Then $\mathcal{C}$ is a base for a Hausdorff topology $\mathcal{V}$ on 2^X.

PROOF: 1. Let $< V_1, \ldots, V_k >, < W_1, \ldots, W_n > \in \mathcal{C}$ with

$$[A] \in < V_1, \ldots, V_k > \cap < W_1, \ldots, W_n > .$$

2. Set $V = \bigcup_{i=1}^k V_i$ and $W = \bigcup_{j=1}^n W_j$.
3. Set $\hat{V}_i = V_i \cap W$ and $\hat{W}_j = V \cap W_j$ for $1 \leq i \leq k$ and $1 \leq j \leq n$.
4. $[A] \in < \hat{V}_1, \ldots, \hat{V}_k, \hat{W}_1, \ldots, \hat{W}_n > \subset < V_1, \ldots, V_k > \cap < W_1, \ldots, W_n >$.
(by 1, 2, 3)
5. Let $[A], [B] \in 2^X$ with $[A] \neq [B]$.
6. We may assume without loss of generality that there exists $a \in A \setminus B$.
(by 5)

7. Let $U, V \subset X$ be open with $a \in U$, $B \subset V$ and $V \cap U = \emptyset$.
8. $[A] \in \,< U, X \setminus \{a\} >$ or $[A] = [\{a\}] \in\, < U >$, and $[B] \in \,< V >$. (by 7)
9. Let $C \subset X$ with $C \subset V$.
10. $C \cap U = \emptyset$. (by 7)
11. $< U, X \setminus \{a\} > \cap < V > = < U > \cap < V > = \emptyset$. (by 9, 10)
12. $(2^X, \mathcal{V})$ is Hausdorff. (by 5, 8, 11)

Proposition 5.A.3 Let X be a compact Hausdorff space. Then 2^X provided with the Vietoris topology $\mathcal{V}$, is compact.

PROOF: 1. Let $\mathcal{F}$ be a collection of closed subsets of 2^X which has the finite intersection property, and let $\mathcal{U}$ be an ultrafilter containing $\mathcal{F}$. (see **1.A.4**)
2. For every collection Φ, of subsets of X, (e.g. every $\Phi \in \mathcal{U}$) set $\hat{\Phi} = \overline{\bigcup\{A \mid [A] \in \Phi\}}$.
3. $\Phi_1 \hat{\cap} \Phi_2 \subset \hat{\Phi}_1 \cap \hat{\Phi}_2$ for any $\Phi_1, \Phi_2 \in \mathcal{U}$. (by 2)
4. $\{\hat{\Phi} \mid \Phi \in \mathcal{U}\}$ is a collection of closed subsets of X which has the finite intersection property. (by 3)
5. Set $A = \bigcap_{\Phi \in \mathcal{U}} \hat{\Phi}$.
6. A is closed and non-empty. (by 4, 5, X is compact)
7. Let $[A] \in \,< V_1, \ldots, V_k > \,\in \mathcal{C}$, and $\Psi_0 \in \mathcal{F}$.
8. $X \setminus (\bigcup_{i=1}^{k} V_i) \subset X \setminus A = \bigcup_{\Phi \in \mathcal{U}} (X \setminus \hat{\Phi})$. (by 5, 7)
9. There exist $\Phi_1, \ldots, \Phi_n \in \mathcal{F}$ with $X \setminus (\bigcup V_i) \subset (X \setminus \hat{\Phi}_1) \cup \cdots \cup (X \setminus \hat{\Phi}_n)$. (by 8, $X \setminus (\bigcup V_i)$ is compact)
10. Set $\Psi = \Psi_0 \cap \Phi_1 \cap \cdots \cap \Phi_n$.
11. $\Psi \in \mathcal{F}$ with $\Psi \subset \Psi_0$ and $\hat{\Psi} \subset \bigcup_{i=1}^{k} V_i$. (by 1, 7, 9, 10)
12. Set $\Psi_i = \{[B] \in \Psi \mid B \cap V_i = \emptyset\}$ for $1 \leq i \leq k$.
13. Assume that $\Psi_1 \cup \cdots \cup \Psi_k = \Psi$.
14. $\Psi_1 \cup \cdots \cup \Psi_k \in \mathcal{F} \subset \mathcal{U}$. (by 11, 13)
15. There exists j with $\Psi_j \in \mathcal{U}$. (by 1, **1.A.18**)
16. $A \subset \hat{\Psi}_j \subset \overline{X \setminus V_j} = X \setminus V_j$. (by 5, 12, 15)
17. $\Psi_1 \cup \cdots \cup \Psi_k \subsetneq \Psi$. (by 12, 13, since 16 contradicts 7)
18. There exists $[B] \in \Psi$ such that $B \cap V_i \neq \emptyset$ for all $1 \leq i \leq k$. (by 12, 17)
19. $\emptyset \neq \Psi \cap < V_1, \ldots, V_k > \subset \Psi_0 \cap < V_1, \ldots, V_k >$. (by 11, 18)
20. $[A] \in \overline{\Psi_0} = \Psi_0$. (by 1, 7, 19)
21. $[A] \in \bigcap \mathcal{F}$. (by 7, 20)

EXERCISES FOR CHAPTER 5

Exercise 5.1 Let X be a set provided with a compact Hausdorff topology $\mathcal{T}$. Let

$$\mathcal{U} = \{\alpha \subset X \times X \mid \alpha \text{ is a neighborhood of } \Delta\}$$

be the unique uniformity on X which induces $\mathcal{T}$. We refer to an element $\alpha \in \mathcal{U}$ as an *index on* X and write

$$x\alpha = \{y \mid (x, y) \in \alpha\} \quad \text{and} \quad A\alpha = \bigcup_{x \in A} x\alpha$$

for any $x \in X$ and $A \subset X$. Given any index $\alpha \in \mathcal{U}$ we set

$$\hat{\alpha} = \{([A], [B]) \mid B \subset A\alpha \text{ and } A \subset B\alpha\} \subset 2^X \times 2^X.$$

We then define

$$\hat{\mathcal{U}} = \{\hat{\alpha} \mid \alpha \in \mathcal{U}\}.$$

Then

(a) $\hat{\mathcal{U}}$ is a base for a uniformity on 2^X.
(b) For any $[A] \in 2^X$ and $\alpha \in \mathcal{U}$, there exists $W \in \mathcal{C}$ with $[A] \in W \subset [A]\hat{\alpha}$.
(c) For any $[A] \in < V_1, \ldots, V_k > \in \mathcal{C}$, there exists $\alpha \in \mathcal{U}$ such that $[A]\hat{\alpha} \subset < V_i, \ldots, V_k >$.
(d) The uniformity $\hat{\mathcal{U}}$ induces the topology $\mathcal{V}$.

Exercise 5.2 (See **5.6**) Let:

(i) X be a compact Hausdorff space, and
(ii) $\sigma\colon X \to 2^X$.
   $x \to [\{x\}]$

Show that σ is continuous.

Exercise 5.3 Let $\pi : X \to Y$ be continuous, open and onto. Then

(a) $2^\pi(< V_1, \ldots, V_k >) =< \pi(V_1), \ldots, \pi(V_k) >$.
(b) $2^\pi : 2^X \to 2^Y$ is open.

Exercise 5.4 (See **5.10**) Let:

(i) X be compact metric space with metric d: $X \times X \to \mathbf{R}$, and
(ii) $h\colon 2^X \times 2^X \to \mathbf{R}$ be defined by

$$h([A], [B]) = \max\big\{\sup\{d(x, B) \mid x \in A\}, \sup\{d(x, A) \mid x \in B\}\big\}.$$

Show that h is a metric on 2^X which induces the Veitoris topology.

Exercise 5.5 Let:

(i) (X, T) be a flow,
(ii) $\emptyset \neq A \subset X$, and
(iii) $p \in \beta T$.

Show that $A \circ p = \bigcap \{\overline{A(N \cap T)} \mid N \text{ is a neighborhood of } p \text{ in } \beta T\}$.

Exercise 5.6 Let:

(i) $\pi : (X, T) \to (Y, T)$ be a homomorphism, and
(ii) (Y, T) be minimal.

Then the following are equivalent:

(a) π is open,
(b) $\pi^{-1}(B \circ p) = \pi^{-1}(B) \circ p$ for all $[B] \in 2^Y$ and $p \in M$,
(c) $\pi^{-1}(yp) = \pi^{-1}(y) \circ p$ for all $y \in Y$ and $p \in M$, and
(d) $\pi^{-1}(y_0 p) = \pi^{-1}(y_0) \circ p$ for some $y_0 \in Y$ and all $p \in M$.

Notice that (i) and (ii) imply that π is onto.

Exercise 5.7 Let:

(i) $R_1 \subset R_2$ be icers on M,
(ii) $X = M/R_1$, $Y = M/R_2$, and
(iii) $\hat{Y}_y = \{[(\pi_{R_2}^{R_1})^{-1}(yp)] \mid p \in M\} \subset 2^X$ for all $y \in Y$.

Then

(a) $\hat{Y}_y = \hat{Y}_z$ for all $y, z \in Y$.
(b) $\hat{Y}_y$ is minimal.
(c) $M/(R_1 \cap S) \to M/(R_2 \cap S)$ is open where $\hat{Y}_y = M/S$.
(d) if N is an icer on M with $M/(R_1 \cap N) \to M/(R_2 \cap N)$ open, then $N \subset S$.

PART II

Equivalence relations and automorphisms of flows

According to **3.16**, every minimal flow is a quotient $(M/R, T)$ of the universal minimal flow (M, T) by some icer R on M. We now proceed to investigate the icers on M using the structure of M and its group of automorphisms. Applying **3.12**, **3.16**, and **3.15**, to M (a fixed minimal ideal in $\beta T \cong E(\beta T, T)$), we conclude that M can be written as a disjoint union of copies of a subgroup $G \subset M$, one for each of the idempotents $u^2 = u \in M$. We show (see **7.4**) that this group G can be identified with the group of automorphisms of the flow (M, T). The icers on M can then be characterized in terms of the idempotents in M and the subgroups of G. This characterization relies on the so-called relative product of two relations which is discussed in section 6. Given two relations R and S on a space X, the relative product $R \circ S$ is a relation on X which contains $R \cup S$. Theorem **7.21** shows that every icer R on M can be written uniquely as a relative product

$$R = (R \cap P_0) \circ gr(G(R)). \qquad (*)$$

Here $P_0 \subset P(M)$ is a an equivalence relation generated by the pairs of idempotents in M, and the subgroup

$$G(R) = \{\alpha \in G \mid gr(\alpha) \subset R\} \subset G$$

consists of the automorphisms of M whose graphs are contained in R. The expression $(*)$ of every icer on M as a relative product plays a fundamental role in our study of minimal flows via the icers on M. Section 7 develops the language and techniques for proving this result and understanding its consequences.

Section 8 is concerned with the so-called regular flows; these are the flows which can be uniquely expressed in the form M/R. We show that the regular flows are those which have as many automorphisms as possible. For a regular flow the pair $\{X, Aut(X)\}$ has properties analogous to those of the pair $\{M, G\}$.

In section 9 we introduce what we call the quasi-relative product $R(S)$ of two icers. $R(S)$ though it need not be closed is always an invariant equivalence relation. When $R \circ S$ is an icer, $R(S) = R \circ S$, but $R(S)$ is an icer in situations when $R \circ S$ is not. We show that when $R(S)$ is an icer the flow $M/R(S)$ is a quasi-factor of X/S (see **9.9**). $R(S)$ arose originally as we searched for a new proof of the Furstenberg structure theorem for distal extensions of compact Hausdorff (not necessarily metric) spaces. The resulting proof (see **9.13**) of a result which is equivalent to the Furstenberg theorem, requires a construction of metric flows via icers in a situation where the relative product alone does not suffice.

6

Quotient spaces and relative products

Any continuous mapping $f\colon X \to Y$ of compact Hausdorff spaces which is onto is a quotient map, and the equivalence relation $R_f = \{(x_1, x_2) \mid f(x_1) = f(x_2)\}$ is closed. Conversely any closed equivalence relation R on X gives rise to a compact Hausdorff space X/R and continuous quotient map $X \to X/R$. Thus continuous mappings can be defined by constructing closed equivalence relations; this motivates the study of the so-called *relative product* of two relations.

In the interest of completeness we include a proof (see also [Massey, (1977)]) of the fact that if R is a closed equivalence relation on a compact Hausdorff space X, then X/R is Hausdorff. We then introduce and investigate some basic properties of the relative product of two relations on a compact Hausdorff space. The techniques developed here are used extensively throughout the remainder of the text; for example the relative product is used in section 7 to construct and characterize the icers on the universal minimal flow M. Another example is seen in **6.18** where we obtain a generalization of the fact that any factor of a distal flow is itself a distal flow; a refinement of this result will be used in section 9.

The following notation will be used to facilitate our discussion of equivalence relations, quotient spaces, and quotient mappings.

Definition and Notation 6.1 Let $R \subset S$ be closed equivalence relations on a compact Hausdorff space X. We define an *R-cell* to be a set of the form:

$$xR = \{y \in X \mid (x, y) \in R\} \subset X.$$

If $A \subset X$ is any subset of X we write

$$AR = \bigcup_{x \in A} xR = \{y \in X \mid (x, y) \in R \text{ for some } x \in A\}$$

for the union of the R-cells corresponding to points of A. We use the notation

$$X/R = \{xR \mid x \in X\}$$

for the quotient space (provided with the quotient topology). The notation π_R and π_S^R will be used for the quotient map, and canonical projection map respectively:

$$\begin{array}{rclcrcl} \pi_R\colon X & \to & X/R & \text{and} & \pi_S^R\colon X/R & \to & X/S \\ x & \to & xR & & xR & \to & xS. \end{array}$$

Thus the composition $\pi_S^R \circ \pi_R = \pi_S$. By "abuse of notation" we will also write:

$$\begin{array}{rcl} \pi_R \equiv \pi_R \times \pi_R\colon X \times X & \to & X/R \times X/R. \\ (x, y) & \to & (xR, yR) \end{array}$$

Proposition 6.2 Let:

(i) X be a compact Hausdorff space, and

(ii) R be a closed equivalence relations on X.

Then X/R is Hausdorff.

PROOF: 1. Let $x, y \in X$.

2. $U \subset X$ be open with $xR \subset U$.

3. $xR \cap (X \setminus U) = \emptyset$. (by 2)

4. $x \notin (X \setminus U)R$ which is closed. (by 3, (ii), X is compact)

5. $x \in F \equiv X \setminus \big((X \setminus U)R\big) = \pi_R^{-1}(\pi_R(X \setminus (X \setminus U)R)) = \pi_R^{-1}(\pi_R(F)) = FR$ which is open. (by 4)

6. $xR \subset \overline{F} \subset \overline{U}$. (by 5)

7. $xR = \bigcap\{\overline{F} \mid x \in F \text{ open with } FR = F\}$. (by 2, 6, (i), (ii))

8. $yR = \bigcap\{\overline{H} \mid y \in H \text{ open with } HR = H\}$. (by 2-7 applied to y)

9. Assume every open neighborhood of $\pi_R(x)$ intersects every open neighborhood of $\pi_R(y)$.

10. Let F, H be open sets with $x \in F$, $y \in H$, $FR = F$, and $HR = H$.

11. $\pi_R(F)$ and $\pi_R(H)$ are open neighborhoods of $\pi_R(x)$ and $\pi_R(y)$ respectively. (by 10)

12. $\emptyset \neq \pi_R^{-1}(\pi_R(F) \cap \pi_R(H)) \subset \pi_R^{-1}(\pi_R(F)) \cap \pi_R^{-1}(\pi_R(H)) = F \cap H$. (by 10, 11)

13. $\{\overline{H} \cap \overline{F} \mid x \in F \text{ open with } FR = F\}$ has the finite intersection property. (by 12)

14. $\overline{H} \cap xR \neq \emptyset$. (by 7, 13, compactness)

15. $\{\overline{H} \cap xR \mid y \in H \text{ open with } HR = H\}$ has the finite intersection property. (by 10, 14)
16. $yR \cap xR \neq \emptyset$. (by 8, 15, compactness)
17. $xR = yR$ and $\pi_R(x) = \pi_R(y)$. (by 16)
18. X/R is Hausdorff. (by 9, 17)

Applying **6.2** allows us to study continuous mappings via closed equivalence relations. In particular if S is another closed equivalence relation on X, one is often interested in spaces Y that fit into a commutative diagram of the form.

$$\begin{array}{ccc} X & \rightarrow & X/S \\ \downarrow & & \downarrow \\ X/R & \rightarrow & Y. \end{array} \qquad (*)$$

Any such Y is a quotient of X, so finding such a Y amounts to finding a closed equivalence relation on X which contains both R and S. This motivates the following definition.

Definition 6.3 Let R, S be any relations on X (a compact Hausdorff space). We define the *relative product*, $R \circ S$, *of* R *and* S by

$$R \circ S = \{(p, q) \in X \times X \mid \text{there exists } r \in X \text{ with } (p, r) \in R \text{ and } (r, q) \in S\}.$$

Using the notation introduced above the cells of the relative product $R \circ S$ of two relations can be written

$$x(R \circ S) = (xR)S;$$

in other words

$$(x, z) \in R \circ S \iff \text{there exists } y \in xR \text{ with } z \in yS.$$

In this sense the relative product is indeed a product.

Our investigation of the properties of the relative product begins with a few preliminary results; many of the proofs are direct consequences of the definitions and are left to the reader. Though we will be primarily interested in the case of closed equivalence relations where the quotient spaces are Hausdorff, often these results hold in somewhat greater generality.

Lemma 6.4 Let:

(i) X be a compact Hausdorff space,
(ii) R, S be closed equivalence relations on X,
(iii) $A \subset X \times X$, and
(iv) $B, C \subset X/R \times X/R$.

Then:

(a) $(\pi_R \times \pi_S)^{-1}((\pi_R \times \pi_S)(A)) = R \circ A \circ S$,
(b) $\pi_R^{-1}(B^{-1}) = \left(\pi_R^{-1}(B)\right)^{-1}$, and
(c) $\pi_R^{-1}(B \circ C) = \pi_R^{-1}(B) \circ \pi_R^{-1}(C)$.

PROOF: We leave this as an exercise for the reader.

Corollary 6.5 Let:

(i) X be a compact Hausdorff space,
(ii) N be a closed equivalence relation on X, and
(iii) $R \subset X/N \times X/N$ be a closed equivalence relation on X/N.

Then $\pi_N^{-1}(R)$ is a closed equivalence relation on X.

PROOF: 1. $\Delta_X \subset N = \pi_N^{-1}(\Delta_{X/N}) \subset \pi_N^{-1}(R)$. (by (ii), (iii))
2. $(\pi_N^{-1}(R))^{-1} = \pi_N^{-1}(R^{-1}) = \pi_N^{-1}(R)$. (by (iii), and **6.4**)
3. $\pi_N^{-1}(R) \circ \pi_N^{-1}(R) = \pi_N^{-1}(R \circ R) = \pi_N^{-1}(R)$. (by (iii), and **6.4**)

Lemma 6.6 Let:

(i) X be a compact Hausdorff space,
(ii) N be a closed equivalence relation on X, and
(iii) $R, S \subset X \times X$ be arbitrary relations on X.

Then:

(a) $\pi_N^{-1}(\pi_N(R)) = N \circ R \circ N$,
(b) $\pi_N(R \circ S) \subset \pi_N(R) \circ \pi_N(S) \subset \pi_N(R \circ N \circ S)$, and
(c) if either $R \circ N$ or $S \circ N$ are closed equivalence relations on X, then

$$\pi_N(R \circ S) = \pi_N(R) \circ \pi_N(S).$$

PROOF: We leave this as an exercise for the reader.

Note that if R and S are closed, then the relative product $R \circ S$ is also closed. Similarly if (X, T) is a flow, and R and S are invariant, then $R \circ S$ is invariant. On the other hand the relative product of two equivalence relations is reflexive, but need not be symmetric or transitive. However in this case if $R \circ S$ is symmetric, then $R \circ S = S \circ R$ and it is also transitive (see **6.9**). It is clear that any equivalence relation which contains both R and S must also contain $R \circ S$. Consequently if $R \circ S$ is an equivalence relation, then it must be the smallest one which contains $R \cup S$. This says that $Y = X/(R \circ S)$ fits into the diagram in $(*)$, and is maximal in the sense that any other Y is a quotient of $X/(R \circ S)$. In this maximal example the diagram in $(*)$ has particularly nice properties which are used throughout the text. Moreover we will see that the relative product can be used to understand these diagrams in general by describing the equivalence relations involved in their construction. We proceed to make these notions precise.

Proposition 6.7 Let:

(i) R_1, R_2 be closed equivalence relations on X, and
(ii) $N = \bigcap\{S \mid S \text{ is a closed equivalence relation with } R_1 \cup R_2 \subset S\}$.

Then:

(a) N is a closed equivalence relation on X, and
(b) if S is a closed equivalence relation on X with $R_1 \cup R_2 \subset S$, then $N \subset S$.

PROOF: This is clear.

Definition 6.8 Let R, S be closed equivalence relations on X. We define the *infimum of R and S*, $\inf(R, S)$ by

$$\inf(R, S) = \bigcap\{N \mid N \text{ is a closed equivalence relation on } X \text{ with } R \cup S \subset N\}.$$

According to proposition **6.7**, $\inf(R, S)$ is a closed equivalence relation on X; we proceed to show that $R \circ S$ is an equivalence relation if and only if $R \circ S = \inf(R, S)$.

Proposition 6.9 Let R, S be closed equivalence relations on X.
Then the following are equivalent:

(a) $S \circ R = R \circ S$,
(b) $R \circ S$ is a closed equivalence relation on X, and
(c) $R \circ S = \inf(R, S)$.

PROOF: (a) $\Rightarrow$ (b)

1. Assume that $R \circ S = S \circ R$.
2. As has been previously remarked $R \circ S$ is closed and reflexive.
3. $(R \circ S)^{-1} = S^{-1} \circ R^{-1} = S \circ R = R \circ S$ so $R \circ S$ is symmetric.
4. $(R \circ S) \circ (R \circ S) = R \circ (R \circ S) \circ S = R \circ S$ so $R \circ S$ is transitive.

(b) $\Rightarrow$ (c)

1. Assume that $R \circ S$ is a closed equivalence relation on X.
2. $R \cup S \subset R \circ S$. (by **6.3**)
3. $\inf(R, S) \subset R \circ S$. (by **6.7**)

4. $R \circ S \subset \inf(R, S) \circ \inf(R, S) = \inf(R, S)$. (by **6.7**)
5. $R \circ S = \inf(R, S)$. (by 3, 4)

(c) $\Rightarrow$ (a)

1. Assume that $R \circ S = \inf(R, S)$.
2. $R \circ S$ is a closed equivalence relation on X. (by 1)
3. $S \circ R = S^{-1} \circ R^{-1} = (R \circ S)^{-1} = R \circ S$. (by 2)

There are many situations where the projection $\pi_R(S)$ is an equivalence relation on X/R even though S is not an equivalence relation on X. For flows (X, T), one such example is when $S = P(X)$ is the proximal relation introduced earlier; we investigate this further in section 13. Another example is the regionally proximal relation introduced in section 15 and studied further in sections 16 and 19. Though the main application of the following proposition is the corollary which follows, where R, S and $R \circ S$ are closed equivalence relations on X, it does give a condition under which the projection of a reflexive symmetric relation is an equivalence relation.

Proposition 6.10 Let:

(i) X be a compact Hausdorff space,
(ii) R be a closed equivalence relation on X, and
(iii) S be a closed reflexive symmetric relation on X.

Then the following are equivalent:

(a) $R \circ S \circ R$ is a closed equivalence relation on X,
(b) $R \circ S \circ R \circ S = R \circ S \circ R$, and
(c) $\pi_R(S)$ is a closed equivalence relation on X/R.

Moreover all of these conditions imply:

(d) $(X/R)/\pi_R(S) \cong X/(R \circ S \circ R)$.

PROOF: (a) $\Rightarrow$ (b)

1. Assume that $R \circ S \circ R$ is a closed equivalence relation on X.
2. $R \circ S \circ R \subset R \circ S \circ R \circ S \subset (R \circ S \circ R) \circ (R \circ S \circ R) \subset R \circ S \circ R$. (by 1, (ii), (iii))

(b) $\Rightarrow$ (c)

1. Assume that $R \circ S \circ R \circ S = R \circ S \circ R$.
2. It suffices to show that $\pi_R(S)$ is transitive. (by (iii))

$$3.\ \pi_R^{-1}(\pi_R(S) \circ \pi_R(S)) \underset{\mathbf{6.4}}{=} \pi_R^{-1}(\pi_R(S)) \circ \pi_R^{-1}(\pi_R(S))$$
$$\underset{\mathbf{6.6}}{=} R \circ S \circ R \circ R \circ S \circ R$$
$$\underset{\text{(ii)}}{=} R \circ S \circ R \circ S \circ R \underset{1}{=} R \circ S \circ R \circ R$$
$$\underset{\text{(ii)}}{=} R \circ S \circ R = \pi_R^{-1}(\pi_R(S)).$$

4. $\pi_R(S) \circ \pi_R(S) = \pi_R(S)$. (by 3)

(c) $\Rightarrow$ (a)

1. Assume that $\pi_R(S)$ is a closed equivalence relation on X/R.
2. $\pi_R^{-1}(\pi_R(S))$ is a closed equivalence relation on X. (by 1, and **6.5**)
3. $\pi_R^{-1}(\pi_R(S)) = R \circ S \circ R$. (by **6.6**)

(c) $\Rightarrow$ (d)

$$(X/R)/\pi_R(S) \cong X/\pi_R^{-1}(\pi_R(S)) \cong X/(R \circ S \circ R).$$

Corollary 6.11 Let R, S be closed equivalence relations on X such that $R \circ S$ is a closed equivalence relation on X. Then:

(a) $\pi_R(S)$ is a closed equivalence relation on X/R,
(b) $\pi_S(R)$ is a closed equivalence relation on X/S, and
(c) $(X/R)/\pi_R(S) \cong X/(R \circ S)$.

PROOF: 1. $S \circ R \circ S = S \circ S \circ R = S \circ R = S \circ R \circ R = R \circ S \circ R$. (by **6.9**)
2. $S \circ R \circ S$ and $R \circ S \circ R$ are closed equivalence relations on X. (by 1)
3. $\pi_S(R)$ and $\pi_R(S)$ are closed equivalence relations on X/S and X/R respectively. (by 2, **6.10**)
Part (c) follows from **6.9** and **6.10**.

We will make extensive use of **6.11** in constructing icers on M and on flows in general; it can be interpreted as saying that when R, S, and $R \circ S$ are closed equivalence relations on X, the following diagram of quotient maps is commutative:

$$\begin{array}{ccc} X & \overset{\pi_S}{\rightarrow} & X/S \\ \downarrow \pi_R & & \downarrow \pi^S_{R \circ S} \\ X/R & \overset{\pi^R_{R \circ S}}{\rightarrow} & X/(R \circ S). \end{array}$$

In fact we now show that in this diagram π_R and π_S map fibers **onto** fibers, a property which characterizes the relative product.

Proposition 6.12 Let:

(i) X be a compact Hausdorff space,
(ii) R, S, N be closed equivalence relations on X, and
(iii) $R \cup S \subset N$.

Then we can form the commutative diagram

$$\begin{array}{ccc} X & \overset{\pi_S}{\to} & X/S \\ \downarrow \pi_R & & \downarrow \pi_N^S \\ X/R & \overset{\pi_N^R}{\to} & X/N \end{array}$$

and the following are equivalent:

(a) π_S maps the fibers of the map π_R onto the fibers of π_N^S; i.e. $\pi_S(\{xR\}) = (\pi_N^S)^{-1}(xN)$ or equivalently $\pi_S(\{xR\}) = \pi_S(\{xN\})$ for all $x \in X$.
(b) $N = R \circ S$.
(c) π_R maps the fibers of the map π_S onto the fibers of π_N^R; i.e. $\pi_R(\{xS\}) = (\pi_N^R)^{-1}(xN)$ or equivalently $\pi_R(\{xS\}) = \pi_R(\{xN\})$ for all $x \in X$.
(d) $\{(xR, yS) \in X/R \times X/S \mid \pi_N^R(xR) = \pi_N^S(yS)\} = (\pi_R \times \pi_S)(\Delta)$.

PROOF: (a) $\Longrightarrow$ (b)

1. Assume that π_S maps the fibers of π_R onto the fibers of π_N^S.
2. Let $(p, q) \in N$.
3. $\pi_N(p) \underset{1}{=} \pi_N(q) \underset{\mathbf{6.1}}{=} \pi_N^S(\pi_S(q))$ so $\pi_S(q)$ is in the fiber over $\pi_N(p)$.
4. There exists $r \in X$ in the fiber over $\pi_R(p)$ with $\pi_S(r) = \pi_S(q)$. (by 1, 3)
5. $(p, r) \in R$, and $(r, q) \in S$. (by 4)
6. $(p, q) \in R \circ S$. (by 5)
7. $N \underset{2,6}{\subset} R \circ S \underset{(iii)}{\subset} N \circ N \underset{(ii)}{=} N$.

(b) $\Longrightarrow$ (a)

1. Assume that $N = R \circ S$.
2. Let $q \in X$ with $\pi_S(q)$ in the fiber over $\pi_N(p)$.
3. $\pi_N(q) = \pi_N^S(\pi_S(q)) = \pi_N(p)$. (by 2)
4. $(p, q) \in N$. (by 3)
5. There exists $r \in X$ with $(p, r) \in R$ and $(r, q) \in S$. (by 1, 4)

6. $\pi_R(r) = \pi_R(p)$ so r is in the fiber over $\pi_R(p)$. (by 5)
7. $\pi_S(r) = \pi_S(q)$. (by 4)

(b) $\Longleftrightarrow$ (c)

Interchange the roles of R and S in the proof that (a) $\Longleftrightarrow$ (b).

(b) $\Longrightarrow$ (d)

1. Assume that $N = R \circ S$.
2. $(\pi_R \times \pi_S)(\Delta) \subset \{(xR, yS) \in X/R \times X/S \mid \pi_N^R(xR) = \pi_N^S(yS)\}$ is immediate.
3. Let (xR, yS) with $\pi_N^R(xR) = \pi_N^S(yS)$.
4. $xN = yN$. (by 3)
5. There exists $z \in X$ with $(x, z) \in R$ and $(z, y) \in S$. (by 1, 4)
6. $(xR, yS) = (zR, zS) = (\pi_R(z), \pi_S(z)) \in (\pi_R \times \pi_S)(\Delta)$. (by 5)

(d) $\Longrightarrow$ (b)

1. Assume $(\pi_R \times \pi_S)(\Delta) = \{(xR, yS) \in X/R \times X/S \mid \pi_N^R(xR) = \pi_N^S(yS)\}$.
2. Let $(x, y) \in N$.
3. $\pi_N^R(xR) = xN = yN = \pi_N^S(yS)$. (by 2)
4. There exists $z \in X$ with $(\pi_R(z), \pi_S(z)) = (xR, yR)$. (by 1, 3)
5. $(x, z) \in R$ and $(z, y) \in S$. (by 4)
6. $(x, y) \in R \circ S$. (by 5)

Another useful property of the relative product is given below.

Proposition 6.13 Let:

(i) X be a compact Hausdorff space,
(ii) R, S, N be closed equivalence relations on X, and
(iii) $R \circ S = N$.

Then as in **6.12** we can form the commutative diagram

$$\begin{array}{ccc} X & \overset{\pi_S}{\rightarrow} & X/S \\ \downarrow \pi_R & & \downarrow \pi_N^S \\ X/R & \overset{\pi_N^R}{\rightarrow} & X/N \end{array}$$

and

(a) π_S open implies that π_N^R is open,
(b) π_R open implies that π_N^S is open.

PROOF: (a) 1. Assume that π_S is open.
2. Let $V \subset X/R$ be open.
3. $\pi_S((\pi_R)^{-1}(V)) = (\pi_N^S)^{-1}(\pi_N^R(V))$ is open in X/S. (by 1, 2, **6.12**)
4. $\pi_N^R(V)$ is open in X/N. (by 3)
(b) This is just (a) with the roles of R and S interchanged.

The basic properties of the relative product outlined above will be used to construct equivalence relations and quotient maps. We will also need a result which says that the relative product construction commutes with the inverse limit construction. First note that any descending chain

$$S_1 \supset S_2 \supset \cdots \supset S_n \supset \cdots$$

of closed equivalence relations on X gives rise to a sequence of continuous maps

$$X/S_1 \leftarrow X/S_2 \leftarrow \cdots \leftarrow X/S_n \leftarrow \cdots$$

whose inverse limit

$$\varprojlim X/S_i \cong X/\bigcap S_i.$$

For completeness sake and for future reference, we make this precise in the slightly more general context where $\{S_i\}$ is a filter base of equivalence relations on X.

Proposition 6.14 Let:

(i) R be a closed equivalence relation on X, and
(ii) $\{S_i \mid i \in I\}$ be a filter base of closed equivalence relations on X.

Then:

(a) $X/\bigcap_{i\in I} S_i \cong \varprojlim X/S_i$ (the inverse limit), and
(b) $x \in U \subset X/\bigcap_{i\in I} S_i$ with U open implies that there exists U_j open in X/S_j such that $x \in \pi_j^{-1}(U_j) \subset U$ where $\pi_j \colon X/\bigcap S_i \to X/S_j$ is the canonical map.

PROOF: Left as an exercise for the reader.

The relative product construction commutes with this inverse limit construction in the following sense. If R is a closed equivalence relation on X such that each of the $R \circ S_i$ is an equivalence relation, then $R \circ \left(\bigcap S_i\right) = \bigcap (R \circ S_i)$ is an equivalence relation and

$$\varprojlim X/(R \circ S_i) \cong X/\left(R \circ \bigcap S_i\right).$$

This result holds when $\{S_i\}$ is a filter base of closed equivalence relations; the details are given in the next proposition and its corollary.

Proposition 6.15 Let:

(i) R be a closed relation on X (a compact Hausdorff space), and
(ii) $\{S_i \mid i \in I\}$ be a filter base (see **1.A.1**) of closed relations on X.

Then:

(a) $R \circ \left(\bigcap_{i\in I} S_i\right) = \bigcap_{i\in I}(R \circ S_i)$, and

(b) $\left(\bigcap_{i\in I} S_i\right) \circ R = \bigcap_{i\in I}(S_i \circ R)$.

PROOF: (a) 1. Let $S = \bigcap_{i\in I} S_i$.

Proof that: $R \circ S \subset \bigcap_{i\in I}(R \circ S_i)$

2. Let $(p, q) \in R \circ S$.
3. There exists $r \in S$ with $(p, r) \in R$ and $(r, q) \in S$. (by 2)
4. $(p, q) \in \bigcap_{i\in I}(R \circ S_i)$. (by 1, 3)

Proof that: $\bigcap_{i\in I}(R \circ S_i) \subset R \circ S$

1. Let C be a closed subset of $X \times X$ with $C \cap S_j \neq \emptyset$ for all $j \in I$.
2. Let $\{i_1, \dots i_n\} \subset I$.
3. There exists $i \in I$ with $S_i \subset S_{i_1} \cap \dots \cap S_{i_n}$. (by (ii))
4. $\emptyset \neq C \cap S_i \subset (C \cap S_{i_1}) \cap \dots \cap (C \cap S_{i_n})$. (by 1, 3)
5. The collection $\{C \cap S_i \mid i \in I\}$ has the finite intersection property. (by 2, 4)
6. $C \cap S = \bigcap_{i\in I}(C \cap S_i) \neq \emptyset$. (by 5, $X \times X$ is compact)
7. Let $(p, q) \in \bigcap_{i\in I}(R \circ S_i)$ and N be an open neighborhood of S in $X \times X$.
8. $((X \times X) \setminus N) \cap S = \emptyset$.
9. $((X \times X) \setminus N) \cap S_j = \emptyset$ for some $j \in I$.
(by 1, 6, 7 with $C = (X \times X) \setminus N$)
10. $S_j \subset N$. (by 9)
11. $(p, q) \in R \circ S_j \subset R \circ N$. (by 7, 10)
12. There exists r_N with $(p, r_N) \in R$ and $(r_N, q) \in N$. (by 11)
13. The set $K_N = \{r \mid (p, r) \in R \quad \text{and} \quad (r, q) \in N\}$ is nonempty. (by 12)
14. $\{\overline{K_N} \mid N$ is an open neighborhood of $S\}$ has the finite intersection property.
(by 8, 13)

15. Let $x \in \bigcap_N \overline{K_N}$ and V be an open neighborhood of x in X.
16. $V \cap K_N \neq \emptyset$ for all N. (by 15)
17. There exists $r \in V$ with $(p, r) \in R$ and $(r, q) \in N$. (by 13, 16)
18. $(p, x) \in \overline{R} = R$ and $(x, q) \in \bigcap_N \overline{N} = S$. (by 15, 17)
19. $(p, q) \in R \circ S$.

(b) Similar argument.

The following result will be used in section 20.

Corollary 6.16 Let:

(i) R be a closed equivalence relation on X,
(ii) $\{S_i \mid i \in I\}$ be a filter base of closed equivalence relations on X, and
(iii) $R \circ S_i$ be an equivalence relation for every i.

Then $R \circ \left(\bigcap_{i \in I} S_i\right)$ is a closed equivalence relation.

PROOF: 1. $\bigcap_{i \in I} S_i$ is a closed equivalence relation. (by (ii))

2. $R \circ \left(\bigcap_{i \in I} S_i\right) \underset{\mathbf{6.15}}{=} \bigcap_{i \in I}(R \circ S_i) \underset{\text{(iii)}}{=} \bigcap_{i \in I}(S_i \circ R) \underset{\mathbf{6.15}}{=} \left(\bigcap_{i \in I} S_i\right) \circ R.$

3. $R \circ \left(\bigcap_{i \in I} S_i\right)$ is an equivalence relation. (by 1, 2, **6.9**)

The results in this section have been stated for closed equivalence relations on any compact Hausdorff space X. If (X, T) is a flow and the equivalence relations are invariant under the action of T, then the results remain valid for icers on X, and the quotient maps are homomorphisms of flows. In particular we recover results on minimal flows by setting $X = M$ and requiring that the equivalence relations be icers.

We end this section with some applications of the relative product to homomorphisms of flows. These results are motivated by considering the following diagram:

$$\begin{array}{ccc} X & \xrightarrow{f} & Y \\ \pi_L \downarrow & & \downarrow \pi \\ X/L & \rightarrow & Z \end{array}$$

where f is an epimorphism of flows, L is an icer on X, and π is a homomorphism of flows. Assuming that π_L is a distal homomorphism, when can we conclude that π is also distal? Note that when $X/L = \{pt\} = Z$, the answer is **always**. (In this case we are just asking whether (X, T) distal implies that (Y, T) is distal, which is the content of **4.10**.) In general, using R_f to denote the icer on X corresponding to f, if $R_{\pi\circ f} \subset R_f \circ L \circ R_f$, and (X, T) is pointwise almost periodic, then π is distal. In the interest of clarity, we state this as the next proposition. If the diagram is a relative product, that is $R_{\pi\circ f} = L \circ R_f$ and $Z = X/(L \circ R_f)$, then π is distal under the weaker assumption that (Y, T) is pointwise almost periodic. We give a proof of this in **6.17**.

Proposition 6.17 Let:

(i) (X, T) be a pointwise almost periodic flow,
(ii) L be an icer on X,
(iii) $\pi_L : X \to X/L$ be distal,
(iv) $f : (X, T) \to (Y, T)$ be an epimorphism of flows,
(v) the following be a commutative diagram of flows and flow homomorphism, and

$$\begin{array}{ccc} X & \xrightarrow{f} & Y \\ \pi_L \downarrow & & \downarrow \pi \\ X/L & \rightarrow & Z \end{array}$$

(vi) $R_{\pi\circ f} \subset R_f \circ L \circ R_f$.

Then $\pi : Y \to Z$ is distal.

PROOF: 1. (L, T) is pointwise almost periodic. (by (i), (ii), (iii), and **4.15**)
2. $f(L) = f(R_f \circ L \circ R_f) \underset{\text{(vi)}}{=} f(R_{\pi\circ f}) = f(f^{-1}(R_\pi)) \underset{\text{(iv)}}{=} R_\pi$.
3. (R_π, T) is pointwise almost periodic. (by 1, 2)
4. π is distal. (by **4.15**)

Lemma 6.18 Let:

(i) (X, T) be a flow,
(ii) L and N be an icers on X,
(iii) $X \to X/L$ be distal,
(iv) $(X/N, T)$ be pointwise almost periodic, and
(v) $L \circ N$ be an icer on X.

In other words assume that the following is a commutative diagram of flows such that the flows in the right hand column are pointwise almost periodic and the left hand column is a distal extension.

$$\begin{array}{ccc} X & \rightarrow & X/N \\ \downarrow & & \downarrow \\ X/L & \rightarrow & X/(L \circ N) \end{array}$$

Then $X/N \to X/(L \circ N)$ is distal.

PROOF: 1. $\pi_N(L \circ N) = \pi_N(L)$ is a closed equivalence relation on X/N with

$$(X/N)/\pi_N(L) = X/(L \circ N). \qquad \text{(by } \mathbf{6.11}\text{)}$$

2. Let $\pi_N(x_1, x_2) \in P(X/N)$ with $(x_1, x_2) \in L \circ N$.
3. There exists a minimal ideal $I \subset E(X, T)$ with $(x_1 p, x_2 p) \in N$ for all $p \in I$. (by 2, **4.4**)
4. There exists a minimal idempotent $u \in I$ with

$$\pi_N(x_1 u) = \pi_N(x_1)u = \pi_N(x_1). \qquad \text{(by 3, (iv), } \mathbf{4.3}\text{)}$$

5. $(x_1 u, x_2) \in N \circ L \circ N = L \circ N$. (by 2, 3, (v))
6. There exists $x_3 \in X$ with $(x_1 u, x_3) \in L$ and $(x_3, x_2) \in N$. (by 5)
7. $(x_1 u, x_3 u) = (x_1 u, x_3)u \in Lu \subset L$. (by 6, (ii))
8. $(x_3, x_3 u) \in L$. (by 6, 7, (ii))
9. $x_3 = x_3 u$. (by 8, (iii))
10. $(x_3, x_2 u) = (x_3 u, x_2 u) \in Nu \subset N$. (by 6, 9)
11. $(x_3, x_1 u) \in N \circ N = N$. (by 3, 5, 10, (ii))
12. $(x_3, x_1) \in N$. (by 4. 11)
13. $(x_2, x_1) \in N$ and so $\pi_N(x_1, x_2) \in \Delta_{X/N}$. (by 6, 12)
14. $P(X/N) \cap \pi_N(L) = \Delta_{X/N}$. (by 2, 13)
15. $X/N \to X/(L \circ N)$ is distal. (by 1, 14, and (v))

A refinement of the preceding result will be needed in section 9.

Corollary 6.19 Let:

(i) (X, T) be a (not necessarily minimal) flow,
(ii) L be an icer on X,
(iii) $X \to X/L$ be distal,
(iv) $H \subset T$ be a subgroup of T,
(v) N be a closed H-invariant equivalence relation on X,

(vi) $(X/N, H)$ be pointwise almost periodic, and
(vii) $L \circ N$ be an H-invariant equivalence relation on X.

In other words assume that the following is a commutative diagram of H-flows such that the flows in the right hand column are H-pointwise almost periodic and the left hand column is a T-distal extension.

$$\begin{array}{ccc} X & \rightarrow & X/N \\ \downarrow & & \downarrow \\ X/L & \rightarrow & X/(L \circ N) \end{array}$$

Then $(\pi_N(L), H)$ is pointwise almost periodic.

PROOF: The assumption that $X \rightarrow X/L$ is distal as a T-homomorphism, implies that it is distal as an H-homomorphism. Thus the result follows immediately by applying **6.18** and **4.15** to the group H.

EXERCISES FOR CHAPTER 6

Exercise 6.1 (See **6.4**) Let:

(i) X be a compact Hausdorff space,
(ii) R, S be icers on X,
(iii) $A \subset X \times X$, and
(iv) $B, C \subset X/R \times X/R$.

Then:

(a) $(\pi_R \times \pi_S)^{-1}((\pi_R \times \pi_S)(A)) = R \circ A \circ S$,
(b) $\pi_R^{-1}(B^{-1}) = \left(\pi_R^{-1}(B)\right)^{-1}$, and
(c) $\pi_R^{-1}(B \circ C) = \pi_R^{-1}(B) \circ \pi_R^{-1}(C)$.

Exercise 6.2 (See **6.9**) Let:

(i) X be a compact Hausdorff space, and
(ii) $R, S \subset X \times X$ be equivalence relations on X.

Show that $R \circ S$ is an equivalence relation on X if and only if $R \circ S \subset S \circ R$.

Exercise 6.3 (See **6.6**) Let:

(i) X be a compact Hausdorff space,
(ii) N be a closed equivalence relation on X, and
(iii) $R, S \subset X \times X$ be arbitrary relations on X.

Show that:

(a) $\pi_N^{-1}(\pi_N(R)) = N \circ R \circ N$,
(b) $\pi_N(R \circ S) \subset \pi_N(R) \circ \pi_N(S) \subset \pi_N(R \circ N \circ S)$, and
(c) if $R \circ N$ or $S \circ N$ is a closed equivalence relations on X, then

$$\pi_N(R \circ S) = \pi_N(R) \circ \pi_N(S).$$

Exercise 6.4 (See **6.14**) Let:

(i) R be a closed equivalence relation on X, and
(ii) $\{S_i \mid i \in I\}$ be a filter base of closed equivalence relations on X.

Show that:

(a) $X/\bigcap_{i\in I} S_i \cong \varprojlim X/S_i$ (the inverse limit), and
(b) $x \in U \subset X/\bigcap_{i\in I} S_i$ with U open implies that there exists U_j open in X/S_j such that $x \in \pi_j^{-1}(U_j) \subset U$ where $\pi_j \colon X/\bigcap S_i \to X/S_j$ is the canonical map.

7

Icers on M and automorphisms of M

For this section, and the remainder of this book, we will use M to denote a fixed minimal ideal in βT. As we saw in **3.16**, given any minimal flow (X, T) there exists an epimorphism of flows $f: M \to X$, so that $M/R_f \cong X$. Here $R_f = \{(p, q) \in M \times M \mid f(p) = f(q)\}$ is an invariant closed equivalence relation on M. We will study minimal flows and their properties by studying invariant closed equivalence relations on the universal minimal set M. We emphasize this point of view by defining a category $\mathcal{M}$.

Definition 7.1 The category $\mathcal{M}$.
We define a category $\mathcal{M}$ whose objects, $obj(\mathcal{M})$ are the *icers* (closed invariant equivalence relations) on M. If R and S are icers on M with $R \subset S$, then π_S^R will denote the canonical map

$$\pi_S^R: M/R \to M/S.$$

We refer to M/S as a *factor* of M/R and to M/R as an *extension* of M/S. We then define

$$morph(R, S) = \begin{cases} \{\pi_S^R\} & \text{if } R \subset S \\ \emptyset & \text{if } R \not\subset S \end{cases}.$$

We write π_R for

$$\pi_R \equiv \pi_R^\Delta: M \to M/R,$$

Note that if $R \subset S$, then $\pi_S = \pi_S^R \circ \pi_R$.

Given an icer R on M, the quotient space is a compact Hausdorff space (by **6.2**) on which T acts. In other words $(M/R, T)$ is a flow; moreover it's minimal since it's an image of M. Conversely any minimal flow (X, T) is represented by the element R_f in the category $\mathcal{M}$, where $f: M \to X$ is a homomorphism of flows. In general however, if g is another such homomorphism, then R_g

need not equal R_f. Thus X may have many distinct representatives in the category $\mathcal{M}$. In other words $obj(\mathcal{M})$ is **not** in one-one correspondence with the collection of minimal flows. We will see in section 8 (see **8.1** and **8.3**) that the regular minimal flows are exactly those minimal flows (X, T), for which there is a unique icer with $X \cong M/R$. Similarly $morph(R, S)$ consists of at most one element and is **not** in one-one correspondence with $Hom(M/R, M/S)$. In particular $morph(R, R)$ consists of a single element, while $Aut(M/R)$, in general, contains many automorphisms of the flow $(M/R, T)$. Nevertheless $\mathcal{M}$ provides a useful context in which to frame many of the questions concerning minimal flows and their extensions. Before addressing some of these questions we state an elementary proposition which will be used frequently in subsequent discussions.

Proposition 7.2 Let:

(i) $X = M/R$ be a flow,
(ii) N be an icer on X, and
(iii) $S = \pi_R^{-1}(N)$.

Then:

(a) S is an icer on M, and
(b) $M/S \cong X/N$.

PROOF: Straightforward.

In section 3 we showed (see **3.12**) that any minimal ideal I in the enveloping semigroup $E(X)$ of a flow (X, T) can be written as a disjoint union of isomorphic groups, one for each idempotent in I. Since $E(\beta T) \cong \beta T$ this holds for M. In fact M can be written as a disjoint union of copies of the group G of automorphisms of M. In particular any element $p \in M$ can be written uniquely in the form $p = \alpha(u)$ where $\alpha \in G$ and u is an idempotent in M. This description, which is a consequence of the proposition below, plays a key role in our study of the category $\mathcal{M}$.

Proposition 7.3 Let:

(i) $p, q \in M$,
(ii) $L_p \colon M \to M$ be defined by $L_p(m) = pm$, and
(iii) $\alpha \colon M \to M$ be any homomorphism of the flow (M, T).

Then:

(a) there exists a unique idempotent $u \in M$ such that $L_p(u) = pu = p$,
(b) L_p is an automorphism of (M, T),
(c) $L_p = L_q$ if and only if $p = qu$, and

(d) $\alpha = L_{\alpha(v)}$ for any idempotent $v \in M$, and hence α is an automorphism of M.

PROOF: (a) This follows immediately from **3.12**.

(b) 1. $L_p(qt) = p(qt) = (pq)t = L_p(q)t$ for all $q \in M$ and $t \in T$.

2. L_p is a homomorphism of flows. (by 1)

3. There exists $q \in M$ with $qp = u = pq$. (by **3.12**)

4. $L_p(L_q(r)) = p(qr) = ur = r = (qp)r = L_q(L_p(r))$ for all $r \in M$. (by **3.12**)

5. L_q is the inverse of L_p. (by 4)

(c) 1. If $L_p = L_q$, then $p = L_p(u) = L_q(u) = qu$. (by part (a))

2. If $p = qu$, then $L_p(r) = pr = (qu)r = q(ur) = qr = L_q(r)$ for all $r \in M$, so $L_p = L_q$. (by **3.12**)

(d) $L_{\alpha(v)}(r) = \alpha(v)r = \alpha(vr) = \alpha(r)$ for all $r \in M$. (by **3.12**)

We state the following corollary for emphasis, and so that it may be referred to later in the text.

Corollary 7.4 Let:

(i) $G = \{\alpha \mid \alpha\colon M \to M\}$ be the group of automorphisms of M, and
(ii) $J = \{u \in M \mid u^2 = u\}$ be the set of idempotents of M.

Then the map $\begin{array}{ccc} G \times J & \to & M \\ (\alpha, u) & \to & \alpha(u) \end{array}$ is bijective.

PROOF: This follows immediately from **7.3** and **3.12**.

Given two minimal flows (X, T) and (Y, T), we would like to study homomorphisms

$$\varphi\colon X \to Y.$$

It is clear that we can find icers R and S on M such that $X \cong M/R$ and $Y \cong M/S$; we would like to do this in such a way that $R \subset S$, and then relate φ to the canonical projection map π_S^R. As a first step we show that there exists an automorphism $\alpha \in G$ such that

$$\begin{array}{ccc} M & \xrightarrow{\alpha} & M \\ \pi_R \downarrow & & \downarrow \pi_S \\ M/R & \xrightarrow{\varphi} & M/S \end{array}$$

is a commutative diagram of flow homomorphisms. This is a consequence of the fact that every minimal subset of $M \times M$ is of the form $gr(\alpha) \equiv \{(p, \alpha(p)) \mid p \in M\}$ for some automorphism $\alpha \in G$. This result, an analog of which holds for regular flows (see **8.9**), is of independent interest and will also be used in section 10 to analyze the τ-topology on G.

Proposition 7.5 A nonvacuous subset $Y \subset M \times M$ is minimal if and only if $Y = gr(\alpha)$ for some $\alpha \in G$ (that is Y is the graph of an automorphism of M).

PROOF: 1. The graph of an automorphism of M is a homomorphic image of M and hence is a minimal subset of $M \times M$. (by **3.5**)
2. Let $Y \subset M \times M$ be a minimal set and $(p, q) \in Y$.
3. There exists an idempotent $u \in M$ with $pu = p$. (by **3.12**)
4. There exists $p' = p'u \in M$ with $pp' = u = p'p$. (by **3.12**)
5. $\alpha = L_{qp'} \in G$. (by **7.3**)
6. $(u, \alpha(u)) = (pp', (qp')u) = (p, q)p' \in Yp' \subset Y$. (by 2, 4, 5)
7. $gr(\alpha) \cap Y \neq \emptyset$. (by 6)
8. $Y = gr(\alpha)$. (by 2, 7, since $gr(\alpha)$ is minimal)

We now use **7.5** to characterize homomorphisms. For this purpose, and for the remainder of the text, it will be convenient to write:

$$\alpha(p, q) \equiv (\alpha \times \alpha)(p, q) = (\alpha(p), \alpha(q)),$$

for any $\alpha \in G$ and $(p, q) \in M \times M$. Thus when R is an icer on M, we will use the notation:

$$\alpha(R) \equiv \{\alpha(p, q) \mid (p, q) \in R\}.$$

Proposition 7.6 Let R, S be icers on M. Then:

(a) if $\varphi \in Hom(M/R, M/S)$, there exists $\alpha \in G$ with $\alpha(R) \subset S$ and $\varphi \circ \pi_R = \pi_S \circ \alpha$,
(b) if $\alpha \in G$ with $\alpha(R) \subset S$, there exists $\varphi \in Hom(M/R, M/S)$ such that $\varphi \circ \pi_R = \pi_S \circ \alpha$,
(c) if $\varphi \in Iso(M/R, M/S)$, there exists $\alpha \in G$ with $\alpha(R) = S$ and $\varphi \circ \pi_R = \pi_S \circ \alpha$, and
(d) if $\alpha \in G$ with $\alpha(R) = S$, there exists $\varphi \in Iso(M/R, M/S)$ such that $\varphi \circ \pi_R = \pi_S \circ \alpha$.

PROOF: (a) 1. Let $\varphi\colon M/R \to M/S$ be a flow homomorphism.
2. $gr(\varphi)$ is a minimal subset of $M/R \times M/S$. (by **3.5**)
3. There exists a minimal set $Y \subset M \times M$ with $(\pi_R \times \pi_S)(Y) = gr(\varphi)$.
(by **3.5**, since $\pi_R \times \pi_S$ is an epimorphism)

4. There exists $\alpha \in G$ with $Y = gr(\alpha)$. (by 3, and **7.5**)
5. Let $p \in M$.
6. There exists $q \in M$ with $(\pi_R(p), \varphi(\pi_R(p))) = (\pi_R \times \pi_S)(q, \alpha(q))$. (by 3, 4)
7. $\varphi(\pi_R(q)) = \varphi(\pi_R(p)) = \pi_S(\alpha(q))$. (by 6)
8. $\varphi \circ \pi_R = \pi_S \circ \alpha$. (7, M is minimal)
9. $(\pi_S \times \pi_S)(\alpha(R)) = (\varphi \times \varphi)(\pi_R \times \pi_R)(R) = \Delta \subset M/S \times M/S$. (by 8)
10. $\alpha(R) \subset S$. (by 9)

(b) 1. Let $\alpha \in G$ with $\alpha(R) \subset S$.
2. Let $(p, q) \in R$.
3. $(\alpha(p), \alpha(q)) \in S$. (by 1, 2)
4. $\pi_S(\alpha(p)) = \pi_S(\alpha(q))$. (by 3)
5. $\varphi: M/R \to M/S$ is a well-defined homomorphism. (by 2, 4)
$\quad pR \to \alpha(p)S$
6. $\varphi \circ \pi_R = \pi_S \circ \alpha$. (by 5)

(c) This follows from (a); we leave the details to the reader.
(d) This follows from (b); we leave the details to the reader.

Corollary 7.7 Let R, S be icers on M. Then $M/R \cong M/S$ if and only if $\alpha(R) = S$ for some $\alpha \in G$.

PROOF: This follows immediately from **7.6**.

As an additional consequence of **7.6** we note that up to pre-composition with an isomorphism, any homomorphism of flows can be represented by the canonical homomorphism $\pi_S^R : M/R \to M/S$, for some choice of $R \subset S$. Thus not only properties of individual flows, but also properties of homomorphisms of flows can be studied using the category $\mathcal{M}$.

Corollary 7.8 Let $\varphi: X \to Y$ be a homomorphism of minimal flows. Then there exist icers $R \subset S$ on M, and an isomorphism $\psi : X \to M/R$ such that $M/S \cong Y$ and $\varphi = \pi_S^R \circ \psi$.

PROOF: 1. There exists an icer S on M such that $Y \cong M/S$. (by **3.16**)
2. There exists an icer R_1 on M such that $X = M/R_1$. (by **3.16**)
3. There exists $\alpha \in G$ such that $R \equiv \alpha(R_1) \subset S$ and $\pi_S \circ \alpha = \varphi \circ \pi_{R_1}$. (by **7.6**)
4. There exists an isomorphism $\psi : X = M/R_1 \to M/R$ such that $\pi_R \circ \alpha = \psi \circ \pi_{R_1}$. (by **7.6**)
5. Let $x \in X$.
6. There exists $p \in M$ with $\pi_{R_1}(p) = x$. (by 2)

7. $\pi_S^R(\psi(x)) \underset{6}{=} \pi_S^R(\psi(\pi_{R_1}(p))) \underset{4}{=} \pi_S^R(\pi_R(\alpha(p))) = \pi_S(\alpha(p)) \underset{3}{=} \varphi(\pi_{R_1}(p))$
$\underset{6}{=} \varphi(x)$.

According to **7.6**, every automorphism $\varphi\colon M/R \to M/R$ fits into a commutative diagram:

$$\begin{array}{ccc} M & \overset{\alpha}{\longleftrightarrow} & M \\ \pi_R \downarrow & & \downarrow \pi_R \\ M/R & \overset{\varphi}{\longleftrightarrow} & M/R \end{array}$$

for some choice of $\alpha \in G$. In this sense every automorphism of M/R is induced by an $\alpha \in G$ with $\alpha(R) = R$. This motivates us to consider the subgroup

$$aut(R) = \{\alpha \in G \mid \alpha(R) = R\} \subset G$$

consisting of the automorphisms of M which descend to automorphisms of M/R. Of special interest are those α which induce the **identity** on M/R; this happens when:

$$\pi_R(\alpha(p)) = \pi_R(p)$$

for all $p \in M$, or equivalently: $gr(\alpha) \subset R$. In other words those α which descend to the identity on M/R are the members of the subgroup $G(R) = \{\alpha \in G \mid gr(\alpha) \subset R\} \subset aut(R) \subset G$. We make these ideas precise and set some notation in the following definition and proposition.

Definition 7.9 Let R be an icer. As indicated above we will use the notation:

$$aut(R) = \{\alpha \in G \mid \alpha(R) = R\} \qquad G(R) = \{\alpha \in G \mid gr(\alpha) \subset R\}$$

and we refer to $G(R)$ as the *group of the flow* M/R. (It will also be denoted by $G(M/R)$.)

It should be noted that the terminology introduced in **7.9** is an abuse of language since $G(M/R)$ depends on the choice of icer R. This terminology is consistent with the literature where the group of a flow is defined for *pointed flows* (where a base point is specified), and it depends on the choice of base point. This abuse of language is often justified by **7.8** since we are most often interested in properties of flows and homomorphisms which are invariant under isomorphisms. It is important to keep in mind however that in general $(M/R, T) \cong (M/\alpha(R), T)$ for every $\alpha \in aut(R)$. We will show (see **7.16**) that $G(\alpha(R)) = \alpha G(R)\alpha^{-1}$ which implies that $G(M/R)$ is well-defined up to conjugacy. (A change of base point in the original definition given in [Ellis, R., 1969] also results in a conjugate subgroup.)

Clearly, if $R \subset S$ are icers on M, then $G(R) \subset G(S)$. Of course this is not necessarily the case for $aut(R)$; indeed we will show that any icer S contains an icer $reg(S)$ (the so-called *regularizer* of S, see **8.4**) with $aut(reg(S)) = G$.

Proposition 7.10 Let:

(i) R be an icer on M, and
(ii) $X = M/R$.

Then:

(a) there exists a group epimorphism $\chi_R\colon aut(R) \to Aut(X)$ such that

$$\pi_R(\alpha(p)) = \chi_R(\alpha)(\pi_R(p)) \quad \text{for all } p \in M \text{ and } \alpha \in aut(R), \qquad (*)$$

(b) $ker(\chi_R) = G(R)$, and
(c) $G(R)$ is a normal subgroup of $aut(R)$.

PROOF: (a) The fact that χ_R is well-defined and onto follows immediately from **7.7**. We leave it to the reader to verify that χ_R is a group homomorphism.

(b) Proof that $ker(\chi_R) \subset G(R)$:

1. Assume that $\chi_R(\alpha) = id$.
2. $\pi_R(\alpha(p)) = \pi_R(p)$. for all $p \in M$. (by 1, and $(*)$)
3. $(p, \alpha(p)) \in R$ for all $p \in M$. (by 2)
4. $gr(\alpha) \subset R$. (by 3)
5. $\alpha \in G(R)$. (by 4)

Proof that $G(R) \subset ker(\chi_R)$:

1. Assume that $gr(\alpha) \subset R$.
2. $(p, \alpha(p)) \in R$ for all $p \in M$. (by 1)
3. $\pi_R(\alpha(p)) = \pi_R(p)$ for all $p \in M$. (by 2)
4. $\chi_R(\alpha) = id$. (by 3 and $(*)$)

(c) This follows immediately from (a) and (b).

When R is an icer on M, the group $G(R)$ plays a fundamental role in understanding the dynamics of the minimal flow M/R. This motivates and serves as one of the main themes of parts IV and V of this text. As a first elementary example of this principle we show that a homomorphism of minimal flows $\varphi\colon X \to Y$ is proximal if and only if the groups $G(X) = G(Y)$ are equal. As we remarked above this is technically an abuse of language. We make things precise by appealing to **7.8** and writing $X = M/R$, $Y = M/S$ with $R \subset S$ icers on M. Then $\varphi = \pi_S^R \circ \psi$ for some isomorphism $\psi\colon X \to M/R$. Now it is clear (see exercise **4.6**) that φ is proximal if and only π_S^R is proximal. Thus in the category $\mathcal{M}$ we will show that $\pi_S^R\colon M/R \to M/S$ is proximal if and only if $G(R) = G(S)$.

Proposition 7.11 Let:

(i) $R \subset S$ icers on M, and
(ii) $X = M/R$, and $Y = M/S$.

Then $\pi_S^R\colon M/R \to M/S$ is proximal if and only if $G(R) = G(S)$.

PROOF: $\Longrightarrow$

1. Assume that X is a proximal extension of Y.
2. Let $\alpha \in G(S)$ and $u \in J$.
3. $\pi_S^R(\pi_R(u)) = \pi_S(u) \underset{2}{=} \pi_S(\alpha(u)) = \pi_S^R(\pi_R(\alpha(u)))$.
4. $(\pi_R(u), \pi_R(\alpha(u))) \in P(X)$. (by 1, 3)
5. There exists $p \in \beta T$ with $\pi_R(up) = \pi_R(\alpha(up))$. (by 4)
6. $\alpha \in G(R)$. (by 5)
7. $G(S) \subset G(R)$. (by 2, 6)
8. $G(R) \subset G(S)$. (since $R \subset S$)

$\Longleftarrow$

1. Assume that $G(S) = G(R)$.
2. Let $x_1, x_2 \in X$ with $\pi_S^R(x_1) = \pi_S^R(x_2)$.
3. There exist $\alpha_1, \alpha_2 \in G$ and $u_1, u_2 \in J$ with $(x_1, x_2) = \pi_R(\alpha_1(u_1), \alpha_2(u_2))$. (by **7.4**)
4. $\pi_S(\alpha_1(u)) = \pi_S^R(\pi_R(\alpha_1(u_1))) = \pi_S^R(x_1) = \pi_S^R(x_2) = \pi_S(\alpha_2(u_2))$. (by 2)
5. $(\alpha_1(u_1), \alpha_2\alpha_1^{-1}(\alpha_1(u_1))) = (\alpha_1(u_1), \alpha_2(u_2))u_1 \in Su_1 \subset S$. (by 4)
6. $\alpha_2\alpha_1^{-1} \in G(S) = G(R)$. (by 1, 4)
7. $(x_1, x_2)u_1 = \pi_R(\alpha_1(u_1), \alpha_2(u_2))u_1 = \pi_R(\alpha_1(u_1), \alpha_2(u_1)) \in \Delta_X$. (by 6)
8. $(x_1, x_2) \in P(X)$. (by 7, and **4.4**)

Corollary 7.12 Let $X = M/R$. Then X is a proximal flow if and only if $G(R) = G$.

PROOF: This is immediate from **7.11**

Returning to our consideration of homomorphisms $\varphi\colon X \to Y$ of minimal flows, the case where there exists a point $y_0 \in Y$ such that the fiber $\varphi^{-1}(y_0)$ over y_0 consists of a single point is referred to as an *almost one-one* extension. This terminology is motivated by the observation that $\varphi^{-1}(y_0)t = \varphi^{-1}(y_0t)$ so that if the fiber over y_0 is a single point then the same is true for every point in the orbit of y_0. Since we are assuming that (Y, T) is minimal this means that the set of y for which $\varphi^{-1}(y)$ is a single point is dense in Y. Just as in the case of proximal extensions φ is almost one-one if and only if $\varphi \circ \psi$ is almost one-one for any isomorphism $\psi: X \to X$. Applying **7.8** we need only focus on the canonical projection $\pi_S^R: M/R \to M/S$ for $R \subset S$. We make this the definition of an almost one-one extension and then show that any almost one-one extension is a proximal extension.

Definition 7.13 Let $X = M/R$, and $Y = M/S$ be flows with $R \subset S$ icers on M. We say that X is an *almost one-one extension* of Y if there exists $p \in M$ such that $pR = pS$. Clearly this is equivalent to the existence of $y_0 \in Y$ such that $(\pi_S^R)^{-1}(y_0)$ consists of a single point.

Proposition 7.14 Let:

(i) $X = M/R$, and $Y = M/S$ be flows with $R \subset S$ icers on M, and
(ii) X be an almost one-one extension of Y.

Then X is a proximal extension of Y.

PROOF: 1. $G(R) \subset G(S)$. (by (i))
2. Let $\alpha \in G(S)$.
3. There exists $p \in M$ with $pR = pS$. (by (ii), **7.13**)
4. $\alpha(p) \in pS = pR$. (by 2, 3)
5. $\alpha \in G(R)$. (by 4)
6. $G(S) \subset G(R)$. (by 2, 5)
7. $G(R) = G(S)$. (by 1, 6)
8. X is a proximal extension of Y. (by 7, **7.11**)

We will be interested in what the group $G(R)$ tells us about an icer R, and to what extent R is determined by $G(R)$. In order to answer these questions we first deduce some elementary properties of $G(R)$. We will use an elementary lemma concerning the graphs of elements of G; it will be convenient to use the notation:

$$gr(A) = \bigcup\{gr(\alpha) \mid \alpha \in A\}$$

for any subset $A \subset G$.

Lemma 7.15 Let:

(i) $\alpha, \gamma \in G$, and
(ii) $\emptyset \neq A \subset G$.

Then:

(a) $\overline{gr(\alpha\gamma\alpha^{-1})} = \alpha(gr(\gamma))$, and
(b) $\overline{gr(\alpha A\alpha^{-1})} = \alpha(\overline{gr(A)})$.

PROOF: (a) 1. Let $p \in M$.
2. $(p, \alpha\gamma\alpha^{-1}(p)) = (\alpha(\alpha^{-1}(p)), \alpha(\gamma(\alpha^{-1}(p))) \in \alpha(gr(\gamma))$.
3. $gr(\alpha\gamma\alpha^{-1}) \subset \alpha(gr(\gamma))$. (by 1, 2)
4. $\alpha(gr(\gamma)) = \alpha(gr(\alpha^{-1}(\alpha\gamma\alpha^{-1})\alpha)) \underset{3}{\subset} \alpha(\alpha^{-1}(gr(\alpha^{-1}\gamma\alpha))) = gr(\alpha\gamma\alpha^{-1})$.
(b) 1. $\overline{gr(\alpha A\alpha^{-1})} = \alpha(\overline{gr(A)})$. (by (a))
2. $\alpha(\overline{gr(A)}) \subset \overline{\alpha(gr(A))} \underset{1}{=} \overline{gr(\alpha A\alpha^{-1})}$. ($\alpha$ is continuous)
3. $\alpha(\overline{gr(A)})$ is compact hence closed.
4. $\overline{gr(\alpha A\alpha^{-1})} \subset \alpha(\overline{gr(A)})$. (by 1, 3)

Proposition 7.16 Let R be an icer on M and $\alpha \in G$. Then:

(a) $\alpha(R)$ is an icer on M, and
(b) $G(\alpha(R)) = \alpha G(R)\alpha^{-1}$.

PROOF: We leave this as an exercise for the reader.

Note that if $\alpha \in aut(R)$, then $\alpha(R) = R$ and applying **7.16** we get

$$\alpha G(R)\alpha^{-1} = G(\alpha(R)) = G(R),$$

which gives an alternate proof that $G(R)$ is a normal subgroup of $aut(R)$. This idea is also used in the following proposition to determine when $G(R)$ is a normal subgroup of G.

Proposition 7.17 Let R be an icer on M. Then the following are equivalent:

(a) $G(R)$ is a normal subgroup of G,
(b) $G(\alpha(R)) = G(R)$ for all $\alpha \in G$,
(c) $\alpha(gr(G(R))) \subset R$ for all $\alpha \in G$, and
(d) $\alpha(\overline{gr(G(R))}) \subset R$ for all $\alpha \in G$.

PROOF: This follows immediately from **7.15** and **7.16**.

Proposition 7.18 Let R and S be icers on M with $M/R \cong M/S$. Then $G(S) = \alpha G(R)\alpha^{-1}$ for some $\alpha \in G$.

PROOF: 1. There exists $\alpha \in G$ with $\alpha(R) = S$. (by **7.7**)
2. $G(S) = G(\alpha(R)) = \alpha G(R)\alpha^{-1}$. (by **7.16**)

Proposition 7.19 Let $\{R_i\}$ be a family of icers on M. Then $G(\bigcap R_i) = \bigcap G(R_i)$.

PROOF:

$$\alpha \in G\left(\bigcap R_i\right) \iff gr(\alpha) \subset \bigcap R_i \iff gr(\alpha) \subset R_i \text{ for all } i$$
$$\iff \alpha \in G(R_i) \text{ for all } i \iff \alpha \in \bigcap G(R_i).$$

An immediate consequence of **7.19** is the statement that the group of an inverse limit of flows is the intersection of the groups of those flows.

We turn now to the task of identifying and constructing icers on M. A key role is played by a certain subset, P_0 of the proximal relation $P(M)$ in M; intuitively P_0 is the subset generated by the pairs of idempotents in M.

Definition 7.20 Let $M \subset \beta T$ be a fixed minimal ideal in βT, $G = aut(M)$, and $J \subset M$ be the set of idempotents in M. Now we define a set $P_0 \subset P(M)$ by:

$$P_0 = \{(\alpha(u), \alpha(v)) \mid u, v \in J \text{ and } \alpha \in G\} = \{(p, pv) \mid v \in J \text{ and } p \in M\}.$$

It is easy to see that P_0 is an equivalence relation on $M \times M$ (though it is neither closed nor invariant). Our plan is to show that every icer on R on M is determined by its intersection with P_0 and its group $G(R)$; in fact we will show that:

$$R = (R \cap P_0) \circ gr(G(R)). \qquad (*)$$

By way of motivation we begin with the icer $M \times M$. It follows from **7.4** that for any $(p, q) \in M \times M$, we can write

$$(p, q) = (\alpha(u), \beta(v)) = (\alpha(u), \beta\alpha^{-1}(\alpha(v))),$$

for some $\alpha, \beta \in G$ and idempotents $u, v \in J$. Now

$$(\alpha(u), \alpha(v)) \in P_0 \qquad \text{and} \qquad (\alpha(v), \beta\alpha^{-1}(\alpha(v))) \in gr(\beta\alpha^{-1}) \subset gr(G),$$

so $(p, q) \in P_0 \circ gr(G)$. Thus we have shown that

$$M \times M = P_0 \circ gr(G);$$

in other words equation $(*)$ holds for the icer $M \times M$. The next theorem shows that $(*)$ holds in general.

Theorem 7.21 Let R be an icer on M. Then

$$R = (R \cap P_0) \circ gr(G(R)) = gr(G(R)) \circ (R \cap P_0).$$

PROOF: 1. $(R \cap P_0) \circ gr(G(R)) \subset R$. ($R$ is transitive)
2. Let $(p, q) = (\alpha(u), \beta(v)) \in R$.

3. $(p,q)\alpha^{-1}(v) = (\alpha(u)\alpha^{-1}(v), \beta(v)\alpha^{-1}(v))$

$$= (\alpha(u\alpha^{-1}(v)), \beta(v\alpha^{-1}(v))) \underset{3.12}{=} (v, \beta\alpha^{-1}(v)) \in R.$$

(R is closed and invariant)

4. $\beta\alpha^{-1} \in G(R)$. (by 3)
5. $(\alpha(u), \alpha(v)) = (p, \alpha\beta^{-1}(q)) \in R \circ R = R$. (by 2, 4, R is transitive)
6. $(\alpha(v), \beta(v)) = (\alpha(v), \beta\alpha^{-1}(\alpha(v))) \in gr(\beta\alpha^{-1}) \subset R$. (by 4)
7. $(p,q) = (\alpha(u), \beta(v)) \in (P_0 \cap R) \circ gr(G(R))$. (by 5, 6)
8. $R = (R \cap P_0) \circ gr(G(R))$. (by 1, 2, 7)
9. $gr(G(R)) \circ (R \cap P_0) = R^{-1} = R = (R \cap P_0) \circ gr(G(R))$.
(by 8, R is symmetric)

Corollary 7.22 Let R, S be icers on M. Then the following are equivalent:

(a) $S \cap P_0 \subset R \cap P_0$,
(b) $S \subset R \circ gr(G(S))$, and
(c) $\pi_R(G(S)(p)) = \pi_R(pS)$ for all $p \in M$.

PROOF: (a) $\Longrightarrow$ (b)

1. Assume that $S \cap P_0 \subset R \cap P_0$.
2. $S = (S \cap P_0) \circ gr(G(S)) \subset R \circ gr(G(S))$. (by 1, **7.21**)

(b) $\Longrightarrow$ (a)

1. Assume that $S \subset R \circ gr(G(S))$.
2. $S \cap P_0 \subset \big(R \circ gr(G(S))\big) \cap P_0 \subset R \cap P_0$. (by 1, **7.21**)

(a) $\Longrightarrow$ (c)

1. Assume that $S \cap P_0 \subset R \cap P_0$ and let $q \in pS$.
2. $(p,q) \in S = (S \cap P_0) \circ gr(G(S))$. (by 1, **7.21**)
3. There exists $v \in J$ and $\alpha \in G(S)$ such that $(p, pv) \in S \cap P_0$ and $q = \alpha(pv)$. (by 2)
4. $(\alpha(p), q) = \alpha(p, pv) \in \alpha(S \cap P_0) = S \cap P_0 \subset R \cap P_0$. (by 1, 3)
5. $\pi_R(q) = \pi_R(\alpha(p)) \in \pi_R(G(S)(p))$. (by 3, 4)
6. $\pi_R(pS) \subset \pi_R(G(S)(p))$. (by 1, 5)

(c) $\Longrightarrow$ (a)

1. Assume that $\pi_R(pS) \subset \pi_R(G(S)(p))$ for all $p \in M$.
2. Let $(p, pv) \in S \cap P_0$.
3. There exists $\alpha \in G(S)$ with $(\alpha(p), pv) \in R$. (by 1, 2)
4. $\alpha \in G(R)$ and $(p, pv) \in R$. (by 3, **7.21**)
5. $S \cap P_0 \subset R \cap P_0$. (by 2, 4)

The characterization of icers given in **7.21** will play a fundamental role in our exposition. We mention one important consequence here; namely that any distal extension (homomorphism) of minimal flows is an open extension.

Proposition 7.23 Let:

(i) $R \subset S$ be icers on M, and
(ii) $\pi \equiv \pi_S^R : M/R \to M/S$ be a distal homomorphism.

Then:

(a) $R \cap P_0 = S \cap P_0$, and
(b) π is an open map.

PROOF: (a) 1. Clearly $R \cap P_0 \subset S \cap P_0$. (by (i))
2. Let $\alpha \in G$ and $u, v \in J$ with $(\alpha(u), \alpha(v)) \in S \cap P_0$.
3. $\pi\ (\pi_R(\alpha(u))) = \pi_S(\alpha(u)) = \pi_S(\alpha(v)) = \pi\ (\pi_R(\alpha(v)))$. (by 2, (ii))
4. $\pi_R(\alpha(u), \alpha(v)) \in \pi_R(P_0) \subset \pi_R(P(M)) = P(M/R)$. (by 2, **4.7**)
5. $\pi_R(\alpha(u)) = \pi_R(\alpha(v))$. (by 3, 4, (ii))
6. $(\alpha(u), \alpha(v)) \in R \cap P_0$. (by 2, 5)
7. $R \cap P_0 = S \cap P_0$. (by 1, 2, 6)
(b) 1. Let U be open in M/R.
2.
$$\begin{aligned}\pi_R^{-1}\left(\pi^{-1}(\pi(U))\right) &= \{p \in M \mid \pi(pR) \in \pi(U)\}\\ &= \{p \in M \mid \pi(pR) = \pi(qR) \text{ for some } qR \in U\}\\ &= \{p \in M \mid (p, q) \in S \text{ for some } qR \in U\}\\ &\underset{\mathbf{7.21}}{=} \{p \in M \mid (\beta(p), q) \in P_0 \cap S\\ &\qquad \text{for some } \beta \in G(S) \text{ and } q \in \pi_R^{-1}(U)\}\\ &\underset{\text{(ii)},\mathbf{7.22}}{=} \{p \in M \mid (\beta(p), q) \in P_0 \cap R\\ &\qquad \text{for some } \beta \in G(S) \text{ and } q \in \pi_R^{-1}(U)\}\\ &= \{p \in M \mid \beta(p) \in \pi_R^{-1}(U) \text{ for some } \beta \in G(S)\}\\ &= \{p \in M \mid p \in \beta(\pi_R^{-1}(U)) \text{ for some } \beta \in G(S)\}\\ &= \bigcup_{\beta \in G(S)} \beta(\pi_R^{-1}(U)).\end{aligned}$$
3. $\pi_R^{-1}\left(\pi^{-1}(\pi(U))\right)$ is open. (by 1, 2)
4. $\pi^{-1}(\pi(U))$ is open. (by 3)
5. $\pi(U)$ is open. (by 4)

We will study distal extensions in more detail in section 18; in particular we will show (see **18.5**) that for $R \subset S$, the extension $M/R \to M/S$ is distal **if**

and only if $R \cap P_0 = S \cap P_0$. The reader may wish to verify this as an exercise now since the proof does not rely on any results not yet discussed.

Theorem **7.21** will often allow us to construct icers and hence minimal flows by manipulating relative products. To facilitate this we present the following lemma.

Lemma 7.24 Let:

(i) R be an equivalence relation on M,
(ii) $A, B \subset G$ be subgroups of G, and
(iii) $A \subset aut(R)$.

Then:

(a) $gr(A) \circ gr(B) = gr(BA)$,
(b) if $AB = BA$, then $gr(AB)$ is an equivalence relation on M,
(c) $R \circ gr(A) = gr(A) \circ R$, and
(d) $R \circ gr(A)$ is an equivalence relation on M.

PROOF: (a) 1. Let $(p, q) \in gr(A) \circ gr(B)$.
2. There exists $r \in M$ with $(p, r) \in gr(A)$ and $(r, q) \in gr(B)$.
3. There exist $\alpha \in A$ and $\beta \in B$ with $r = \alpha(p)$ and $\beta(r) = q$. (by 2)
4. $(p, q) = (p, \beta(\alpha(p))) \in gr(BA)$. (by 3)
5. $gr(A) \circ gr(B) \subset gr(BA)$. (by 1, 4)
6. Let $(p, \beta(\alpha(p))) \in gr(BA)$ with $\alpha \in A$ and $\beta \in B$.
7. $(p, \alpha(p)) \in gr(A)$ and $(\alpha(p), \beta(\alpha(p))) \in gr(B)$. (by 6)
8. $(p, \beta(\alpha(p))) \in gr(A) \circ gr(B)$. (by 7)
9. $gr(BA) \subset gr(A) \circ gr(B)$. (by 6, 8)
(b) 1. Assume that $AB = BA$.
2. AB is a group. (by 1)
3. $gr(AB)$ is an equivalence relation on M. (by 2)
(c) 1. Let $(p, r) \in R \circ gr(A)$.
2. There exists $q \in M$ with $(p, q) \in R$ and $\alpha(q) = r$ for some $\alpha \in A$. (by 1)
3. $(\alpha(p), r) = \alpha(p, q) \in R$. ($\alpha \in A \subset aut(R)$ by 2, (iii))
4. $(p, \alpha(p))) \in gr(A)$ so $(p, r) \in gr(A) \circ R$. (by 3)
5. $R \circ gr(A) \subset gr(A) \circ R$. (by 1, 4)
6. $gr(A) \circ R = (R \circ gr(A))^{-1} \subset (gr(A) \circ R)^{-1} = R \circ gr(A)$. (by 5)
7. $R \circ gr(A) = gr(A) \circ R$. (by 5, 6)
(d) follows immediately from part (c). (by **Ex. 6.2**)

It is clear from **7.21** that an icer R on M is completely determined by $R \cap P_0$ and $G(R)$; we make this precise for future reference in the following proposition.

Proposition 7.25 Let N, S be icers on M such that:

(i) $P_0 \cap S \subset P_0 \cap N$, and
(ii) $G(S) \subset G(N)$.

Then $S \subset N$.

PROOF: $S \underset{\mathbf{7.21}}{=} (S \cap P_0) \circ gr(G(S)) \underset{\text{(i),(ii)}}{\subset} (N \cap P_0) \circ gr(G(N)) \underset{\mathbf{7.21}}{=} N.$

Clearly we will be interested in how $G(R \circ S)$ is related to $G(R)$ and $G(S)$ when R, S and $R \circ S$ are icers on M. This question is answered by the following proposition.

Proposition 7.26 Let R, S be icers on M such that $R \circ S$ is an icer on M. Then $G(R \circ S) = G(R)G(S)$.

PROOF: 1. $R \subset R \circ S$ and $S \subset R \circ S$.
2. $G(R) \subset G(R \circ S)$ and $G(S) \subset G(R \circ S)$. (by 1)
3. $G(R)G(S) \subset G(R \circ S)$. (by 2)
4. Let $\alpha \in G(R \circ S)$ and $u^2 = u \in M$.
5. $(u, \alpha(u)) \in R \circ S = S \circ R$. (by 4)
6. There exists $\beta(v) \in M$ with $(u, \beta(v)) \in S$ and $(\beta(v), \alpha(u)) \in R$. (by 5, **7.4**)
7. $(u, \beta(u)) = (u, \beta(v))u \in Su \subset S$ and $(\beta\alpha^{-1}(\alpha(v)), \alpha(v)) = (\beta(v), \alpha(u))$ $v \in Rv \subset R$. (by 6)
8. $\beta \in G(S)$ and $\alpha\beta^{-1} = (\beta\alpha^{-1})^{-1} \in G(R)$. (by 7)
9. $\alpha = (\alpha\beta^{-1})\beta \in G(R)G(S)$. (by 8)

When R and S are icers on M **7.26** can be thought of as giving an obstruction to $R \circ S$ being an icer on M. In order for $R \circ S$ to be an icer, the product $G(R)G(S) = G(R \circ S)$ must be a group. This motivates our next result which gives a sufficient condition for $R \circ S$ to be an icer.

Proposition 7.27 Let:

(i) R, S be icers on M,
(ii) $G(R)G(S) = G(S)G(R)$, and
(iii) $R \cap P_0 = S \cap P_0$.

Then $R \circ S$ is an icer on M.

PROOF: Since $R \circ S$ is reflexive, closed and invariant it suffices to show that $R \circ S = S \circ R$.

$$R \circ S \underset{\mathbf{7.21}}{=} (R \cap P_0) \circ gr(G(R)) \circ (S \cap P_0) \circ gr(G(S))$$

$$\underset{\text{(iii)}}{=} (R \cap P_0) \circ gr(G(R)) \circ (R \cap P_0) \circ gr(G(S))$$

$$\underset{\text{(i)}}{=} (R \cap P_0) \circ gr(G(R)) \circ gr(G(S))$$

$$\underset{\mathbf{7.24}}{=} (R \cap P_0) \circ gr(G(S)G(R))$$

$$\underset{\text{(ii),(iii)}}{=} (S \cap P_0) \circ gr(G(R)G(S))$$

$$\underset{\mathbf{7.24}}{=} (S \cap P_0) \circ gr(G(S)) \circ gr(G(R))$$

$$\underset{\text{(i)}}{=} (S \cap P_0) \circ gr(G(S)) \circ (S \cap P_0) \circ gr(G(R))$$

$$\underset{\text{(iii)}}{=} (S \cap P_0) \circ gr(G(S)) \circ (R \cap P_0) \circ gr(G(R))$$

$$\underset{\mathbf{7.21}}{=} (S \circ R).$$

We end this section with another sufficient condition for $R \circ S$ to be an icer; we will use this result in sections 15, 18, and 19.

Proposition 7.28 Let:

(i) R, S be icers on M,
(ii) $G(S) \subset aut(R)$, and
(iii) $S \cap P_0 \subset R \cap P_0$.

Then:

(a) $R \circ S$ is an icer on M,
(b) $\pi_R(S)$ is an icer on M/R, and
(c) $\pi_S(R)$ is an icer on M/S.

PROOF: (a) It suffices to show that $R \circ S = S \circ R$.

1. $R \circ S \underset{\mathbf{7.21}}{=} gr(G(R)) \circ (R \cap P_0) \circ (S \cap P_0) \circ gr(G(S)).$

$$\underset{\text{(iii)}}{=} gr(G(R)) \circ (R \cap P_0) \circ gr(G(S)).$$

$$\underset{\text{(ii)},\mathbf{7.24}}{=} gr(G(R)) \circ gr(G(S)) \circ (R \cap P_0).$$

$$\underset{\mathbf{7.24}}{=} gr(G(S)G(R)) \circ (R \cap P_0).$$

2. $S \circ R \underset{\mathbf{7.21}}{=} gr(G(S)) \circ (S \cap P_0) \circ (R \cap P_0) \circ gr(G(R))$

$\underset{\text{(iii)}}{=} gr(G(S)) \circ (R \cap P_0) \circ gr(G(R))$

$\underset{\mathbf{7.21}}{=} gr(G(S)) \circ gr(G(R)) \circ (R \cap P_0)$

$\underset{\mathbf{7.24}}{=} gr(G(R)G(S)) \circ (R \cap P_0)$.

3. $G(R)$ is a normal subgroup of $aut(R)$. (by **7.10**)
4. $G(R)G(S) = G(S)G(R)$. (by 3, (ii))
5. $R \circ S = S \circ R$. (by 1, 2, 4)

Parts (b) and (c) follow from part (a) and **6.11**.

NOTES ON SECTION 7

In the body of this section we have identified a minimal flow with an object in the category $\mathcal{M}$, of icers on M. Following the treatment in [Ellis (1969)], one can instead identify a minimal flow with an object in the category $al(M)$, of T-invariant uniformly closed subalgebras of $\mathcal{C}(M)$ where

$$\mathcal{C}(M) = \{f{:}M \to \mathbf{R} \mid f \text{ is continuous}\}.$$

We now examine a few of the key ideas in this section from this point of view. The objects in the two categories are related by the following proposition which is an immediate consequence of the Stone-Weierstrass theorem. (see [Rudin, Walter (1953)])

Note 7.N.1 For every $R \in obj(\mathcal{M})$ and $\mathcal{A} \in al(M)$ let

(i) $al(R) = \{f \in \mathcal{C}(M) \mid f(x) = f(y) \text{ for all } (x, y) \in R\}$, and
(ii) $R(\mathcal{A}) = \{(x, y) \in M \times M \mid f(x) = f(y) \text{ for all } f \in \mathcal{A}\}$.

Then the map $\begin{matrix} obj(\mathcal{M}) \to al(M) \\ R \to al(R) \end{matrix}$ is bijective, with inverse $\begin{matrix} al(M) \to obj(\mathcal{M}) \\ \mathcal{A} \to R(\mathcal{A}) \end{matrix}$.

We now describe the group $G(R)$ of an icer on M in terms of the subalgebra $al(R) \subset \mathcal{C}(M)$. We begin with a preliminary lemma.

Note 7.N.2 Let $f, g \in Hom(M, X)$ with X minimal. Then:

(a) there exists $\alpha \in G$ with $f = g \circ \alpha$,
(b) $G(R_f) = \{\beta \in G \mid f\beta = f\}$, and
(c) $G(R_f) = \alpha^{-1}G(R_g)\alpha$.

PROOF: (a) 1. Let S be an icer on M with $X = M/S$.
2. There exist $\beta, \gamma \in G$ with $\beta(\Delta) \subset S$, $\gamma(\Delta) \subset S$ and $f \circ \pi_\Delta = \pi_S \circ \beta$, $g \circ \pi_\Delta = \pi_S \circ \gamma$. (by **7.6**)
3. $f = \pi_S \circ \beta = (\pi_S \circ \gamma) \circ (\gamma^{-1}\beta) = g \circ (\gamma^{-1}\beta)$. (by 2)
(b) 1. Let $\gamma \in G(R_f)$.
2. $gr(\gamma) \subset R_f$.
3. $f(p) = f(\gamma(p))$ for all $p \in M$. (by 2)
4. $f = f \circ \gamma$. (by 3)
5. Let $\gamma \in G$ with $f\gamma = f$.
6. $f(p) = f(\gamma(p))$ for all $p \in M$. (by 5)
7. $(p, \gamma(p)) \in R_f$. (by 6)
(c)

$$\gamma \in G(R_f) \iff f\gamma = f \iff g \circ \alpha \circ \gamma = g \circ \alpha$$
$$\iff g \circ (\alpha \circ \gamma \circ \alpha^{-1}) = g \iff \alpha \circ \gamma \circ \alpha^{-1} \in G(R_g)$$
$$\iff \gamma \in \alpha^{-1} G(R_g)\alpha.$$

Note 7.N.3 Let R be an icer on M and $\mathcal{A} = al(R)$. Then

$$G(R) = \{\alpha \mid f\alpha = f \text{ for all } f \in \mathcal{A}\} \equiv B.$$

PROOF: 1. Let $\alpha \in G(R)$, and $f \in \mathcal{A}$.
2. $gr(\alpha) \subset R$. (by 1)
3. $f(x) = f(\alpha(x))$ for all $f \in \mathcal{A}$ and $x \in M$. (by 2, $\mathcal{A} = al(R)$)
4. $\alpha \in B$. (by 3)
5. Now let $\alpha \in B$, and $x \in M$.
6. $f(\alpha(x)) = f(x)$ for all $f \in \mathcal{A}$. (by 5)
7. $(x, \alpha(x)) \in R$. (by 6)
8. $gr(\alpha) \subset R$. (by 5, 7)
9. $\alpha \in G(R)$. (by 8)

The preceding result shows that the definition of the group $G(R)$ associated to the flow $(M/R, T)$ coincides with the group of the flow associated to the corresponding algebra $\mathcal{A}$ as defined in [Ellis, R., (1969)]. We end the notes on this section with the observation that the inf of two icers (which is their relative product when it's an icer) corresponds to the intersection of the corresponding subalgebras of $\mathcal{C}(M)$.

Note 7.N.4 Let:

(i) R, S be icers on M, and
(ii) $\mathcal{A} = al(R)$ and $\mathcal{B} = al(S)$.

Then $al(\inf(R, S)) = \mathcal{A} \cap \mathcal{B}$.

PROOF: 1. Let $\mathcal{C} \equiv al(\inf(R, S))$.
2. $R, S \subset \inf(R, S)$.
3. $\mathcal{C} \subset \mathcal{A} \cap \mathcal{B}$. (by 1, 2, and **7.N.1**)
4. $R = R(\mathcal{A}) \subset R(\mathcal{A} \cap \mathcal{B})$ and $S = R(\mathcal{B}) \subset R(\mathcal{A} \cap \mathcal{B})$. (by **7.N.1**)
5. $\inf(R, S) \subset R(\mathcal{A} \cap \mathcal{B})$. (by 4)
6. $\mathcal{A} \cap \mathcal{B} \subset al(\inf(R, S)) = \mathcal{C}$. (by 5)

EXERCISES FOR CHAPTER 7

Exercise 7.1 (See **7.16**) Let R be an icer on M and $\alpha \in G$. Show that

(a) $\alpha(R)$ is an icer on M, and
(b) $G(\alpha(R)) = \alpha G(R)\alpha^{-1}$.

Exercise 7.2 Let R be an icer. Then

$$G(R) = \{\alpha \in G \mid (1 \times \alpha)(R) \subset R\} = \{\alpha \in G \mid (1 \times \alpha)(R) = R\}.$$

Exercise 7.3 Let R be an icer on M. Then

$$R = \{(\alpha(u), \beta(v)) \mid \alpha\beta^{-1} \in G(R), (u, v) \in (J \times J) \cap \alpha^{-1}(R)\}.$$

Exercise 7.4 Let R be an icer on M. Then $R = \bigcup_{\alpha \in G(R)} (1 \times \alpha)(R \cap P_0)$.

Exercise 7.5 Suppose that R is an icer on M; given the results of **Ex. 7.3** and **Ex. 7.4** it is natural to consider the relations

$$R_H = \{(\alpha(u), \beta(v)) \mid \alpha\beta^{-1} \in H, (u, v) \in (J \times J) \cap \alpha^{-1}(R) \cap \beta^{-1}(R)\}$$

and

$$R^H = \bigcup_{\alpha \in H} (1 \times \alpha)(R \cap P_0)$$

where H is a subgroup of G. Note that if $H = G(R)$, then **Ex. 7.3** and **Ex. 7.4** imply that $R_H = R = R^H$. In general R_H and R^H need not be equal, and neither are icers. Prove the following:

(a) $R_H \subset R^H$,
(b) if $H \subset aut(R)$, then $R_H = R^H$ is an equivalence relation, and
(c) if $G(R) \subset H$, then $R^H = \bigcup_{\alpha \in H} (1 \times \alpha)(R)$.

Exercise 7.6 Complete the proof of proposition **7.10**.

Exercise 7.7 Let R, S be icers on M. Then:

(a) $\bigcup_{\alpha \in G(S)} (1 \times \alpha)(R) \subset R \circ S$, and

(b) if $S \cap P_0 \subset R \cap P_0$, then $R \circ S = \bigcup_{\alpha \in G(S)} (1 \times \alpha)(R)$.

Exercise 7.8 (A partial converse to **7.24**) Let:

(i) $R \subset P_0$ be an equivalence relation on M,
(ii) A be a subgroup of G, and
(iii) $R \circ gr(A) = gr(A) \circ R$.

Then $A \subset aut(R)$. (In particular $G(R) \subset aut(R \cap P_0)$ for any icer R.)

Exercise 7.9 Let:

(i) R, S be icers on M, and
(ii) N be an icer on M with $R \cup S \subset N$.

Then we can form the commutative diagram

$$\begin{array}{cccc} (\pi_R \times \pi_S)(N) \subset & M/R \times M/S & \rightarrow & M/S \\ & \downarrow & & \downarrow \pi_N^S \\ & M/R & \overset{\pi_N^R}{\rightarrow} & M/N \end{array}$$

and $(\pi_R \times \pi_S)(N)$ is minimal if and only if $N = R \circ S$.

Exercise 7.10 Let:

(i) R, S be icers on M, and
(ii) $N = \inf(R, S)$.

Then $G(N) = G(R)G(S)$ if and only if $(\pi_R \times \pi_S)(N)$ contains only one minimal set.

8

Regular flows

Regular flows were introduced in [Auslander, J., *Regular minimal sets*, 1966]. The original definition is motivated by the idea that regular flows are those which admit as many automorphisms as possible. Indeed for a regular flow, its group of automorphisms, $Aut(X)$, acts almost transitively in the sense that for any $p, q \in X$ there exists an $\alpha \in Aut(X)$ such that $\alpha(p)$ is proximal to q. Here we focus on the icer R on the universal minimal ideal M, writing $X = M/R$. In this context it is natural to consider those icers for which $aut(R) = G$, which motivates our definition of a *regular flow*.

Definition 8.1 Let R be an icer on M. We say that R is *regular* if $\alpha(R) = R$ for every $\alpha \in G$. Thus R is regular if and only if $aut(R) = G$ (see **7.9**). We also refer to the flow $X = M/R$ as a *regular flow*.

We begin with a few immediate consequences of the definition.

Proposition 8.2 Let:

(i) R be an icer on M, and
(ii) $\alpha(R) \subset R$ for all $\alpha \in G$.

Then:

(a) R is regular, and
(b) $G(R)$ is a normal subgroup of G.

PROOF: (a) 1. $R = \alpha(\alpha^{-1}(R)) \subset \alpha(R)$ for all $\alpha \in G$. (by (ii))
2. $R = \alpha(R)$ for all $\alpha \in G$. (by 1, (ii))

(b) R is regular so $aut(R) = G$, and hence $G(R)$ is a normal subgroup of G. (by **7.10**)

For a given minimal flow (X, T), there may be many icers R on M with $M/R \cong X$. The flows for which R is unique are exactly the regular flows, as shown in the next proposition.

Proposition 8.3 Let:

(i) R and S be icers on M,
(ii) R be regular, and
(iii) $\varphi : M/R \to M/S$ be a homomorphism.

Then:

(a) $R \subset S$, and
(b) if φ is one-one, then $R = S$.

PROOF: (a) 1. There exists $\alpha \in G$ with $\alpha(R) \subset S$. (by (ii), **7.6**)
2. $R \subset S$. (by 1, (iii))
(b) 1. Assume that φ is one-one.
2. $\varphi^{-1}:M/S \to M/R$ is a homomorphism. (by 1)
3. There exists $\alpha \in G$ with $\alpha(S) \subset R$. (by **7.6**)
4. $R \underset{(a)}{\subset} S \underset{3}{\subset} \alpha^{-1}(R) \underset{(iii)}{=} R$.

Given any icer R on M we construct the largest regular icer $reg(R) \subset R$, the so-called *regularizer* of R.

Definition 8.4 Let R be an icer on M. The *regularizer of* R is defined by

$$reg(R) = \bigcap_{\alpha \in G} \alpha(R).$$

When $X = M/R$ we will write $reg(X) = M/reg(R)$.

It is clear from the definition that $aut(reg(R)) = G$ so that $reg(R)$ is a regular icer on M. We now show that when R is an icer the flow $reg(M/R)$ can be identified with a minimal ideal in the enveloping semigroup of M/R.

Proposition 8.5 Let:

(i) R be an icer and $X = M/R$,
(ii) $\Phi_X\colon \beta T \to E(X)$ the canonical map,
(iii) $I(X) = \Phi_X(M)$, and
(iv) $N = R_{\Phi_X} \equiv \{(p, q) \in M \times M \mid \Phi_X(p) = \Phi_X(q)\}$.

Then $N = reg(R)$ and hence $I(X) = reg(X)$.

PROOF: Proof that $N \subset reg(R)$:
1. Let $(p, q) \in N$, $\alpha \in G$, and $u \in J$.
2. $\pi_R(\alpha(p)) = \pi_R(\alpha(up)) = \pi_R(\alpha(u)p) \underset{\mathbf{2.9}}{=} \pi_R(\alpha(u))\Phi_X(p)$

$$= \pi_R(\alpha(u))\Phi_X(q) = \pi_R(\alpha(u)q) = \pi_R(\alpha(uq)) = \pi_R(\alpha(q)).$$

3. $(\alpha(p), \alpha(q)) \in R$. (by 2)
4. $(p, q) \in reg(R)$. (by 1, 3)

Proof that $reg(R) \subset N$:

1. Let $(p, q) \in reg(R)$ and $x \in X$.
2. There exists $\alpha \in G$ and $u \in J$ with $\pi_R(\alpha(u)) = x$.
3. $x\Phi_X(p) = \pi_R(\alpha(u))\Phi_X(p) = \pi_R(\alpha(p)) \underset{1}{=} \pi_R(\alpha(q)) = x\Phi_X(q)$.
4. $\Phi_X(p) = \Phi_X(q)$. (by 1, 3)
5. $(p, q) \in N$. (by 1, 4)

Applying **8.5** in the case where R is a regular icer on M, so that $R = reg(R)$, we see that $X \cong I(X) = \Phi_X(M)$. Here $X = M/R$ and $\Phi_X: \beta T \to E(X)$ is the canonical map. As a result X inherits a semigroup structure which is given by:

$$\pi_R(p)\pi_R(q) = \pi_R(pq),$$

for all $p, q \in M$. The semigroup X has a structure analogous to that of the universal minimal flow M; we state this precisely in the following propositions.

Proposition 8.6 Let:

(i) R be a regular icer on M, and
(ii) $X = M/R$.

Then:

(a) X has a semigroup structure for which $\pi_R: M \to X$ is both a flow and a semigroup homomorphism, in particular $\pi_R(p)\pi_R(q) = \pi_R(pq)$, for all $p, q \in M$.
(b) the maps $\begin{array}{rcl} L_x : X & \to & X \\ y & \to & xy \end{array}$ are flow homomorphisms for all $x \in X$.
In particular X is an E-semigroup (see **Ex. 2.3** and **Ex. 3.4**).

PROOF: This follows immediately from **8.5** and **2.9**.

We saw in **7.4** that the elements of M can be written uniquely in the form $\alpha(u)$ where $\alpha \in G$ and u is an idempotent in M; the analogous statement holds for the elements of a regular flow X. In order to state this result and some of its consequences we introduce some notation.

Notation 8.7 Let R be an icer on M and $X = M/R$. When X is regular, the group $Aut(X)$ of automorphisms of X will often be denoted G_X. The collection of idempotents in X will be denoted $J_X = \{u \in X \mid u^2 = u\}$.

We proceed with a series of results which show that when X is a regular minimal flow, the pair $\{X, G_X\}$ has properties analogous to those of the pair $\{M, G\}$.

Proposition 8.8 Let:

(i) $X = M/R$ be a regular flow, and
(ii) $J_X = \{u \in X \mid u^2 = u\}$.

Then:

(a) the map $\begin{matrix} G_X & \to & Xu \\ \alpha & \to & \alpha(u) \end{matrix}$ is an isomorphism of groups for every $u \in J_X$, and
(b) the map $\begin{matrix} \varphi: G_X \times J_X & \to & X \\ (\alpha, u) & \to & \alpha(u) \end{matrix}$ is bijective. Thus any element of X can be written uniquely in the form $\alpha(v)$ for some $\alpha \in G_X$ and $v \in J_X$.

PROOF: We leave this to the reader (compare **7.4**).

Proposition 8.9 Let X be a minimal flow. Then the following are equivalent:

(a) X is regular,
(b) $Y \subset X \times X$ is minimal if and only if $Y = gr(\beta)$ for some $\beta \in Aut(X)$ (compare **7.5**), and
(c) for any $x, y \in X$ there exists $\beta \in Aut(X)$ such that $\beta(x)$ is proximal to y.

PROOF: We leave the proof as an exercise for the reader.

In section 7 we developed the machinery which allows us to analyze the factors of M (all minimal flows) using the icers on M and the subgroups of the group G of automorphisms of M. The structure of any regular flow X allows for the same treatment of the factors of X in terms of the icers on X and the subgroups of the group G_X of automorphisms of X. In order to outline these ideas we introduce some additional notation.

Definition 8.10 Let $X = M/R$ be a regular flow. Let N be an icer on X. We define the *X-group of N* by

$$G_X(N) = \{\alpha \in G_X \mid gr(\alpha) \subset N\}.$$

We will also use the notation

$$aut_X(N) \equiv \{\alpha \in G_X \mid \alpha(N) = N\}.$$

As in **7.10**, $G_X(N)$ is a normal subgroup of $aut_X(N)$. For reference we restate proposition **7.10** here using the current notation.

Proposition 8.11 Let $X = M/R$ be a regular flow. Then there exists a group epimorphism

$$\chi_R : G = aut(R) \to G_X$$

such that:

(a) $\pi_R(\alpha(p)) = \chi_R(\alpha)(\pi_R(p))$ for all $p \in M$ and $\alpha \in aut(R)$,
(b) $ker(\chi_R) = G(R)$, and
(c) $G_X \cong G/G(R)$.

The next three results explore the naturality of these constructions with respect to the map χ_R. The following definition will be used in later sections; the naturality results can easily be extended to the map χ_S^R.

Definition 8.12 Let $X = M/R$ and $Y = M/S$ be regular flows with $R \subset S$. Then the (canonical) map of $G_X \cong G/G(R)$ onto $G_Y \cong G/G(S)$ induced by the inclusion $G(R) \subset G(S)$ will be denoted χ_S^R. Thus $\chi_S^R \circ \chi_R = \chi_S$.

Proposition 8.13 Let:

(i) $X = M/R$ be a regular flow,
(ii) N be an icer on X, and
(iii) $S = \pi_R^{-1}(N)$.

Then:

(a) S is an icer on M,
(b) $M/S \cong X/N$, and
(c) $\chi_R(G(S)) = \{\alpha \in G_X \mid gr(\alpha) \subset N\} \equiv G_X(N)$.

PROOF: (a) and (b) are simply **7.2** in the case where R is regular.

(c) Proof that $\chi_R(G(S)) \subset G_X(N)$:

1. Let $\alpha \in G(S)$ and $x \in X$.
2. There exists $p \in M$ with $\pi_R(p) = x$.
3. $(p, \alpha(p)) \in S$. (by 1)
4. $(x, \chi_R(\alpha)(x)) \underset{2}{=} (\pi_R(p), \chi_R(\alpha)(\pi_R(p)))$

$$\underset{\mathbf{8.11}}{=} (\pi_R(p), \pi_R(\alpha(p))) = \pi_R(p, \alpha(p)) \underset{3}{\in} \pi_R(S) \underset{\text{(iii)}}{=} N.$$

5. $\chi_R(\alpha) \in G_X(N)$. (by 1, 4)

Proof that $G_X(N) \subset \chi_R(G(S))$:

1. Let $\gamma \in G_X(N)$ and $p \in M$.
2. There exists $\alpha \in G$ with $\chi_R(\alpha) = \gamma$. (by **8.11**)

3. $\pi_R(p, \alpha(p)) = (\pi_R(p), \pi_R(\alpha(p))) \underset{2,\mathbf{8.11}}{=} (\pi_R(p), \gamma(\pi_R(p))) \underset{1}{\in} N$.

4. $(p, \alpha(p)) \underset{3}{\in} \pi_R^{-1}(N) \underset{\text{(iii)}}{=} S$.

5. $\gamma \underset{2}{=} \chi_R(\alpha) \underset{4}{\in} \chi_R(G(S))$.

Corollary 8.14 Let:

(i) $X = M/R$ be a regular flow,
(ii) N be an icer on X,
(iii) $S = \pi_R^{-1}(N)$, and
(iv) $aut_X(N) = \{\alpha \in G_X \mid \alpha(N) = N\}$.

Then $\chi_R(aut(S)) = aut_X(N)$.

PROOF: Similar to **8.13**.

Corollary 8.15 Let:

(i) $X = M/R$ be a regular flow,
(ii) N be an icer on M, and
(iii) $\pi_R(N)$ be an icer on X.

Then $\chi_R(G(N)) = G_X(\pi_R(N)) \equiv \{\alpha \in G_X \mid gr(\alpha) \subset \pi_R(N)\}$.

PROOF: Set $S = \pi_R^{-1}(\pi_R(N)) \underset{\mathbf{6.4}}{=} R \circ N \circ R$.

Proof that $\chi_R(G(N)) \subset G_X(\pi_R(N))$:

1. $N \subset S$.
2. $G(N) \subset G(S)$. (by 1)
3. $\chi_R(G(N)) \subset \chi_R(G(S)) = G_X(\pi_R(N))$. (by 2, **8.13**)

Proof that $G_X(\pi_R(N)) \subset \chi_R(G(N))$:

1. $G_X(\pi_R(N)) \underset{\mathbf{8.13}}{=} \chi_R(G(S)) = \chi_R(G(R \circ N \circ R)) \subset \chi_R(G(R)G(N)G(R)) = \chi_R(G(N))$.

(We leave it to the reader to check that $G(R \circ N \circ R) \subset G(R)G(N)G(R)$.)

We end this section with a generalization of the notion of regular flows to homomorphisms (extensions) of minimal flows which will be useful in part V.

Definition 8.16 Let $\pi_S^R : M/R \to M/S$ be the canonical homomorphism of minimal flows where $R \subset S$ are icers on M. We say that π_S^R is *regular homomorphism* and that $R \subset S$ is a *regular extension* if $G(S) \subset aut(R)$.

Note that if $S = M \times M$, then $G(S) = G$, so that M/R is a regular extension of the point flow if and only if R is a regular icer. A construction analogous to that of the regularizer associates to any extension a corresponding regular extension.

Proposition 8.17 Let:

(i) $N \subset S$ be icers on M, and
(ii) $R = \bigcap_{\alpha \in G(S)} \alpha(N)$.

Then π_S^R is regular ($R \subset S$ is a regular extension).

PROOF: Straighforward.

Proposition 8.18 Let:

(i) $R \subset S$ be icers on M, and
(ii) $\pi \equiv \pi_S^R : X \equiv M/R \to M/S$.

Then the following are equivalent:

(a) $R \subset S$ is regular,
(b) $Y \subset \pi_R(S)$ is minimal if and only if $Y = gr(\beta)$ for some $\beta \in Aut(X)$, and
(c) for any $y \in \pi^{-1}\pi(x) \subset X$ there exists $\beta \in Aut(X)$ such that $\beta(x)$ is proximal to y.

PROOF: We leave the proof as an exercise for the reader.

NOTES ON SECTION 8

Note 8.N.1 A close reading of section 7 reveals that none of the results depend on any special property of M other than regularity. This suggests that they remain valid when M is replaced by an arbitrary regular minimal flow. We have touched on this in **8.8**, **8.9**, and **8.10**. For emphasis and to be more specific: Let Z be a regular minimal flow. Then mimicking **7.1** we define the category $\mathcal{Z}$:

$$obj(\mathcal{Z}) \equiv \text{icers on } Z$$

$$morph(\mathcal{Z}) = \begin{cases} \{\pi_S^R\} & \text{if } R \subset S \text{ are icers on } Z, \\ \emptyset & \text{otherwise.} \end{cases}$$

Here π_S^R is the canonical map $Z/R \to Z/S$ when $R \subset S$.
Other relevant definitions and notation:

1. $G_Z \equiv$ the set of automorphisms of Z.
2. Let R be an icer on Z. Then:

2.1. the Z-group of R is defined by $G_Z(R) = \{\alpha \in G_Z \mid gr(\alpha) \subset R\}$. (as in **8.10**)
2.2. $aut_Z(R) \equiv \{\alpha \in G_Z \mid \alpha(R) = R\}$. (again as in **8.10**)
2.3. R is Z-regular if $aut_Z(R) = G_Z$.
2.4. $\pi_R: Z \to Z/R$ denotes the canonical map.

Finally all the results of sections 8 and 9 remain valid if the category $\mathcal{M}$ and related concepts are replaced be the category $\mathcal{Z}$ and the concepts defined above. Moreover these constructions behave naturally with respect to the canonical projection map (see for example **8.13**, **8.14**, and **8.15**).

EXERCISES FOR CHAPTER 8

Exercise 8.1 Let R be an icer on M and $\alpha \in G$. Then $\alpha(pR) = \alpha(p)(\alpha(R))$. In particular if R is regular, then $\alpha(pR) = \alpha(p)R$.

Exercise 8.2 (See **8.9**) Let X be a minimal flow. Then the following are equivalent:

(a) X is regular,
(b) $Y \subset X \times X$ is minimal if and only if $Y = gr(\beta)$ for some $\beta \in Aut(X)$, and
(c) for any $x, y \in X$ there exists $\beta \in Aut(X)$ such that $\beta(x)$ is proximal to y.

Exercise 8.3 (See **8.15**) Let:

(i) $X = M/R$ be a regular flow,
(ii) N be an icer on M, and
(iii) $\pi_R(N)$ be an icer on X.

Show that $G(R \circ N \circ R) \subset G(R)G(N)G(R)$.

Exercise 8.4 (See **8.18**) Let:

(i) $R \subset S$ be icers on M, and
(ii) $\pi \equiv \pi_S^R: X \equiv M/R \to M/S$.

Then the following are equivalent:

(a) $R \subset S$ is regular,
(b) $Y \subset \pi_R(S)$ is minimal if and only if $Y = gr(\beta)$ for some $\beta \in Aut(X)$, and
(c) for any $y \in \pi^{-1}\pi(x)(\pi\ (x)) \subset X$ there exists $\beta \in Aut(X)$ such that $\beta(x)$ is proximal to y.

9

The quasi-relative product

The definition of the quasi-relative product given below is motivated by the fact that when R and S are closed equivalence relations on X, the relative product $R \circ S$, though closed, may not be an equivalence relation. The quasi-relative product $R(S)$, while not always closed, is the largest equivalence relation N with $R \subset N \subset R \circ S$. In **9.8**, we show that when the projection map π_R is open, $R(S)$ is closed. Under the same assumption we show in **9.9**, that $(X/R(S), T)$ is a quasi-factor of X/S. That is, $(X/R(S), T)$ is isomorphic to a sub-flow of $(2^{X/S}, T)$. This motivates the use of the term quasi-relative product.

At the end of this section we use the quasi-relative product to give a proof that if R is an icer on a minimal flow such that (R, T) is both pointwise almost periodic and topologically transitive, then $R = \Delta$. This result (which is equivalent to the generalized Furstenberg structure theorem for distal extensions) was proven for metric flows in **4.19**, and will be discussed further in section 20. We will also use the quasi-relative product in section 17 as a means of studying so-called RIC extensions of minimal flows.

We begin this section by deriving some properties of the quasi-relative product and using them to give conditions under which the relative product of two equivalence relations is an equivalence relation. As in section 6 which dealt with the relative product, many results in this section are stated for equivalence relations on any compact Hausdorff space X. If (X, T) is a flow and the equivalence relations are invariant under the action of T, then the results remain valid.

In order to state the definition of the quasi-relative product we recall the notation introduced in **6.1**; namely if R is a relation on X, then

$$xR = \{y \in X \mid (x, y) \in R\},$$

denotes the R-cell containing x.

Definition 9.1 Let R and S be any relations on X. We define the *quasi-relative product of R and S*

$$R(S) = \{(p, q) \in X \times X \mid p(R \circ S) = q(R \circ S)\}.$$

It is immediate from the definition that even when neither R nor S is an equivalence relation, $R(S)$ is an equivalence relation. When S is an equivalence relation on X note that

$$z \in p(R \circ S) \iff \text{there exists } x \in pR \text{ with } (x, z) \in S \iff \pi_S(z) \in \pi_S(pR).$$

Thus in this case $p(R \circ S) = \pi_S^{-1}(\pi_S(pR))$, which proves the following lemma.

Lemma 9.2 Let:

(i) X be a compact Hausdorff space,
(ii) R be any relation on X, and
(ii) S be an equivalence relation on X.

Then $R(S) = \{(p, q) \mid \pi_S(pR) = \pi_S(qR)\}$.

The elementary properties of the quasi-relative product are readily deduced from those of the relative product. In section 6 (see **6.3**) we observed that for any two relations, $x(R \circ S) = (xR)S$; since we will refer to this fact frequently we state it explicitly as part of the next lemma.

Lemma 9.3 Let:

(i) (X, T) be a flow,
(ii) R, S be subsets of $X \times X$,
(iii) $t \in T$, and
(iv) $z \in X$.

Then:

(a) $(zR)t = (zt)(Rt)$, and
(b) $(zR)S = z(R \circ S)$.

PROOF: We leave this as an exercise for the reader.

The quasi-relative product $R(S)$ is the largest equivalence relation which contains R and is contained in $R \circ S$. We deduce this and a couple of elementary consequences before examining the question of when $R(S)$ is closed.

Proposition 9.4 Let:

(i) (X, T) be a flow,
(ii) $H \subset T$ be a subgroup of T, and
(iii) R, S be H-invariant equivalence relations on X.

Then:

(a) $R(S)$ is an H-invariant equivalence relation on X,
(b) $R \subset R(S) \subset R \circ S$, and
(c) $R(S) = R \circ (R(S) \cap S) = (R(S) \cap S) \circ R$.

PROOF: (a) 1. $R(S)$ is clearly reflexive, symmetric, and transitive.
2. Let $(p, q) \in R(S)$ and $t \in H$.
3. $(pt)(R \circ S) = (pt)(Rt \circ St) \underset{\mathbf{9.3}}{=} ((pt)(Rt))(St) \underset{\mathbf{9.3}}{=} ((pR)t)(St)$

$$\underset{\mathbf{9.3}}{=} ((pR)S)t = (p(R \circ S))t \underset{2}{=} (q(R \circ S))t \underset{\mathbf{9.3},\text{(iii)}}{=} (qt)(R \circ S).$$

4. $(pt, qt) \in R(S)$. (by 3)
(b) 1. $(p, q) \in R(S) \Rightarrow p(R \circ S) = q(R \circ S) \Rightarrow q \in p(R \circ S) \Rightarrow (p, q) \in R \circ S$.
2. Let $(p, q) \in R$.
3. $pR = qR$. (by 2, R is transitive)
4. $p(R \circ S) = (pR)S = (qR)S = q(R \circ S)$. (by 3, **9.3**)
5. $(p, q) \in R(S)$. (by 4)
(c) 1. The fact that $R \circ (R(S) \cap S) \subset R(S)$ follows from parts (a) and (b).
2. Let $(p, q) \in R(S)$.
3. $p(R \circ S) = q(R \circ S)$. (by 2)
4. $q \in p(R \circ S)$.
5. There exists $x \in X$ with $(p, x) \in R$ and $(x, q) \in S$. (by 4)
6. $x(R \circ S) \underset{5}{=} p(R \circ S) \underset{3}{=} q(R \circ S)$.
7. $(x, q) \in R(S) \cap S$. (by 5, 6)
8. $(p, q) \in R \circ (R(S) \cap S)$. (by 5, 7)
9. $R(S) = R \circ (R(S) \cap S)$. (by 1, 2, 8)
10. $R \circ (R(S) \cap S) = (R(S) \cap S) \circ R$. (by 9, and part (a))

Proposition 9.5 Let:

(i) X be a compact Hausdorff space,
(ii) R, S, and K be equivalence relations on X, and
(iii) $R \subset K \subset R \circ S$.

Then $K \subset R(S)$.

PROOF: 1. $K \circ R \circ S \underset{\text{(ii)},\text{(iii)}}{=} K \circ S \underset{\text{(iii)}}{\subset} R \circ S \circ S \underset{\text{(ii)}}{=} R \circ S$.
2. Let $(p, q) \in K$.

3. $p(R \circ S) \underset{1}{=} p(K \circ (R \circ S)) \underset{\mathbf{9.3}}{=} (pK)(R \circ S) \underset{2,(\mathrm{i})}{=} (qK)(R \circ S)$

$$\underset{\mathbf{9.3}}{=} q(K \circ (R \circ S)) \underset{2}{=} q(R \circ S).$$

4. $(p, q) \in R(S)$. (by 3)

Corollary 9.6 Let:

(i) X be a compact Hausdorff space,
(ii) R, S, and K be equivalence relations on X,
(iii) $K \subset S$, and
(iv) $R \circ K$ be an equivalence relation.

Then $K \subset R(S)$.

PROOF: Applying **9.5** to $R \circ K$ we obtain $K \subset R \circ K \subset R(S)$.

Corollary 9.7 Let R and S be equivalence relations on X. Then the following are equivalent:

(a) $S \subset R(S)$,
(b) $R(S) = R \circ S$, and
(d) $R \circ S$ is an equivalence relation.

PROOF: (a) $\Rightarrow$ (b)

1. Assume that $S \subset R(S)$.
2. $R \circ S \subset R(S) \subset R \circ S$. (by 1, **9.4**)

(b) $\Rightarrow$ (c)

1. Assume that $R(S) = R \circ S$.
2. $R \circ S$ is an equivalence relation. (by **9.4**)

(c) $\Rightarrow$ (a)

1. Assume that $R \circ S$ is an equivalence relation.
2. $S \subset R(S)$. (by 1, and **9.6** with $K = S$)

As we have seen the quasi-relative product of two equivalence relations is an equivalence relation. In general $R(S)$ need not be closed even when both R and S are closed. We will often make use of the fact, proven in the following proposition, that if the canonical projection map associated to R is an open mapping, then $R(S)$ is closed.

Proposition 9.8 Let:

(i) R and S be closed equivalence relations on X, and
(ii) $\pi : X \to X/R$ (the canonical projection) be an open map.

Then $R(S)$ is closed.

PROOF: 1. Let $(x, z) \in \overline{R(S)}$.
2. Let $y \in x(R \circ S)$.
3. There exists $p \in X$ with $(x, p) \in R$ and $(p, y) \in S$. (by 2)
4. Let V and W be open neighborhoods of p and z respectively.
5. $\pi(V)$ is an open neighborhood of $\pi(p) = \pi(x)$ in X/R. (by 3, 4, (ii))
6. $\pi^{-1}(\pi(V))$ is an open neighborhood of x in X. (by 5)
7. There exists an open neighborhood U of x with $\pi(U) \subset \pi(V)$. (by 6)
8. There exists $(x_U, z_W) \in (U \times W) \cap R(S)$. (by 1, 4, 7)
9. There exists $x_V \in V$ with $\pi(x_V) = \pi(x_U)$. (by 7, 8)
10. $x_V(R \circ S) = (x_V R)S \underset{9}{=} (x_U R)S = x_U(R \circ S) \underset{8}{=} z_W(R \circ S)$.
11. $(z_W, x_V) \in R \circ S$. (by 10, (i))
12. $(z, p) \in \overline{R \circ S} = R \circ S$. (by 4, 8, 9, 11, (i))
13. $(z, y) \in R \circ S \circ S = R \circ S$. (by 3, 12, (i))
14. $y \in z(R \circ S)$. (by 13)
15. $x(R \circ S) \subset z(R \circ S)$. (by 2, 14)
16. $z(R \circ S) \subset x(R \circ S)$. (by 1-15 with the roles of x and z interchanged)
17. $(x, z) \in R(S)$. (by 15, 16)

Our use of the terminology quasi-relative product for $R(S)$ is motivated by the fact that under certain assumptions the flow $X/R(S)$ is a quasi-factor of the flow X/S. Keeping in mind that $R(S)$ is invariant if R and S are invariant this amounts to showing that there is a homomorphism of flows $\psi: X \to 2^{X/S}$ with $R(S) = \{(x, y) \mid \psi(x) = \psi(y)\}$.

Proposition 9.9 Let:

(i) (X, T) be a flow,
(ii) R and S be icers on X,
(iii) $\pi_R : X \to X/R$ and $\pi_S: X \to X/S$ be the canonical maps, and
(iv) π_R be open.

Then the map

$$\begin{aligned} \psi: X &\to 2^{X/S} \\ x &\to [\pi_S(xR)] \end{aligned}$$

is a flow homomorphism which induces an isomorphism of $X/R(S)$ onto a subflow of $2^{X/S}$.

PROOF: 1. The map $\begin{aligned} \sigma: X &\to 2^{X/R} \\ x &\to [\{\pi_R(x)\}] \end{aligned}$ is continuous. (by **5.6**)

2. π_R^* is continuous. (by (iv), **5.7**)

3. $\psi = 2^{\pi_S} \circ \pi_R^* \circ \sigma$ is a homomorphism of flows. (by 2, **5.5**)
4. $R(S) \underset{\mathbf{9.2}}{=} \{(x, y) \mid \pi_S(xR) = \pi_S(yR)\} = \{(x, y) \mid \psi(x) = \psi(y)\}$.

Note that the argument given in **9.9** applies when there is no T-action on X, producing a continuous map $\psi: X \to 2^{X/S}$ with $R(S) = \{(x, y) \mid \psi(x) = \psi(y)\}$. In particular this shows that $R(S)$ is closed, giving an alternative proof of **9.8**.

We will need to use the quasi-relative product to construct metrizable flows. As long as X/S is metrizable it follows from **9.9** that $X/R(S)$ is metrizable.

Corollary 9.10 Let:

(i) X be a compact Hausdorff space,
(ii) R and S be icers on X,
(iii) $X \to X/R$ be an open map, and
(iii) X/S be metrizable.

Then $X/R(S)$ is metrizable.

PROOF: This follows immediately from **9.9** and the fact that $2^{X/S}$ is metrizable. (by **Ex. 5.4**)

We saw in section 6 that the relative product construction commutes with this inverse limit construction. We wish to prove that the same is true for the quasi-relative product. Recall from **6.15** that

$$\bigcap_{i \in I}(R \circ S_i) = R \circ \left(\bigcap_{i \in I} S_i\right)$$

from which we deduced that under appropriate assumptions:

$$\lim_{\leftarrow} X/(R \circ S_i) \cong X/\left(R \circ \bigcap S_i\right).$$

Here we prove the corresponding results for the quasi-relative product.

Proposition 9.11 Let:

(i) R be a closed equivalence relation on X, and
(ii) $\{S_i \mid i \in I\}$ be a filter base of closed equivalence relations on X.

Then:

(a) $R\left(\bigcap_{i \in I} S_i\right) = \bigcap_{i \in I} R(S_i)$, and
(b) $\lim\limits_{\leftarrow} X/R(S_i) \cong X/R(\bigcap S_i)$.

PROOF: (a) We leave it to the reader to check that $S_1 \subset S_2 \Longrightarrow R(S_1) \subset R(S_2)$, from which it follows immediately that $R\left(\bigcap_{i\in I} S_i\right) \subset \bigcap_{i\in I} R(S_i)$.

Proof that $\bigcap_{i\in I} R(S_i) \subset R\left(\bigcap_{i\in I} S_i\right)$:

1. $R(S_i)$ is an equivalence relation with $R \subset R(S_i) \subset R \circ S_i$. (by **9.4**)
2. $\bigcap_{i\in I} R(S_i)$ is an equivalence relation with $R \subset \bigcap_{i\in I} R(S_i) \subset \bigcap_{i\in I} R \circ S_i = R \circ \left(\bigcap_{i\in I} S_i\right)$. (by 1, **6.15**)
3. $\bigcap_{i\in I} R(S_i) \subset R\left(\bigcap_{i\in I} S_i\right)$. (by 2, **9.5**)

(b) This follows immediately from part (a) and **6.14**.

We end this section with a proof that the diagonal is the only icer on a minimal flow which is both pointwise almost periodic and topologically transitive. As we have seen this follows immediately from **4.19** in the case of a metric flow. The result in the general case also uses **4.19**, but since **4.19** only applies to metric flows, the proof requires a technical lemma relating the general case to the metric case. The idea behind this lemma is that if L is a topologically transitive pointwise almost periodic icer on X, then the extension $X \to X/L$ can be "shadowed" by a metric extension of the same type, though we may have to restrict ourselves to a countable subgroup of T. In fact if we are given homomorphisms of minimal T-flows with X/N metrizable and (L, T) topologically transitive:

$$\begin{array}{ccc} X & \to & X/N \\ \downarrow & & \\ X/L & & \end{array}$$

then there exists a countable subgroup $H \subset T$, an icer N_∞ on X, and a commutative diagram of minimal H-flows such that X/N_∞ is metrizable and $(\pi_{N_\infty}(L), H)$ is topologically transitive:

$$\begin{array}{ccccc} X & \to & X/N_\infty & \to & X/N \\ \downarrow & & \downarrow & & \\ X/L & \to & X/(L \circ N_\infty). & & \end{array}$$

The extension $X \to X/L$ can be shadowed closely in the sense that an N_∞ with these properties can be found for any N with X/N metrizable. The proof of this lemma relies on a technical construction involving inverse limits, relative products, and quasi-relative products. The reader interested primarily in the metric case may wish to skip the technical details of **9.12** and **9.13**. The details of the shadowing lemma are laid out below; a close inspection reveals that the assumption that $L \subset X \times X$ is pointwise almost periodic (equivalently, by **4.15**, $X \to X/L$ is a distal homomorphism) can be weakened slightly to the assumption that $X \to X/L$ is an open map.

Lemma 9.12 Let:

(i) X be a minimal flow,
(ii) L be an icer on X,
(iii) $X \to X/L$ be an open mapping,
(iv) L be topologically transitive, and
(v) N be a closed equivalence relation on X with X/N metrizable.

Then there exists a countable subgroup $H \subset T$ and a closed equivalence relation N_∞ on X such that:

(a) $N_\infty \subset N$,
(b) $N_\infty H = N_\infty$,
(c) $(X/N_\infty, H)$ is minimal and metrizable,
(d) $L \circ N_\infty$ is a H-invariant closed equivalence relation on X, and
(e) the flow $(\pi_{N_\infty}(L), H)$ is topologically transitive.

In other words assume that we are given homomorphisms of minimal T-flows with X/N metrizable and (L, T) topologically transitive:

$$\begin{array}{ccc} X & \to & X/N \\ \downarrow & & \\ X/L & & \end{array}$$

then there exists a countable subgroup $H \subset T$ and a commutative diagram of minimal H-flows such that X/N_∞ is metrizable and $(\pi_{N_\infty}(L), H)$ is topologically transitive:

$$\begin{array}{ccccc} X & \to & X/N_\infty & \to & X/N \\ \downarrow & & \downarrow & & \\ X/L & \to & X/(L \circ N_\infty). & & \end{array}$$

PROOF: (a), (b), and (c) 1. Let $\mathcal{U}_0$ and $\mathcal{V}_0$ be countable bases for the topologies on X/N and $\pi_N(L) \subset X/N \times X/N$ respectively.
2. $\pi_N^{-1}(U)$ is open in X for every $U \in \mathcal{U}_0$.
3. For every $U \in \mathcal{U}_0$ there exists a finite set $F_U \subset T$ with $\pi_N^{-1}(U)F_U = X$.
(by 2, (i))

4. For every pair $V_1, V_2 \in \mathcal{V}_0$ there exists $t_{(V_1,V_2)} \in T$ with

$$\pi_N^{-1}(V_1)t_{(V_1,V_2)} \cap \pi_N^{-1}(V_2) \cap L \neq \emptyset. \qquad \text{(by 1, (iv))}$$

5. Let T_1 be the subgroup of T generated by

$$\bigcup\{F_U \mid U \in \mathcal{U}_0\} \cup \{t_{(V_1,V_2)} \mid V_1, V_2 \in \mathcal{V}_0\}.$$

6. Then:
6.1 T_1 is countable. (by 1, 3, 4, 5)
6.2 $\pi_N^{-1}(W)T_1 = X$ for all open sets $\emptyset \neq W \subset X/N$. (by 1, 3, 5)
6.3 $\pi_N^{-1}(V_1)T_1 \cap \pi_N^{-1}(V_2) \cap L \neq \emptyset$ for any pair of nonvacuous open sets $V_1, V_2 \subset \pi_N(L)$. (by 1, 4, 5)
7. Let $N_1' = \bigcap_{t \in T_1} Nt$.
8. Set $N_1 = L(N_1') \cap N_1' \subset N_1' \subset N$.
9. N_1 is a closed T_1-invariant equivalence relation on X.
(by 7, 8, (iii), **9.4**, and **9.8**)
10. X/N_1' is metrizable. (by 6, 7, (v))
11. $X/L(N_1')$ is metrizable. (by 10, (iii), and **9.10**)
12. X/N_1 is metrizable. (by 8, 10, 11)
13. $L \circ N_1 = L(N_1')$ is a T_1-invariant closed equivalence relation. (by 8, **9.4**)
14. Assume that

$$T_1 \subset T_2 \subset \cdots \subset T_n \subset T \quad \text{and} \quad N \supset N_1 \supset \cdots \supset N_n$$

have been constructed so that for all $1 \leq i < n$:
14.1 T_i is countable,
14.2 N_i is a closed T_i-invariant equivalence relation with X/N_i metrizable,
14.3 $L \circ N_i$ is a closed T_i-invariant equivalence relation on X,
14.4 $\pi_{N_i}^{-1}(W)T_{i+1} = X$ for all open sets $\emptyset \neq W \subset X/N_i$, and
14.5 $\pi_{N_i}^{-1}(V_1)T_{i+1} \cap \pi_{N_i}^{-1}(V_2) \cap L \neq \emptyset$ for any pair of open sets $V_1 \neq \emptyset \neq V_2$ in $\pi_{N_i}(L)$.
15. Let $\mathcal{U}_n$ and $\mathcal{V}_n$ be countable bases for the topologies on X/N_n and $\pi_{N_n}(L)$ respectively.
16. For every $U \in \mathcal{U}_n$ there exists a finite set $F_U \subset T$ with $\pi_{N_n}^{-1}(U)F_U = X$.
(by (i))
17. For every pair $V_1, V_2 \in \mathcal{V}_n$ there exists $t_{(V_1,V_2)} \in T$ with

$$\pi_{N_n}^{-1}(V_1)t_{(V_1,V_2)} \cap \pi_{N_n}^{-1}(V_2) \cap L \neq \emptyset. \qquad \text{(by 15, (iv))}$$

18. Let T_{n+1} be the subgroup of T generated by

$$T_n \bigcup\{F_U \mid U \in \mathcal{U}_n\} \cup \{t_{(V_1,V_2)} \mid V_1, V_2 \in \mathcal{V}_n\}.$$

19. Then: (by 15-18)
19.1 T_{n+1} is countable and contains T_n,
19.2 $\pi_{N_n}^{-1}(W)T_{n+1} = X$ for all open sets $\emptyset \neq W \subset X/N_n$, and

19.3 $\pi_{N_n}^{-1}(V_1)T_{n+1} \cap \pi_{N_n}^{-1}(V_2) \cap L \neq \emptyset$ for any pair of open sets $V_1 \neq \emptyset \neq V_2$ in $\pi_{N_n}(L)$.
20. Let $N'_{n+1} = \bigcap_{t \in T_{n+1}} N_n t$.
21. Set $N_{n+1} = L(N'_{n+1}) \cap N'_{n+1} \subset N'_{n+1} \subset N_n$.
22. N_{n+1} is a closed T_{n+1}-invariant equivalence relation on X.
(by 20, 21, (iii), **9.4**, and **9.8**)
23. X/N'_{n+1} is metrizable. (by 19, 20, and (v))
24. $X/L(N'_{n+1})$ is metrizable. (by 23, (iii), and **9.10**)
25. X/N_{n+1} is metrizable. (by 21, 23, 24)
26. $L \circ N_{n+1} = L(N'_{n+1})$ is a T_{n+1}-invariant closed equivalence relation.
(by 21, **9.4**)
27. There exist

$$T_1 \subset T_2 \subset \cdots \subset T_n \subset \cdots \subset T \quad \text{and} \quad N \supset N_1 \supset \cdots \supset N_n \supset \cdots$$

such that:
27.1 T_i is countable,
27.2 N_i is a closed T_i-invariant equivalence relation on X with X/N_i metrizable,
27.3 $L \circ N_i$ is a closed T_i-invariant equivalence relation on X,
27.4 $\pi_{N_i}^{-1}(W)T_{i+1} = X$ for all open sets $\emptyset \neq W \subset X/N_i$, and
27.5 $\pi_{N_i}^{-1}(V_1)T_{i+1} \cap \pi_{N_i}^{-1}(V_2) \cap L \neq \emptyset$ for any pair of open sets $V_1 \neq \emptyset \neq V_2$ in $\pi_{N_i}(L)$. (by induction)
28. Let $H = \bigcup T_i$ and $N_\infty = \bigcap N_i$.
29. N_∞ is a closed H-invariant equivalence relation on X. (by 27, 28)
30. X/N_∞ is metrizable. (by 27, 28)
31. Let $U_0 \subset X/N_\infty$ be open.
32. There exist i and $U \subset X/N_i$ open with $\pi_{N_i}^{-1}(U) \subset U_0$. (by 28, 31, **6.14**)
33. $U_0 H \underset{32}{\supset} \pi_{N_i}^{-1}(U)H \underset{28}{\supset} \pi_{N_i}^{-1}(U)T_{i+1} \underset{27,32}{=} X$.
34. X/N_∞ is minimal. (by 31, 33)
(d) 1. $L \circ N_\infty = L \circ \left(\bigcap N_i\right) = \bigcap (L \circ N_i)$ is a closed equivalence relation on X. (by **6.16**, since $L \circ N_i$ is a closed equivalence relation for every i)
2. $L \circ N_\infty$ is H-invariant. (L and N_∞ are both H-invariant)
(e) 1. Let $V_0 \neq \emptyset \neq W_0$ be open subsets of $\pi_{N_\infty}(L) \subset X/N_\infty \times X/N_\infty$.
2. There exist i, j, V and W be open subsets of $X/N_i \times X/N_i$ and $X/N_j \times X/N_j$ respectively with

$$\emptyset \neq (\pi_{N_i}^{N_\infty})^{-1}(V) \cap \pi_{N_\infty}(L) \subset V_0 \quad \text{and}$$

$$\emptyset \neq (\pi_{N_j}^{N_\infty})^{-1}(W) \cap \pi_{N_\infty}(L) \subset W_0.$$
(by 2, **6.14**)

3. We may assume without loss of generality that $i < j$.
4. $\emptyset \neq L \cap \pi_{N_j}^{-1}\left(\left(\pi_{N_i}^{N_j}\right)^{-1}(V)\right) T_j \cap \pi_{N_j}^{-1}(W) \equiv Y$. (by 2, 3, and 27 above)

5. $\emptyset \neq \pi_{N_\infty}(Y) \subset \pi_{N_\infty}(L) \cap \left(\pi_{N_i}^{N_\infty}\right)^{-1}(V)T_j \cap \left(\pi_{N_j}^{N_\infty}\right)^{-1}(W) \subset \pi_{N_\infty}(L) \cap V_0 H \cap W_0$. (by 2, 4)
6. $(\pi_{N_\infty}(L), H)$ is topologically transitive. (by 1, 5)

Theorem 9.13 Let:

(i) (X, T) be a minimal flow,
(ii) L be an icer on X, and
(iii) (L, T) be topologically transitive and pointwise almost periodic.

Then $L = \Delta_X$.

PROOF: 1. Assume that $(x, y) \in X \times X$ with $x \neq y$.
2. There exists a continuous function $f : X \to \mathbf{R}$ with $f(x) \neq f(y)$. (by 1, X is compact Hausdorff)
3. Let $N = \{(p, q) \in X \times X \mid f(p) = f(q)\}$.
4. N is a closed equivalence relation on X, X/N is metrtizable and $(x, y) \notin N$. (by 1, 2)
5. $X \to X/L$ is open. (by (iii), **4.15** and **7.23**)
6. By 5 and **9.12** there exists a countable subgroup $H \subset T$ and a closed equivalence relation N_∞ on X such that:
(a) $N_\infty \subset N$,
(b) $N_\infty H = N_\infty$,
(c) $(X/N_\infty, H)$ is metrizable and minimal,
(d) $L \circ N_\infty$ is a H-invariant equivalence relation on X, and
(e) The flow $(\pi_{N_\infty}(L), H)$ is topologically transitive.
7. In the diagram

$$\begin{array}{ccc} X & \to & X/N_\infty \\ \downarrow & & \downarrow \\ X/L & \to & X/(L \circ N_\infty), \end{array}$$

the first column is a minimal distal T-extension and the second column is a homomorphism of minimal H-flows. (by 6, (iii), and **4.15**)
8. $(\pi_{N_\infty}(L), H)$ is pointwise almost periodic. (by 7, **6.19**)
9. $(\pi_{N_\infty}(L), H)$ is minimal. (by 6(c), 6(e), 8, and **4.19**)
10. $\pi_{N_\infty}(L) = \Delta_{X/N_\infty}$. (by 9)
11. $\pi_N(L) \underset{6(a)}{=} \pi_N^{N_\infty}(\pi_{N_\infty}(L)) \underset{10}{=} \pi_N^{N_\infty}(\Delta_{X/N_\infty}) = \Delta_{X/N}$.
12. $L \subset N$. (by 11)

13. $(x, y) \notin L$. (4, 12)
14. $L \subset \Delta_X$. (by 1, 13)

Corollary 9.14 Let:

(i) (X, T), (Y, T) be minimal flows,
(ii) $f : (X, T) \to (Y, T)$ be a distal homomorphism, and
(iii) $R_f \equiv \{(x, y) \in X \times X \mid f(x) = f(y)\}$ be topologically transitive.

Then $X = Y$.

PROOF: 1.. R_f is pointwise almost periodic. (by (ii) and **4.15**)
2.. $R_f = \Delta_X$ and hence $X = Y$. (by 1, (iii), and **9.13**)

Restated in terms of icers on M, **9.14** says that if $R \subset S$ is a distal extension and $(\pi_R(S), T)$ is topologically transitive, then $R = S$. Recalling the terminology of **4.16**, this says that if $R \subset S$ is both distal and weak mixing, then $R = S$. When $S = M \times M$, this reduces to saying that if $(M/R, T)$ is a weak mixing flow which is also distal, then $M/R = \{pt\}$; thus **9.14** generalizes **4.25**. We will discuss **9.13** and its consequences further in section 20.

EXERCISES FOR CHAPTER 9

Exercise 9.1 (See **9.3**) Let:

(i) (X, T) be a flow,
(ii) R, S be subsets of $X \times X$,
(iii) $t \in T$, and
(iv) $z \in X$.

Show that:

(a) $(zR)t = (zt)(Rt)$, and
(b) $(zR)S = z(R \circ S)$.

Exercise 9.2 (See **9.11**) Let:

(i) X be a compact Hausdorff space,
(ii) R, S_1, S_2 be subsets of $X \times X$, and
(iii) $S_1 \subset S_2$.

Show that $R(S_1) \subset R(S_2)$.

Exercise 9.3 Let R and S be regular icers on M. Show that $\alpha(R(S)) = R(S)$ for all $\alpha \in G$. (Thus if $R(S)$ is closed, then it is a regular icer on M.)

Exercise 9.4 Let:

(i) R and S be icers on M, and
(ii) $R(S)$ be closed.

Then $aut(R) \cap G(S) \subset G(R(S))$.

Exercise 9.5 Let:

(i) R and S be icers on M, and
(ii) $R(S)$ be closed.

Then $G(R(S)) = \{\alpha \in G \mid (\alpha \times 1)(R \circ S) = R \circ S\}$.

Exercise 9.6 Let:

(i) R and S be icers on M, and
(ii) $R(S)$ be closed.

Then $G(R(S)) = G(R)G(S)$ if and only if $G(S) \subset aut(R \circ S)$. (Note: we do not assume that $R \circ S$ is an equivalence relation.)

Exercise 9.7 Let:

(i) R and S be icers on M,
(ii) $G(S) \subset aut(R)$, and
(iii) $R(S)$ be closed.

Then $G(R(S)) = G(R)G(S)$.

PART III

The τ-topology

In this section we introduce the τ-topology on the group G of automorphisms of the universal minimal set M. Though (G, τ) is not a topological group it is a compact space in which points are closed, inversion is continuous, and multiplication is unilaterally continuous. Most importantly for our purposes the properties of (G, τ) reflect the structure of the category $\mathcal{M}$ of minimal flows. There are several approaches to the construction of this topology. (See for example the books [Auslander, (1988)] [Ellis, (1969)] and [Glasner, (1976)]). We will begin by giving a new approach to defining a τ-topology on the group $Aut(X)$ of automorphisms of any minimal flow X. Our approach was motivated by an observation of J. Auslander's (personal communication) that the τ- topology on G could be obtained from the graphs of the left multiplication maps in M. In the present context this observation amounts to the statement (see **10.7**) that the τ-topology on G is characterized by the fact that $\alpha \in G$, is an element of the τ-closure of $A \subset G$, if and only if, $gr(\alpha) \subset \overline{gr(A)}$.

In section 10 we explicitly construct a base for a topology on $Aut(X)$ for a minimal flow (X, T). In general this topology is T_1 (points are closed), multiplication is unilaterally continuous, and inversion is continuous. When (X, T) is a regular flow, $Aut(X)$ is also compact. In particular taking $X = M$, we obtain a compact T_1 topology on G (the τ-topology).

The construction of the so-called *derived group* is the subject of section 11. When $F \subset G$ is a closed subgroup of G, the derived group $F' \subset F$ is a closed normal subgroup of F which measures the degree to which F fails to be Hausdorff. Indeed, for any closed subgroup $H \subset F$, the quotient space F/H is Hausdorff if and only if $F' \subset H$ (see **11.10**). The derived group $G' \subset G$ plays a particularly important role in analyzing equicontinuous flows and their relationship to distal flows; this is discussed in section 15.

It follows immediately from the characterization of the τ-topology mentioned above that if R is an icer on M, then the group $G(R)$ is a τ-closed subgroup of G. We exploit the interplay between quasi-factors and the τ-topology in section 12 to show that a subgroup A of G is the group of some icer on M if and only if A is τ-closed. In particular we show in **12.2** that if A is τ-closed, then $R = \overline{gr(A)}$ is an icer with $G(R) = A$, and M/R is a quasi-factor of M.

The icer $\overline{gr(A)}$ is clearly the smallest icer with group A, and as such is a proximal extension of any icer with group A. We show in **12.5** that $(\overline{gr(A)}, T)$ is topologically transitive if and only if $A = A'$. This result is used in sections 14 and 20.

10

The τ-topology on $Aut(X)$

For a minimal flow (X, T), we explicitly construct a base for a topology on $Aut(X)$. In general this topology is T_1 (points are closed), multiplication is unilaterally continuous, and inversion is continuous. When (X, T) is a regular flow, $Aut(X)$ which we denote by G_X, is also compact (see **10.6**).

Definition and Notation 10.1 In this section the following will be in force.

(i) (X, T) will denote a minimal flow,
(ii) $Aut(X)$ will denote the group of automorphisms of X, and
(iii) for nonempty open sets $U, V \subset X$, we will write:

$$< U, V > = \{\alpha \in Aut(X) \mid \alpha(U) \cap V \neq \emptyset\}.$$

We will show that the collection of unions of sets of the form $< U, V >$ forms a topology on $Aut(X)$. This will follow once we show that for every pair $< U_1, V_1 >$, $< U_2, V_2 >$ and $\alpha \in < U_1, V_1 > \cap < U_2, V_2 >$, there exist U and V with

$$\alpha \in < U, V > \subset < U_1, V_1 > \cap < U_2, V_2 >;$$

in other words we need to prove the following lemma.

Lemma 10.2 The collection $\mathcal{B}_\tau(X) = \{< U, V > \mid U, V \subset X$ are nonempty open sets$\}$ forms a basis for a topology on $Aut(X)$.

PROOF: 1. Assume that $\alpha \in < U_1, V_1 > \cap < U_2, V_2 >$.
2. There exist $p_1, p_2 \in X$ with $(p_1, \alpha(p_1)) \in U_1 \times V_1$ and $(p_2, \alpha(p_2)) \in U_2 \times V_2$. (by 1)
3. $p_2 \in \alpha^{-1}(V_2) \cap U_2 \neq \emptyset$. (by 2)
4. There exists $t \in T$ such that $p_1 t \in \alpha^{-1}(V_2) \cap U_2$. (by 2, 3, α is continuous, X is minimal)
5. $p_1 \in U_1 \cap U_2 t^{-1}$ and $\alpha(p_1) \in V_1 \cap V_2 t^{-1}$. (by 2, 4)

6. Set $U = U_1 \cap U_2 t^{-1}$ and $V = V_1 \cap V_2 t^{-1}$.

7. $\alpha \underset{5,6}{\in} < U, V > \underset{6}{\subset} < U_1, V_1 > \cap < U_2 t^{-1}, V_2 t^{-1} > = < U_1, V_1 > \cap < U_2, V_2 >$.

Definition and Notation 10.3 The topology on $Aut(X)$ generated by the basis $\mathcal{B}_\tau(X)$ will be denoted τ_X and referred to as the *τ-topology on $Aut(X)$*. Thus we have the topological space $(Aut(X), \tau_X)$. When X is regular, following **8.7**, we write $G_X = Aut(X)$, obtaining the topological space (G_X, τ_X). In particular for $X = M$, the universal minimal set, we simply write (G, τ).

Note that for any $\lambda \in Aut(X)$ and $\emptyset \neq U \subset X$, the set $\lambda(U)$ is open in X and hence $< U, \lambda(U) >$ is an open neighborhood of λ in $Aut(X)$. In fact the collection of all such sets forms a neighborhood base for λ.

Corollary 10.4 Let:

(i) $\lambda \in Aut(X)$,
(ii) $p \in X$, and
(iii) $\mathcal{N}_p = \{V \subset X | V \text{ is open and } p \in V\}$.

Then $\{< U, \lambda(U) > \mid U \in \mathcal{N}_p\}$ is a neighborhood base for λ in the τ-topology on $Aut(X)$.

PROOF: This follows immediately from **10.2**.

The next lemma will be used to show that inversion, left multiplication, and right multiplication are continuous in $Aut(X)$.

Lemma 10.5 Let:

(i) $\lambda, \gamma \in Aut(X)$, and
(ii) $\emptyset \neq V \subset X$ be open.

Then

(a) $< V, \lambda(V) >^{-1} = < \lambda(V), V >$,
(b) $\gamma < V, \lambda(V) > = < V, \gamma\lambda(V) >$, and
(c) $< V, \lambda(V) > \gamma = < \gamma^{-1}V, \lambda(V) >$.

PROOF: (a) $\alpha \in < V, \lambda(V) >^{-1} \Longleftrightarrow \alpha^{-1}(V) \cap \lambda(V) \neq \emptyset$
$\Longleftrightarrow V \cap \alpha\lambda(V) \neq \emptyset$
$\Longleftrightarrow \alpha \in < \lambda(V), V >$.

(b) $\alpha \in \gamma < V, \lambda(V) > \Longleftrightarrow \gamma^{-1}\alpha(V) \cap \lambda(V) \neq \emptyset$
$\Longleftrightarrow \alpha(V) \cap \gamma\lambda(V) \neq \emptyset$
$\Longleftrightarrow \alpha \in < V, \gamma\lambda(V) >$.

(c) $\alpha \in < V, \lambda(V) > \gamma \iff \alpha\gamma^{-1}(V) \cap \lambda(V) \neq \emptyset \iff \alpha \in < \gamma^{-1}(V), \lambda(V) >$.

Proposition 10.6 (a) $(Aut(X), \tau_X)$ is T_1 (points are closed),
(b) multiplication in $(Aut(X), \tau_X)$ is unilaterally continuous,
(c) inversion is continuous in $(Aut(X), \tau_X)$, and
(d) if X is regular, then $(Aut(X), \tau_X) = (G_X, \tau_X)$ is compact.

PROOF: (a) 1. Let $\alpha \neq \lambda \in Aut(X)$.
2. There exists $p \in X$ such that $\alpha(p) \neq \lambda(p)$. (by 1)
3. There exist disjoint open sets $W_1, W_2 \subset X$ such that $\alpha(p) \in W_1$ and $\lambda(p) \in W_2$. (by 2, X is T_2)
4. There exists an open set $V \subset X$ such that $p \in V$, $\alpha(V) \subset W_1$ and $\lambda(V) \subset W_2$. (by 3, α and λ are continuous)
5. $\alpha(V) \cap \lambda(V) = \emptyset$. (by 3, 4)
6. $\alpha \notin < V, \lambda(V) >$. (by 5)
7. $\alpha \notin \overline{\{\lambda\}}$. (by 6)
8. $\overline{\{\lambda\}} = \{\lambda\}$. (by 1, 7)
(b) This follows immediately from **10.5**(b) and **10.5**(c).
(c) This follows immediately from **10.5**(a).
(d) 1. Assume that X is regular and let $\{\mathcal{U}_i \mid i \in I\}$ be an open cover of $Aut(X) = G_X$.
2. For every $\lambda \in G_X$ there exists a nonempty open set $V_\lambda \subset X$ and $i_\lambda \in I$ with $< V_\lambda, \lambda(V_\lambda) > \subset \mathcal{U}_{i_\lambda}$. (by 1, **10.4**)
3. Let $U = \bigcup_{\lambda \in G_X} (V_\lambda \times \lambda(V_\lambda))T$.
4. $U \subset X \times X$ is open and invariant. (by 3)
5. $gr(\lambda) \subset U$ for all $\lambda \in G_X$. (by 3, $gr(\lambda)$ is mimimal)
6. $(X \times X) \setminus U$ is a closed invariant set containing no minimal sets. (by 1, 4, 5, and **8.9**)
7. $U = X \times X$. (by 6, and **3.4**)
8. There exist $\lambda_1, \ldots, \lambda_n \in G_X$ with $X \times X = \bigcup_{i=1}^{n} (V_{\lambda_i} \times \lambda_i(V_{\lambda_i}))T$. (by 3, 7, X is compact)
9. For any $\alpha \in G_X$ we have:

$$\begin{aligned} & gr(\alpha) \cap (V_{\lambda_i} \times \lambda_i(V_{\lambda_i})T \underset{8}{\neq} \emptyset && \text{for some } 1 \leq i \leq n, \\ \Longrightarrow \quad & \alpha(V_{\lambda_i}) \cap \lambda_i(V_{\lambda_i}) \neq \emptyset && \text{for some } 1 \leq i \leq n, \\ \Longrightarrow \quad & \alpha \in < V_{\lambda_i}, \lambda_i(V_{\lambda_i}) > && \text{for some } 1 \leq i \leq n. \end{aligned}$$

10. $G_X \subset \bigcup_{i=1}^{n} < V_{\lambda_i}, \lambda_i(V_{\lambda_i}) > \subset \bigcup_{i=1}^{n} \mathcal{U}_{\lambda_i}$. (by 2, 9)
11. $\{\mathcal{U}_{\lambda_i} \mid 1 \leq i \leq n\}$ is a finite subcover of $\{\mathcal{U}_i \mid i \in I\}$. (by 10)

Let $A \subset Aut(X)$ where (X, T) is a minimal flow. Then $gr(A) = \bigcup\{gr(\alpha) \mid \alpha \in A\}$ need not be a closed subset of $X \times X$, indeed $\overline{gr(A)}$ may contain $gr(\beta)$ for some $\beta \notin A$. When A is τ-closed, however $gr(\beta) \subset \overline{gr(A)}$ implies $\beta \in A$, in fact we will prove that

$$\overline{A} = \{\alpha \mid gr(\alpha) \subset \overline{gr(A)}\}. \qquad (*)$$

It should be noted that as an alternate approach to the τ-topology one can show that the equation above defines a closure operator on subsets of $Aut(X)$, and hence generates a topology on $Aut(X)$ (which coincides with the topology we have defined above). Hence the τ-topology is completely characterized by equation $(*)$.

Proposition 10.7 Let $\emptyset \neq A \subset Aut(X)$, and $\alpha \in Aut(X)$. Then

$$\alpha \in \overline{A} \iff gr(\alpha) \subset \overline{gr(A)}.$$

(Here of course $\overline{A}$ denotes the τ_X-closure of A in $Aut(X)$.)

PROOF: $\Longrightarrow$

1. Assume that $\alpha \in \overline{A}$.
2. Let $p \in X$ and $U \times W$ be any open neighborhood of $(p, \alpha(p))$.
3. There exists an open neighborhood V of p with $V \times \alpha(V) \subset U \times W$. (by 2, α is continuous)
4. There exists $\beta \in A \cap < V, \alpha(V) >$. (by 1, 3, and **10.4**)
5. $\emptyset \neq (V \times \alpha(V)) \cap gr(\beta) \subset (U \times W) \cap gr(A)$. (by 3, 4)
6. $(p, \alpha(p)) \in \overline{gr(A)}$. (by 2, 5)

$\Longleftarrow$

1. Assume that $gr(\alpha) \subset \overline{gr(A)}$.
2. Let $p \in X$ and V be any open neighborhood of p.
3. $gr(A) \cap (V \times \alpha(V)) \neq \emptyset$. (by 1, 2)
4. There exists $\beta \in A$ with $\beta(V) \cap \alpha(V) \neq \emptyset$. (by 3)
5. There exists $\beta \in A \cap < V, \alpha(V) >$. (by 4)
6. $\alpha \in \overline{A}$. (by 2, 5, **10.4**)

It follows immediately from **10.7** that the group $G(R)$ of any icer R on M is a τ-closed subgroup of G; for emphasis the explicit details are given in the corollary below. We will see in section 12 that every closed subgroup of G is of the form $G(R)$ for some icer on M.

Corollary 10.8 Let:

(i) R be an icer on M, and
(i) $G(R) = \{\alpha \in G = Aut(M) \mid gr(\alpha) \subset R\}$.

Then $G(R)$ is τ-closed.

PROOF: 1. Let $\alpha \in \overline{G(R)}$.
2. $gr(\alpha) \subset \overline{gr(G(R))} \subset R$. (by 1, (i), and **10.7**)
3. $\alpha \in G(R)$. (by 2)
4. $G(R)$ is closed. (by 1, 3)

Using **10.7** we give a description of the τ-closure of any subset $A \subset Aut(X)$ by characterizing the elements of $\overline{gr(A)}$.

Proposition 10.9 Let:

(i) (X, T) be a minimal flow,
(ii) $\emptyset \neq A \subset Aut(X)$, and
(iii) $\alpha \in Aut(X)$.

Then:

(a) $\overline{gr(A)} = \{(p, q) \mid q \in \bigcap\{\overline{A(U)} \mid U$ an open neighborhood of p in $X\}\}$, and
(b) $\alpha \in \overline{A}$ if and only if there exists a proximal pair $(x, y) \in X \times X$ with

$$\alpha(y) \in \bigcap\{\overline{A(U)} \mid U \text{ an open neighborhood of } x \text{ in } X\}.$$

(Here of course $\overline{A}$ denotes the τ-closure of A in $Aut(X)$, and $\overline{A(U)}$ denotes the closure of $A(U) = \{\beta(z) \mid \beta \in A, z \in U\}$ in X.)

PROOF: (a) 1. Let $(p, q) \in \overline{gr(A)}$.
2. There exists $\alpha \in A$ such that

$$\emptyset \neq gr(\alpha) \cap (U \times V)$$

for all $U, V \subset X$ open neighborhoods of p and q respectively. (by 1)
3. There exists $z \in U$ with $\alpha(z) \in V$ for all $U \in \mathcal{N}_p$ and $V \in \mathcal{N}_q$. (by 2)
4. $A(U) \cap V \neq \emptyset$ for all $U \in \mathcal{N}_p$ and $V \in \mathcal{N}_q$. (by 3)
5. $q \in \bigcap\{\overline{A(U)} \mid U \in \mathcal{N}_p\}$. (by 4)
6. $\overline{gr(A)} \subset \{(p, q) \mid q \in \bigcap\{\overline{A(U)} \mid U$ an open neighborhood of $p\}\}$. (by 1, 5)
7. $\overline{gr(A)} \supset \{(p, q) \mid q \in \bigcap\{\overline{A(U)} \mid U$ an open neighborhood of $p\}\}$. (read 1-5 in reverse)

(b) 1. Let $\alpha \in \overline{A}$, and $x \in X$.
2. $(x, \alpha(x)) \in \overline{gr(A)}$. (by 1, **10.7**)
3. $\alpha(x) \in \bigcap\{\overline{A(U)} \mid U \in \mathcal{N}_x\}$. (by 2, part (a))
4. Assume that (x, y) is a proximal pair such that $\alpha(y) \in \bigcap\{\overline{A(U)} \mid U \in \mathcal{N}_x\}$.
5. There exists $z \in X$ and $p \in \beta T$ such that $(x, y)p = (z, z)$. (by 4)
6. $(x, \alpha(y)) \in \overline{gr(A)}$. (by 4, part (a))
7. $gr(\alpha) \subset \overline{(z, \alpha(z))T} \underset{5}{\subset} \overline{(x, \alpha(y))T} \underset{6}{\subset} \overline{gr(A)}$.
8. $\alpha \in \overline{A}$. (by 7, **10.7**)

As we will see in section 12, the collection of τ-closed subgroups of the group G of automorphisms of the universal minimal set M is exactly the collection $\{G(R) \mid R \text{ is an icer on } M\}$ of groups of minimal flows. Thus any result concerning the closed subsets of G is of potential interest in studying minimal flows. The description of the closure of a subset $A \subset Aut(X)$ given in **10.9** allows us, in the case where X is regular, to prove that the product of two τ-closed subsets of G_X is τ-closed. This result will be used in sections 11, 12, 18, 19, and 20.

Theorem 10.10 Let:

(i) X be regular, and
(ii) A, B be non-empty closed subsets of G_X.

Then AB is also closed.

PROOF: 1. Let $\gamma \in \overline{AB}$, $u \in J_X$ be an idempotent in X, and $U \in \mathcal{N}_u$.
2. $AB \cap < U, \gamma(U) > \neq \emptyset$. (by 1)
3. $AB(U) \cap \gamma(U) \neq \emptyset$. (by 2)
4. $B(U) \cap A^{-1}\gamma(U) \neq \emptyset$. (by 3)
5. $\{\overline{B(U) \cap A^{-1}\gamma(U)} \mid U \in \mathcal{N}_u\}$ is a filter base of closed subsets of X. (by 1, 4)
6. There exists $x \in \bigcap_{N \in \mathcal{N}_u} \overline{B(U) \cap A^{-1}\gamma(U)}$. (by 5, X is compact)
7. There exist $\beta \in G_X$ and $v \in J_X$ such that $x = \beta(v)$. (by 6, (i), and **8.8**)
8. $\beta(v) \in \bigcap_{\mathcal{N}_u} \overline{B(U)} \cap \bigcap_{\mathcal{N}_u} \overline{A^{-1}\gamma(U)}$. (by 6, 7)
9. (u, v) is a proximal pair. (by 1, 8)
10. $\beta \in \overline{B} \cap \overline{A^{-1}\gamma}$. (by 8, 9, **10.9**)
11. $\beta \in B \cap A^{-1}\gamma$. (by 10, (ii), **10.6**)
12. $\gamma \in A\beta \subset AB$. (by 11)

Corollary 10.11 Let:

(i) X be regular, and
(ii) A, B be non-empty subsets of G_X.

Then $\overline{AB} = \bar{A}\bar{B}$.

PROOF: This follows from **10.10** and **10.6**. We leave the details as an exercise for the reader.

Another result concerning the closed subsets of G which will be of use to us is the following.

Lemma 10.12 Let:

(i) X be regular,
(ii) A be a closed subset of G_X, and
(iii) $\mathcal{B}$ be a filter base of closed non-empty subsets of G_X, (i. e. $B_1, \ldots, B_n \in \mathcal{B}$ implies there exists $B \in \mathcal{B}$ with $B \subset B_1 \cap \cdots \cap B_n$).

Then:

(a) $A(\bigcap \mathcal{B}) = \bigcap_{B \in \mathcal{B}} AB$, and
(b) $(\bigcap \mathcal{B})A = \bigcap_{B \in \mathcal{B}} BA$.

PROOF: (a) Proof that $A\left(\bigcap_{B\in\mathcal{B}} B\right) \subset \bigcap_{B\in\mathcal{B}} AB$:
1. $A(\bigcap \mathcal{B}) \subset AB$ for all $B \in \mathcal{B}$.
2. $A(\bigcap \mathcal{B}) \subset \bigcap_{B\in\mathcal{B}} AB$.

Proof that $\bigcap_{B\in\mathcal{B}} AB \subset A\left(\bigcap_{B\in\mathcal{B}} B\right)$:

1. Let $\alpha \in \bigcap_{B\in\mathcal{B}} AB$.
2. $\{A^{-1}\alpha \cap B \mid B \in \mathcal{B}\}$ is a filter base of non-empty closed subsets of G_X. (by 1, (iii), **10.6**)
3. There exists $\beta \in \bigcap_{B\in\mathcal{B}} B$ with $\beta \in A^{-1}\alpha$. (by 2, and compactness from **10.6**)
4. $\alpha \in A\left(\bigcap_{B\in\mathcal{B}} B\right)$. (by 3)

(b) The proof is completely analogous to the proof of part (a).

The fact that for $A \subset Aut(X)$, we have $\overline{A} = \{\alpha \mid gr(\alpha) \subset \overline{gr(A)}\}$ can be thought of as giving conditions under which an $\alpha \in Aut(X)$ can be obtained as a limit of elements of A. Another approach to this is via nets. Namely, given $\alpha \in Aut(X)$, under what conditions does a net (α_i) converge to α? We now gve a few results in this direction which can be thought of as refinements of **10.7**.

Proposition 10.13 Let:

(i) (p_i), (α_i) be nets in X and $Aut(X)$ respectively,
(ii) $p_i \to p$, and
(iii) $\alpha_i(p_i) \to \alpha(pv)$ for some $v \in J \equiv \{u \in M \mid u^2 = u\}$.

Then $\alpha_i \to \alpha$.

PROOF: 1. Let $\emptyset \neq W \subset X$ be open.
2. There exists $q \in M$ with $pq \in W$. (X is minimal so $X = pM$)
3. $\alpha(pv)q = \alpha(pq) \in \alpha(W)$. (by 2, **3.12**)
4. There exists $t \in T$ with $pt \in W$ and $\alpha(pv)t \in \alpha(W)$.
(by 2, 3, L_p, $L_{\alpha(pv)}$ are continuous)
5. $p_i t \to pt$ and $\alpha_i(p_i t) \to \alpha(pv)t$. (by 4, (ii), (iii), R_t is continuous)
6. There exists i_0 such that if $i > i_0$, then $p_i t \in W$ and $\alpha_i(p_i t) \in \alpha(W)$.
(by 1, 4, 5)
7. There exists i_0 such that if $i > i_0$, then $\alpha_i \in < W, \alpha(W) >$. (by 6)
8. $\alpha_i \to \alpha$. (by 1, 7, **10.4**)

Corollary 10.14 Let:

(i) X be regular,
(ii) (p_i), (α_i) be nets in X and G_X respectively,
(iii) $p_i \to \beta(u)$, where $\beta \in G_X$ and $u \in J_X$, and
(iv) $\alpha_i(p_i) \to \gamma(v)$, where $\gamma \in G_X$ and $v \in J_X$.

Then $\alpha_i \to \gamma\beta^{-1}$.

PROOF: 1. $\alpha_i(p_i) \to \gamma(\beta^{-1}(\beta(v))) = \gamma\beta^{-1}(\beta(u)v)$. (by (iv))
2. $\alpha_i \to \gamma\beta^{-1}$. (by 1, (iii), **10.13**)

Proposition 10.15 Let:

(i) $\alpha_i \to \alpha$ in $Aut(X)$, and
(ii) $p \in X$.

Then there exists a subnet $\{\alpha_{i_V}\} \subset \{\alpha_i\}$, and a net $\{p_{i_V}\} \subset X$, such that

$$p_{i_V} \to p \quad \text{and} \quad \alpha_{i_V}(p_{i_V}) \to \alpha(p).$$

PROOF: We leave the proof as an exercise for the reader.

When $X = M/R$ is a minimal regular flow we have defined a topology τ_X on the group G_X of automorphisms of X; in particular this gives us the τ-topology on the group G of automorphisms of M. On the other hand we saw in **8.11** and **7.10**, that the map $\chi_R \colon G \to G_X$ has kernel $G(R)$, and hence induces an isomorphism of $G/G(R)$ onto G_X. In the next proposition we prove that this isomorphism is a homeomorphism when $G/G(R)$ is provided the quotient topology. For simplicity we denote the maps π_R and χ_R by π and χ respectively.

Proposition 10.16 Let:

(i) $(X, T) = (M/R, T)$ be a regular minimal flow, and
(ii) U and V be nonempty open subsets of X.

Then:

(a) $< \pi^{-1}(U), \pi^{-1}(V) > = \chi^{-1}(< U, V >)$,

(b) the canonical map $\chi \equiv \chi_R : (G, \tau) \to (G_X, \tau_X)$ is continuous,

(c) the map $\chi : (G, \tau) \to (G_X, \tau_X)$ is closed, and

(d) χ induces a homeomorphism $(G/G(R), \tau) \to (G_X, \tau_X)$.

PROOF: (a) 1. $\pi^{-1}(U)$ and $\pi^{-1}(V)$ are nonempty open subsets of M. (by (i), $\pi \equiv \pi_R : M \to M/R \equiv X$ is continuous)

2. Let $\alpha \in < \pi^{-1}(U), \pi^{-1}(V) >$.

3. There exists $p \in M$ with $\pi(p) \in U$ and $\pi(\alpha(p)) \in V$. (by 2)

4. $\chi(\alpha)(\pi(p)) = \pi(\alpha(p))$. (by **8.11**)

5. $\chi(\alpha) \in < U, V >$. (by 3, 4)

6. $< \pi^{-1}(U), \pi^{-1}(V) > \subset \chi^{-1}(< U, V >)$. (by 2, 5)

7. Now let $\beta \in \chi^{-1}(< U, V >)$.

8. There exists $x \in U$ with $\chi(\beta)(x) \in V$. (by 7)

9. There exists $p \in M$ with $\pi(p) = x$. (by (i))

10. $p \in \pi^{-1}(U)$ and $\pi(\beta(p)) = \chi(\beta)(\pi(p)) = \chi(\beta)(x) \in V$. (8, 9)

11. $\beta \in < \pi^{-1}(U), \pi^{-1}(V) >$. (by 10)

12. $\chi^{-1}(< U, V >) \subset < \pi^{-1}(U), \pi^{-1}(V) >$. (by 7, 11)

(b) This follows immediately from part (a) and **10.3**.

(c) 1. Let $\emptyset \neq K$ be a closed subset of G and let $\eta \in \overline{\chi(K)}$.

2. $\pi(gr(K)) = gr(\chi(K))$. (by **8.11**)

3. $\pi(\overline{gr(K)}) = \overline{gr(\chi(K))}$. (by 2)

4. $gr(\eta) \subset \pi(\overline{gr(K)})$. (by 1, 3, **10.7**)

5. There exists a minimal subset $Y \subset \overline{gr(K)}$ with $\pi(Y) = gr(\eta)$. (by 4)

6. There exists $\alpha \in G$ with $gr(\alpha) = Y$. (by (i), 5, **7.5**)

7. $\alpha \in K$. (by 1, 5, 6, **10.7**)

8. $\chi(\alpha) = \eta$. (by 5, 6)

9. $\eta \in \chi(K)$. (by 7, 8)

10. $\chi(K)$ is closed. (by 1, 9)

(d) This follows immediately from parts (b), (c), and **7.10**.

NOTES ON SECTION 10

According to part (d) of **10.6**, when X is regular, $(Aut(X), \tau_X)$ is compact. The key to the proof is **8.9** (when X is regular all the minimal subsets of $X \times X$ are graphs of automorphisms of X). In general it is possible to define a τ-topology on the set of minimal subsets of $X \times X$ making this collection into a compact space. Identifying each automorphism α of X with the minimal subset $gr(\alpha) \subset X \times X$, the resulting subspace topology on $Aut(X)$ is the topology τ_X we have defined above. In the regular case $G_X = Aut(X)$ is the whole space and the two topologies coincide.

In these notes we will outline the construction of a τ-topology on the collection of minimal subsets of any flow (the case $X \times X$ is the one alluded to above), leaving the proofs as exercises for the reader.

Notation 10.N.1 In these notes the following will be in force.

(i) X will denote a not necessarily minimal flow.
(ii) $\mathcal{Q} \equiv \mathcal{Q}(X)$ will denote the collection of minimal subsets of X.
(iii) $\mathcal{U}$ will denote the collection of open invariant subsets of X, here $\emptyset \in \mathcal{U}$.
(iv) When $U \in \mathcal{U}$, we will write $H(U) = \{Y \in \mathcal{Q} \mid Y \subset U\}$.
(v) $\tau = \{H(U) \mid U \in \mathcal{U}\}$.

The following lemma is used to show that τ is a topology on $\mathcal{Q}$.

Lemma 10.N.2 Let $Y \in \mathcal{Q}$ and $U \in \mathcal{U}$. Then $Y \cap U = \emptyset$ or $Y \subset U$.

Proposition 10.N.3 (a) Let $(U_i \mid i \in I)$ be a family of elements of $\mathcal{U}$ and $U = \bigcup U_i$. Then $U \in \mathcal{U}$ and $H(U) = \bigcup H(U_i)$.
(b) Let $U_1, \dots, U_k \in \mathcal{U}$. Then $\bigcap U_i \in \mathcal{U}$, and $H(\bigcap U_i) = \bigcap H(U_i)$.
(c) τ is a topology on $\mathcal{Q}$.
The analog of **10.7** in this context identifies the τ-closure of a subset of $\mathcal{Q}$ as the closure in X of the union of its elements.

Proposition 10.N.4 Let $\Gamma \subset \mathcal{Q}$, $Y \in \mathcal{Q}$. Then $Y \in cls_\tau \Gamma \iff Y \subset \overline{\bigcup \Gamma}$.

Proposition 10.N.5 $(\mathcal{Q}, \tau)$ is compact T_1.

The construction of a topology on $\mathcal{Q}(X)$ for a flow (X, T) is natural in the sense that a homomorphism of flows gives rise to a continuous map of the corresponding topological spaces.

Proposition 10.N.6 Let $\phi\colon W_1 \to W_2$ be a homomorphism of flows. Then

$$\begin{aligned} \Phi : \mathcal{Q}(W_1) &\to \mathcal{Q}(W_2) \\ Y &\to \phi(Y) \end{aligned}$$

is continuous.

Let W be a minimal set, Z a flow, and $f\colon W \to Z$ a homomorphism. Then the graph of f, $gr(f)$ is a minimal subset of $W \times Z$, and the map

$$f \to gr(f) \colon Hom(W, Z) \to \mathcal{Q}(W \times Z)$$

is injective. Thus the τ-topology on $\mathcal{Q}(W \times Z)$ induces a topology on $Hom(W, Z)$ which will also be called the τ-topology. Once again this construction is natural in the sense that for fixed homomorphisms the left and right composition operations are continuous. In particular this approach yields the full strength of **10.6** in this context.

Corollary 10.N.7 Let X, Y be minimal and $g \in Hom(Y, Z)$. Then the map

$$Hom(X, Y) \to Hom(X, Z)$$
$$f \to g \circ f$$

is continuous.

Lemma 10.N.8 Let X, Y be minimal and $h \in Hom(X, Y)$. Then the map

$$Hom(Y, Z) \to Hom(X, Z)$$
$$f \to f \circ h$$

is continuous.

Proposition 10.N.9 Let X be minimal. Then:

(a) The map $(f, g) \to f \circ g \colon Hom(X, X) \times Hom(X, X) \to Hom(X, X)$ is unilaterally continuous.

(b) The map $f \to f^{-1} \colon Aut(X) \to Aut(X)$ is continuous. (Here $Aut(X)$ is the set of invertible elements of $Hom(X, X)$.)

EXERCISES FOR CHAPTER 10

Exercise 10.1 (See **10.15**) Let:

(i) $\alpha_i \to \alpha$ in $Aut(X)$, and

(ii) $p \in X$.

Then there exists a subnet $\{\alpha_{i_\nu}\} \subset \{\alpha_i\}$, and a net $\{p_{i_\nu}\} \subset X$, such that

$$p_{i_\nu} \to p \quad \text{and} \quad \alpha_{i_\nu}(p_{i_\nu}) \to \alpha(p).$$

Exercise 10.2 Let:

(i) X be regular, and

(ii) A, B be non-empty subsets of G_X.

Then $\overline{AB} = \bar{A}\bar{B}$.

11
The derived group

If $F \subset Aut(X)$ is a closed subgroup of the group of automorphisms of a minimal flow (X, T), then the derived group $F' \subset F$ is a closed normal subgroup of F which measures the degree to which the τ-topology on F fails to be Hausdorff. The most interesting case is when (X, T) is regular so that $Aut(X) = G_X$ is compact; the case $X = M$ with $Aut(X) = G$ being of particular interest. The key result in this case is that for a closed normal subgroup $H \subset F$, the quotient F/H is a compact Hausdorff topological group if and only if $F' \subset H$ (see **11.11**) This result along with a few more technical results such as **11.14, 11.15**, will play an important role in the study of equicontinuous flows and almost periodic extensions of minimal flows in sections 15 and 19 respectively.

Definition 11.1 Let X be a minimal flow and F be a closed subgroup of $Aut(X)$. Then the *derived group* F' of F is the intersection of the closed neighborhoods of the identity in F. More precisely

$$F' = \bigcap_{V \in \mathcal{N}_p} \overline{< V, V > \cap F},$$

where $p \in X$ is any element of X and as in **10.4** we are using the following notation:

$$\mathcal{N}_p = \{U \mid p \in U \subset X, \text{ and } U \text{ is open}\}.$$

This definition is independent of the choice of p since for any p the collection

$$\{< V, V > \cap F \mid V \in \mathcal{N}_p\}$$

is a neighborhood base at $1_F \in F$ (see **10.4**). The fact that $\overline{< V, V > \cap F} \subset F$ follows from the assumption that F is closed.

It will be convenient to reformulate the definition of F' using the fact that the collection

$$\{ < V, \alpha(V) > \cap F \mid V \in \mathcal{N}_p \}$$

is a neighborhood base at $\alpha \in F$.

Proposition 11.2 Let:

(i) (X, T) be a minimal flow,
(ii) F is a closed subgroup of $Aut(X)$,
(iii) $\alpha, \beta \in F$, and
(iv) $p \in X$.

Then

(a) $\alpha \in F' \iff < W, \alpha(W) > \cap < V, V > \neq \emptyset$ for all $V, W \in \mathcal{N}_p$, and
(b) $\beta^{-1}\alpha \in F' \iff < W, \alpha(W) > \cap < V, \beta(V) > \neq \emptyset$ for all $V, W \in \mathcal{N}_p$.

PROOF: (a) This follows immediately from **11.1** and **10.4**.
(b) 1. $\beta^{-1}(< W, \alpha(W) > \cap < V, \beta(V) >)$
$= \beta^{-1}(< W, \alpha(W) >) \cap \beta^{-1}(< V, \beta(V) >)$
$=< W, \beta^{-1}\alpha(W) > \cap < V, V >$. (by 10.5)
2. $\beta^{-1}\alpha \in F' \iff < W, \alpha(W) > \cap < V, \beta(V) > \neq \emptyset$ for all $V, W \in \mathcal{N}_p$. (by 1, part (a))

In **11.1** F' is referred to as the derived **group** of F. This terminology will be justified when we show that F' is indeed a group. This requires two preliminary lemmas.

Lemma 11.3 Let:

(i) (X, T) be a minimal flow,
(ii) F be a closed subgroup of $Aut(X)$, and
(iii) U be a non-vacuous open subset of F.

Then $\overline{U}F' = \overline{U} = F'\overline{U}$.

PROOF: 1. Let $\alpha \in F'$ and $\beta \in U$.
2. $\beta^{-1}U$ is open in F, and $1_F \in \beta^{-1}U$. (by 1, and **10.6**)
3. $\alpha \in \overline{\beta^{-1}U} = \beta^{-1}\overline{U}$. (by 1, 2, **10.6**)
4. $\beta\alpha \in \overline{U}$. (by 3)
5. $\overline{U}\alpha = \overline{U\alpha} \subset \overline{U}$. (by 1, 4, **10.6**)
6. $\overline{U}F' \subset \overline{U}$. (by 1, 5)

7. $\overline{U}F' = \overline{U}$. (by 6, $1_F \in F'$)
8. $F'\overline{U} = \overline{U}$. (by an analogous argument)

Lemma 11.4 Let:

(i) (X, T) be a minimal flow,
(ii) F, H be closed subgroups of $Aut(X)$,
(iii) $\varphi : F \to H$ be continuous, and
(iv) $\varphi(1_F) = 1_H$.

Then $\varphi(F') \subset H'$.

PROOF: 1. Let V be a neighborhood of 1_H.
2. There exists a neighborhood U of 1_F with $\varphi(U) \subset V$. (by 1, (ii), (iii))
3. $\varphi(\overline{U}) \subset \overline{V}$. (by 2, (iii))
4. $\varphi(F') \subset \varphi(\overline{U}) \subset \overline{V}$. (by 2, 3)
5. $\varphi(F') \subset \bigcap \overline{V} = H'$.

Proposition 11.5 Let:

(i) (X, T) be a minimal flow, and
(ii) F be a closed subgroup of $Aut(X)$.

Then F' is a subgroup of F.

PROOF: 1. Let U be an open neighborhood of 1_F in F.
2. $F'\alpha \subset \overline{U}\alpha \subset \overline{U}F' \underset{\mathbf{11.3}}{=} \overline{U}$ for all $\alpha \in F'$.
3. $F'\alpha \subset F'$ for all $\alpha \in F'$. (by 1, 2)
4. F' is a closed semigroup. (by 3)
5. The map $\begin{aligned} F &\to F \\ \alpha &\to \alpha^{-1} \end{aligned}$ is continuous. (by **10.6**)
6. $(F')^{-1} = F'$. (by 5, **11.4**)
7. F' is a subgroup of F. (by 4, 6)

Corollary 11.6 Let:

(i) (X, T) be a minimal flow,
(ii) F be a closed subgroup of $Aut(X)$, and
(iii) $\alpha \in Aut(X)$.

Then $\alpha(F')\alpha^{-1} = (\alpha F \alpha^{-1})'$.

PROOF: Apply **11.4** to the map $\varphi : \begin{aligned} F &\to \alpha F \alpha^{-1} \\ \beta &\to \alpha\beta\alpha^{-1} \end{aligned}$ and to its inverse.

Corollary 11.7 Let:

(i) (X, T) be a minimal flow,
(ii) $F \subset B$ be closed subgroups of $Aut(X)$, and
(iii) F be normal in B.

Then F' is a normal subgroup of B.

PROOF: For any $\alpha \in B, \alpha(F')\alpha^{-1} = (\alpha F \alpha^{-1})' = F'$. (by **11.6**)

According to **11.6**, when F is a closed subgroup of $Aut(X)$, F' is a normal subgroup of F so that F/F' is group. If the flow (X, T) is regular, then $G_X = Aut(X)$ and hence F is compact. Thus F/F' is a compact Hausdorff group in which multiplication is unilaterally continuous by **10.6**. We will see in the appendix to section 15 that any compact Hausdorff group in which multiplication is unilaterally continuous is a topological group. In this section we use the next few results to prove directly that F/F' is a topological group. These results can be proved in a slightly more general context (see [Ellis, R., (1969)]), but we wish to prove them using the techniques developed herein. Moreover, these proofs are much simpler than the ones given in the reference above.

Proposition 11.8 Let:

(i) (X, T) be a regular minimal flow,
(ii) F be a closed subgroup of G_X,
(iii) $\emptyset \neq V \subset F$ be open, and
(iv) $F' \subset V$.

Then there exists an open set W with $1_F \in W$ and $F'WW^{-1} \subset V$.

PROOF: 1. Let $\mathcal{N}$ be the collection of open neighborhoods of 1_F in F.
2. $\bigcap_{U \in \mathcal{N}} (F \setminus V) \cap \overline{U} = (F \setminus V) \cap \bigcap_{N \in \mathcal{N}} \overline{U} = (F \setminus V) \cap F' = \emptyset$. (by (iii), (iv))
3. There exists $U_0 \in \mathcal{N}$ with $(F \setminus V) \cap \overline{U_0} = \emptyset$. (by 2, G_X is compact T_1)
4. Set $U = U_0 \cap U_0^{-1}$.
5. $\overline{U} \subset \overline{U_0} \subset V$. (by 3, 4)
6. $\bigcap_{N \in \mathcal{N}} \overline{N}\,\overline{U} \underset{\mathbf{10.12}}{=} \left(\bigcap_{N \in \mathcal{N}} \overline{N} \right) \overline{U} = F'\overline{U} \underset{\mathbf{11.3}}{=} \overline{U} \underset{5}{\subset} V.$
7. $\bigcap_{N \in \mathcal{N}} (F \setminus V) \cap \overline{N}\,\overline{U} = \emptyset$. (by 6)
8. There exists $N_0 \in \mathcal{N}$ with $(F \setminus V) \cap \overline{N_0}\,\overline{U} = \emptyset$. (by 7, and compactness)
9. Set $N = N_0 \cap N_0^{-1}$ and $W = N \cap U$.
10. $F'WW^{-1} \underset{9}{\subset} F'\overline{N}\,\overline{U} \underset{\mathbf{11.3}}{=} \overline{N}\,\overline{U} \underset{8,9}{\subset} V$.

Proposition 11.9 Let:

(i) (X, T) be a regular minimal flow, and
(ii) F be a closed subgroup of G_X.

Then F/F' is a compact topological group.

PROOF: 1. F/F' is a compact group in which multiplication is unilaterally continuous. (by (i), **10.6, 11.7**)
2. Let $\pi : F \to F/F'$ be the canonical map.
3. Let $V \subset F/F'$ be an open neighborhood of the identity.
4. $F' \subset \pi^{-1}(V)$ is open in F. (by 3)
5. There exists an open neighborhood W of 1_F with $F'WW^{-1} \subset \pi^{-1}(V)$. (by 4, **11.8**)
6. $\pi(W)$ is an open neighborhood of the identity in F/F'. (by 5, π is open)
7. $\pi(W)\pi(W)^{-1} = \pi(WW^{-1}) \underset{5}{\subset} \pi(\pi^{-1}(V)) = V$.
8. F/F' is a topological group. (by 1, 3, 7)

Theorem 11.10 Let:

(i) X be regular,
(ii) F be a closed subgroup of G_X, and
(iii) H be a closed subgroup of F.

Then F/H is Hausdorff if and only if $F' \subset H$.

PROOF: $\Longrightarrow$
1. Let $\pi : F \to F/H$ be the canonical map, and $\alpha \notin H$.
2. There exists an open set V with $\pi(1_F) \in V$, and $\pi(\alpha) \notin \overline{V}$. ($F/H$ is Hausdorff)
3. $1_F \in \pi^{-1}(V) \subset \overline{\pi^{-1}(V)} \subset \pi^{-1}(\overline{V})$.
4. $\alpha \notin \overline{\pi^{-1}(V)}$. (by 2, 3)
5. $\alpha \notin F'$. (by 4)

$\Longleftarrow$

1. Let $F' \subset H$, and $\alpha \notin H$.
2. $F' \subset H \subset F \setminus \alpha H$. (by 1)
3. There exists an open neighborhood W of 1_F with $W = W^{-1}$ and $F'WW^{-1} \subset F \setminus \alpha H$. (by 2, (iii), **10.6, 11.8**)
4. $W^2 \subset F'W^2 \subset F \setminus \alpha H$. (by 3)
5. $W^2 \cap \alpha H = \emptyset$. (by 4)
6. $W \cap W\alpha H = \emptyset$. (by 3, 5)

7. $WH \cap W\alpha H = \emptyset$. (by 6, (iii))
8. F/H is Hausdorff. (by 1, 7)

Corollary 11.11 Let:

(i) X be regular,
(ii) F be a closed subgroup of G_X, and
(iii) H be a closed normal subgroup of F.

Then F/H is a compact (Hausdorff) topological group if and only if $F' \subset H$.

PROOF: 

1. Assume that F/H is a topological group.
2. F/H is T_1. (by **10.6**)
3. F/H is Hausdorff. (by 1, 2, and **Ex. 3.5**)
4. $F' \subset H$. (by 3, **11.10**)

$\Longleftarrow$

1. Assume that $F' \subset H$.
2. Let $\pi : F/F' \to F/H$ be the canonical projection.
3. $\pi^{-1}(\pi(V)) = \bigcup\{V\alpha \mid \alpha \in H/F'\}$ for any subset $V \subset F/F'$. (by 2, (iii))
4. π is an open map. (by 2, 3)
5. Let $W \subset F/H$ be an open neighborhood of the identity.
6. $\pi^{-1}(W) \subset F/F'$ is an open neighborhood of the identity. (by 2, 5)
7. There exists an open neighborhood V of the identity in F/F' with $VV^{-1} \subset \pi^{-1}(W)$. (by 6, **11.9**)
8. $\pi(V)$ is an open neighborhood of the identity in F/H with $\pi(V)(\pi(V))^{-1} = \pi(VV^{-1}) \subset \pi(\pi^{-1}(W)) = W$. (by 4, 7)
9. F/H is a topological group. (by 5, 8)

The next two technical lemmas are used to prove **11.14**, **11.15**, and **11.16**, which are needed in future sections.

Lemma 11.12 Let:

(i) X be regular,
(ii) F be a closed subgroup of G_X,
(iii) A, B be closed subgroups of F,
(iv) W be a B-open set with $1_F \in W$, and
(v) $K = B\backslash AW$.

Then:

(a) K is a closed subset of F,
(b) $1_F \notin AK$, and
(c) there exists an F-open set N with $1_F \in N$ and $B \cap AN \subset B \cap AW$.

PROOF: (a) 1. Let $b \in B \cap AW$.
2. $b = aw$ for some $a \in A$ and $w \in W$. (by 1 and (v))
3. $a \in B$. (by (iii), (iv), and 2)
4. $b \in (A \cap B)W$. (by 2, 3)
5. $B \cap AW \subset B \cap (A \cap B)W \subset B \cap AW$. (by 1, 4)
6. $B \setminus AW = B \setminus (A \cap B)W$. (by 5)
7. $(A \cap B)W$ is open in B. (by (iii), (iv))
8. $K = B \setminus AW = B \setminus (A \cap B)W$ is closed in B. (by 7)
9. K is closed. (by 8, (iii))
(b) 1. $A = A\{1_F\} \subset AW$. (by (iii), (iv))
2. $A \cap K = \emptyset$. (by 1)
3. $A \cap AK = \emptyset$. (by 2, (iii))
4. $1_F \notin AK$. (by 3, (iii))
(c) 1. AK is closed in F. (by (iii), (a), **10.10**)
2. $N = F \setminus AK$ is an F-open set with $1_F \in N$. (by 1, (b))
3. Let $b \in B \cap AN$.
4. $b = an$ with $a \in A$ and $n \in N$. (by 3)
5. $a^{-1}b = n \notin AK$. (by 2, 4)
6. $b \notin K$. (by 5)
7. $b \in AW$. (by 3, 6, (v))
8. $B \cap AN \subset B \cap AW$. (by 3, 7)

Lemma 11.13 Let:

(i) X be regular,
(ii) F be a closed subgroup of G_X,
(iii) A, B be closed subgroups of F,
(iv) $\mathcal{N}(F) = \{N \mid N$ is an open neighborhood of 1_F in $F\}$,
(v) $L_0 = \bigcap\{\overline{B \cap AU} \mid U \in \mathcal{N}(F)\}$, and
(vi) $L = \bigcap\{\overline{AB \cap U} \mid U \in \mathcal{N}(F)\}$.

Then:

(a) $L_0 \subset AB'$,
(b) $B' \subset L \subset AB'$, and
(c) $AL = AB'$.

PROOF: (a) 1. Let $\mathcal{N}(B) = \{N \mid N \text{ is an open neighborhood of } 1_B \text{ in } B\}$.

2. $L_0 \underset{(v)}{=} \bigcap\{\overline{B \cap AU} \mid U \in \mathcal{N}(F)\} \underset{1,\mathbf{11},\mathbf{12}}{\subset} \bigcap\{\overline{B \cap AW} \mid W \in \mathcal{N}(B)\}$

$$\subset \bigcap\{\overline{AW} \mid W \in \mathcal{N}(B)\} \underset{\mathbf{10},\mathbf{12}}{=} A\bigcap\{\overline{W} \mid W \in \mathcal{N}(B)\} = AB'.$$

(b) 1. It is clear that $B' \subset L$ since $B \cap U \subset AB \cap U$ for all $U \in \mathcal{N}(F)$.

2. $AB \cap U \subset A(B \cap AU)$ for all $U \in \mathcal{N}(F)$.

$(\alpha\beta \in AB \cap U \Rightarrow \beta \in AU \cap B \Rightarrow \alpha\beta \in A(B \cap AU))$

3. $\overline{AB \cap U} \subset \overline{A(B \cap AU)} = \overline{A}\,\overline{(B \cap AU)} = A\overline{(B \cap AU)}$ for all $U \in \mathcal{N}(F)$.

(by 2, (iii), **10.11**)

4. $L \underset{(vi)}{=} \bigcap_{U \in \mathcal{N}} \overline{AB \cap U} \underset{3}{\subset} \bigcap_{U \in \mathcal{N}} A\overline{(B \cap AU)}$

$$\underset{\mathbf{10.12}}{=} A\left(\bigcap_{U \in \mathcal{N}} \overline{B \cap AU}\right) \underset{(v)}{=} AL_0 \underset{(a)}{\subset} A(AB') \underset{(iii)}{=} AB'.$$

(c) 1. $AB' \subset AL \subset A(AB') = AB'$. (by part (b), (iii))

Corollary 11.14 Let:

(i) X be regular,
(ii) F be a closed subgroup of G_X,
(iii) A, B be closed subgroups of F, and
(iv) $AB = BA$.

Then:

(a) $B' \subset (AB)' \subset AB'$, and
(b) $A(AB)' = AB'$.

PROOF: This follows immediately from **11.13**. (In this case $L = (AB)'$.)

Let $p \in X$ with (X, T) a regular minimal flow. For any closed subgroup $F \subset G_X$ and open neighborhood V of p in X we consider the set $A_p(V) = \{\alpha \in F \mid \alpha(p) \in V\}$. We would like to characterize F' using the collection $\{A_p(V) \mid V \in \mathcal{N}_p\}$. Clearly

$$A_p(V) \subset \{\alpha \in F \mid \alpha(V) \cap V \neq \emptyset\} = F \cap < V, (V) >,$$

so $F' \subset \bigcap\{\overline{A_p(V)} \mid V \in \mathcal{N}_p\}$. On the other hand in general, $\overline{A_p(V)}$ may have empty interior. When p is an almost periodic point of the flow (F, X) where F acts on X on the left, $int\left(\overline{A_p(V)}\right) \neq \emptyset$, and the following proposition gives a complete description of F'.

Proposition 11.15 Let:

(i) (X, T) be a regular minimal flow,
(ii) F be a closed subgroup of G_X,

(iii) p be an almost periodic point of the flow (F, X) (F acts on X on the **left**), and

(iv) $A_p(V) = \{\alpha \in F \mid \alpha(p) \in V\}$ for any $\emptyset \neq V \subset X$.

Then $F' = \bigcap \left\{ \overline{int(\overline{A_p(V)})} \mid V \in \mathcal{N}_p \right\}$.

We write $cic(A_p(V))$ for $\overline{int(\overline{A_p(V)})}$ and D_p for $\bigcap\{cic(A_p(V)) \mid V \in \mathcal{N}_p\}$. Note that all topological references are to the space (F, τ).

PROOF: 1. Let $V \in \mathcal{N}_p$.
2. There exists a finite subset K of F such that $F \subset K(A_p(V))$. (by (iii), and **4.2**)
3. $F = K\overline{A_p(V)}$. (by 2, (ii))
4. $int(\overline{A_p(V)}) \neq \emptyset$. (by 3, K is finite)
5. $D_p \neq \emptyset$. (by 4, (F, τ) is compact by (ii), **10.6**)
6. $F'cic(A_p(V)) \subset cic(A_p(V))$ for all $V \in \mathcal{N}_p$. (by **11.3**)
7. $F'D_p \subset D_p$. (by 6, **10.12**)
8. Let Z be a neighborhood of 1_F.
9. There exists $V \in \mathcal{N}_p$ such that $1_F \in \, < V, V > \, \subset Z$. (by 8, **10.4**)
10. $cic(A_p(V)) \subset cic(< V, V >) \subset \overline{< V, V >} \subset \overline{Z}$. (by 9, (iii))
11. $D_p \subset \bigcap \overline{Z} = F'$. (by 8 and 10)
12. $F' \underset{11}{=} F'D_p \underset{7}{\subset} D_p$. ($F'$ is a group)
13. $F' = D_p$. (by 11, 12)

Proposition 11.16 Let:

(i) (X, T) be a regular minimal flow,
(ii) F be a closed subgroup of G_X,
(iii) p be an almost periodic point of the flow (F, X), and
(iv) $\alpha \in F'$.

Then there exists a net (α_i) in F with $\alpha_i(p) \to \alpha(p)$ and $\alpha_i \to 1_F$.

PROOF: 1. Let $A_p(V) = \{\beta \in F \mid \beta(p) \in V\}$.
2. $\alpha^{-1} \in F' \subset cic(A_p(V))$ for all $V \in \mathcal{N}_p$. (by **11.15**)
3. Let $\mathcal{T}$ denote the set of neighborhoods of α^{-1}.
4. $Z \cap A_p(V) \neq \emptyset$ for all $Z \in \mathcal{T}$, and $V \in \mathcal{N}_p$. (by 1 and 2)
5. Let $\beta(Z, V) \in Z \cap A_p(V)$ for all $Z \in \mathcal{T}$, and $V \in \mathcal{N}_p$.
6. For $Z_1, Z_2 \in \mathcal{T}$, $V_1, V_2 \in \mathcal{N}_p$ set $(Z_1, V_1) \geq (Z_2, V_2)$ if $Z_1 \subset Z_2$ and $V_1 \subset V_2$.
7. Then $(\beta(Z, V) \mid Z \in \mathcal{T},\ V \in \mathcal{N}_p) \to \alpha^{-1}$ and $(\beta(Z, V)(p) \mid Z \in \mathcal{T},\ V \in \mathcal{N}_p) \to p$.
8. Finally $(\alpha\beta(Z, V) \mid Z \in \mathcal{T},\ V \in N_p)$ is the required net.

We end this section with an iteration of the derived group construction which will be used in later sections.

Definition 11.17 Let (X, T) be a regular minimal flow and A a closed subgroup of $G_X = Aut(X)$. Then we associate with A, a closed subgroup $A^\infty \subset A$, by transfinite induction, as follows:

$$\text{set } A^0 = A, A^{\alpha+1} = (A^\alpha)' \text{and} A^\beta = \bigcap_{\alpha<\beta} A^\alpha \text{ if } \beta \text{ is a limit ordinal.}$$

Now if the cardinal number of the set of ordinals less than or equal to ν is greater than the cardinal number of the set A, there must exist an ordinal $\alpha < \nu$ with $A^\alpha = A^\beta$ for all β with $\alpha \leq \beta \leq \nu$. We define $A^\infty = A^\alpha$ where α is the least such ordinal. Note that it follows from **11.7** that A^∞ is a normal subgroup of A.

Corollary 11.18 Let:

(i) (X, T) be a regular minimal flow,
(ii) $B \subset A$ be closed subgroups of G_X, and
(iii) $A'B = A$.

Then $A^\infty B = A$.

PROOF: This follows from **11.14** and **10.12**. We leave the details as an exercise for the reader.

EXERCISES FOR CHAPTER 11

Exercise 11.1 Let:

(i) X be minimal flow,
(ii) F be a closed subgroup of $Aut(X)$, and
(iii) $\alpha, \beta \in F'$.

Then every open neighborhood of α in F intersects every open neighborhood of β in F.

Exercise 11.2 Let:

(i) X be a regular minimal flow,
(ii) F be a closed subgroup of G_X, and
(iii) $\alpha \in F'$.

Then there exists a net (α_i) in F with $\alpha_i \to \alpha$ and $\alpha_i \to 1_F$.

Exercise 11.3 Let:

(i) X be a minimal flow,
(ii) F be a closed subgroup of G_X, and
(iii) A be a closed subgroup of F.

Then $A' \subset A \cap F'$.

Exercise 11.4 Let R, S be regular icers with $R \subset S$. Then the natural map $(G_R, \tau) \to (G_S, \tau)$ is continuous, open, and closed.

Exercise 11.5 Let:

(i) R, S be regular icers with $R \subset S$,
(ii) $A \subset G_R$ be a closed subgroup, and
(iii) $\kappa : A \to G_S$ be the restriction of the natural map to A.

Then:

(a) κ is continuous.
(b) κ is closed.
(c) κ induces a homeomorphism $A/ker(\kappa) \to \chi_S^R(A)$.
(d) κ is open.

Exercise 11.6 Let:

(i) (X, T) be a regular minimal flow,
(ii) A be a closed subgroup of G, and
(iii) $\chi_X : G \to G_X$ be the natural map.

Then $\chi_X(A') = (\chi_X(A))'$.

Exercise 11.7 Give the details of the proof of **11.18**.

12

Quasi-factors and the τ-topology

The study of the transformation group $(2^M, T)$ provides another approach to the τ-topology on G. We use this approach to prove that given any closed subgroup, $A \subset G$ there exists an icer R on M with $G(R) = A$. In fact we show that $R = \overline{gr(A)}$ is such an icer. The key tool is the action of βT on 2^M via the circle operator which was introduced in section 5.

The notation and definitions of **5.1** will be in force throughout this section. In particular all the topological spaces that occur (other than G and its subgroups) are assumed to be compact Hausdorff.

Let $A \subset G$ be a subgroup; we would like to associate with A a quasi-factor of M, that is a minimal sub-flow of $(2^M, T)$. Let $u \in M$ be an idempotent, then the element $[\overline{A(u)}]u = [A(u) \circ u] \in 2^M$ is an almost periodic point of the flow $(2^M, T)$. Thus the orbit closure of $[A(u) \circ u]$ is a minimal flow, and hence a quasi-factor of M. We show in **12.2** that this orbit closure is isomorphic to the flow $M/\overline{gr(A)}$; the proof relies on the following lemma.

Lemma 12.1 Let:

(i) $p, q \in M, u \in J$, and
(ii) $A \subset G$ be a subgroup of G.

Then:

(a) $A(u) \circ p \subset p\overline{gr(A)}$.
(b) $A(u) \circ p = A(u) \circ \alpha(p)$ for all $\alpha \in A$.
(c) $[\overline{A(u)}]p = [\overline{A(u)}]\alpha(p)$ for all $\alpha \in A$.

PROOF: (a) 1. Let $q \in A(u) \circ p$.
2. There exist $\alpha_i \in A$ and $t_i \in T$ with $t_i \to p$ and $\alpha_i(u)t_i \to q$. (by 1)
3. $(p, q) \underset{2}{=} \lim(ut_i, \alpha_i(ut_i)) \in \overline{gr(A)}$.
4. $q \in p\overline{gr(A)}$.
(b) 1. Let $z \in A(u) \circ p$.

2. There exist $\alpha_i \in A$ and $t_i \to p$ with $\alpha_i(u)t_i \to z$. (by 1)
3. $\alpha(ut_i) \to \alpha(p)$. (by 2)
4. $\alpha_i\alpha^{-1}(u)\alpha(ut_i) = \alpha_i(\alpha^{-1}\alpha(u))t_i \to z$. (by 2)
5. Let $U, Z \subset \beta T$ be open neighborhoods of $\alpha(p)$ and z respectively.
6. There exists i such that $\alpha(ut_i) \in U$ and $(\alpha_i\alpha^{-1})(u)(\alpha(ut_i)) \in Z$. (by 3, 4, 6)
7. There exists $t(U, Z) \in T$ with $t(U, Z) \in U$ and $(\alpha_i\alpha^{-1})(u)$ $t(U, Z) \in Z$. ($L_{\alpha_i\alpha^{-1}(u)}$ is continuous)
8. There exist nets $t_{U,Z} \in T$ and $\alpha_{U,Z} \in A$ with $t_{U,Z} \to \alpha(p)$ and $\alpha_{U,Z}(u)$ $t_{U,Z} \to z$. (by 7)
9. $z \in A(u) \circ \alpha(p)$. (by 8)
10. $A(u) \circ p \subset A(u) \circ \alpha(p)$. (by 1, 9)
11. $A(u) \circ \alpha(p) \subset A(u) \circ \alpha^{-1}(\alpha(p)) = A(u) \circ p$. (by 10 applied to α^{-1})

(c) This follows immediately from **5.10** and part (b).

Theorem 12.2 Let:

(i) A be a subgroup of G,
(ii) $u \in M$ be an idempotent,
(iii) $\pi : M \to 2^M$ be defined by $\pi(p) = [A(u) \circ p]$ for all $p \in M$,
(iv) $R = \{(p, q) \mid \pi(p) = \pi(q)\}$.

Then:

(a) π is a homomorphism,
(b) R is an icer on M,
(c) $R = \overline{gr(A)}$,
(d) $pR = A(u) \circ p = p\overline{gr(A)}$,
(e) π is open, and
(f) $G(R) = \overline{A}$.

PROOF: (a) follows from the fact that $[A(u)\circ p] = [\overline{A(u)}]p$ and (b) follows immediately from (a).

(c) Proof that $R \subset \overline{gr(A)}$:

1. Let $(p, q) \in R$.
2. $q = uq \in A(u)q \subset A(u) \circ q = A(u) \circ p \subset p\overline{gr(A)}$. (by 1, (iii), (iv), and **12.1**)

Proof that $\overline{gr(A)} \subset R$:

1. $A(u) \circ p = A(u) \circ \alpha(p)$ for all $\alpha \in A$ and $p \in M$. (by **12.1**)
2. $gr(A) \subset R$. (by 1)

3. $\overline{gr(A)} \subset R$. (by 2, and part (b))

(d) 1. Let $q \in pR$.

2. $q = uq \in A(u)q \subset A(u) \circ q = A(u) \circ p$. (by 1, (iii), (iv))

3. $pR \subset A(u) \circ p$. (by 1, 2)

4. $A(u) \circ p \subset p\overline{gr(A)}$. (by **12.1**)

5. $pR = A(u) \circ p = p\overline{gr(A)}$. (by 3, 4, and part (c))

(e) 1. Let $\varphi : \pi(M) \to 2^M$ be defined by $\varphi(y) = [\pi^{-1}(y)]$ for all $y \in \pi(M)$.

2. Let $y = [A(u) \circ p] = \pi(p) \in \pi(M)$.

3. $\pi^{-1}(y) = yR = A(u) \circ p$. (by 1, (iv) and part (d))

4. $\varphi(y) = [A(u) \circ p] = y$. (by 2, 3)

5. φ is the identity map and hence is continuous. (by 2, 4)

6. π is open. (by 1, 5, and **5.7**)

(f) 1. $\alpha \in G(R) \Leftrightarrow gr(\alpha) \subset R \Leftrightarrow gr(\alpha) \subset \overline{gr(A)} \Leftrightarrow \alpha \in \overline{A}$.

(by (c), **10.7**)

One immediate consequence of theorem **12.2** is a characterization of the τ-closure of a subgroup of G in terms of the circle operator on M.

Corollary 12.3 Let:

(i) $u \in J$, and
(ii) A be a subgroup of G.

Then $\overline{A} = \{\alpha \in G \mid \alpha(u) \in A(u) \circ u\}$.

PROOF: $\alpha \in \overline{A} \underset{\mathbf{10.7}}{\Longleftrightarrow} gr(\alpha) \subset \overline{gr(A)} \Longleftrightarrow \alpha(u) \in u\overline{gr(A)} \underset{\mathbf{12.2}}{=} A(u) \circ u$.

We now prove a theorem which gives conditions under which the $(\overline{gr(A)}, T)$ is topologically transitive. This proof uses a brief technical lemma.

Lemma 12.4 Let:

(i) $A \subset G$ be a closed subgroup,
(ii) $\alpha \in A'$,
(iii) $U \subset M \times M$ be open and invariant, and
(iv) $gr(\alpha) \cap U \neq \emptyset$.

Then $gr(A') \subset \overline{U \cap gr(A)}$.

PROOF: 1. There exists $V \subset M$ open with $V \times \alpha(V) \subset U$. (by (iii), (iv))

2. $A' \subset \overline{< V, V > \cap A}$. (by 1)

3. $gr(A') \underset{2,\mathbf{10.7}}{\subset} \overline{gr(< V, V > \cap A)} \subset \overline{(V \times V)T \cap gr(A)}$

$\underset{1,(iii)}{\subset} \overline{(1 \times \alpha^{-1})(U) \cap gr(A)}$

$= \overline{(1 \times \alpha^{-1})(U \cap gr(A))} = (1 \times \alpha^{-1}) \left(\overline{U \cap gr(A)}\right).$

4. $gr(A') = gr(\alpha A') = (1 \times \alpha)(gr(A')) \underset{3}{\subset} \overline{U \cap gr(A)}.$

Theorem 12.5 Let $A \subset G$ be a closed subgroup. Then $(\overline{gr(A)}, T)$ is topologically transitive if and only if $A = A'$.

PROOF: $\Longleftarrow$

1. Assume that $A = A'$ and let $W \subset M \times M$ be open with $W \cap \overline{gr(A)} \neq \emptyset$.
2. There exists $\alpha \in A'$ with $gr(\alpha) \cap W \neq \emptyset$. (by 1)
3. $gr(A') \subset \overline{WT \cap gr(A)} = \overline{(W \cap gr(A))T} \subset \overline{(W \cap \overline{gr(A)})T}$. (by 2, **12.4**)
4. $\overline{gr(A)} = \overline{gr(A')} \subset \overline{(W \cap \overline{gr(A)})T} \subset \overline{gr(A)}$. (by 1, 3)
5. $\overline{(W \cap \overline{gr(A)})T} = \overline{gr(A)}$. (by 4)
6. $\overline{gr(A)}$ is topologically transitive. (by 1, 5)

$\Longrightarrow$

1. Assume that $\overline{gr(A)}$ is topologically transitive.
2. Let $\alpha \in A$ and $V, W \subset M$ be open sets.
3. $\left((V \times V) \cap \overline{gr(A)}\right) T \cap (W \times \alpha(W)) \cap \overline{gr(A)} \neq \emptyset$. (by 1, 2)
4. There exists $\beta \in A$ with $gr(\beta) \cap (V \times V) \neq \emptyset \neq gr(\beta) \cap (W \times \alpha(W))$. (by 3)
5. $\beta \in A \cap < V, V > \cap < W, \alpha(W) > \neq \emptyset$. (by 4)
6. $\alpha \in \overline{A \cap < V, V >}$. (by 2, 5)
7. $\alpha \in A'$. (by 2, 6)
8. $A = A'$. (by 2, 7)

EXERCISES FOR CHAPTER 12

Exercise 12.1 Let:

(i) $X = M/R$ be regular,
(ii) K be a subgroup of G_X, and
(iii) $u \in J_X$, and $w \in J$ with $\pi_R(w) = u$.

Then $\alpha \in \overline{K}$ if and only if $\alpha(u) \in K(u) \circ w$.

Exercise 12.2 Let:

(i) $X = M/R$ be regular,
(ii) $\emptyset \neq A \subset G_X$, and
(iii) $B = \chi_R^{-1}(A) \subset G$.

Then:

(a) $\pi_R(B(u) \circ p) = A(\pi_R(u)) \circ p$ for all $u \in J$ and $p \in M$.
(b) if π_R is open, then $\pi_R(p\overline{gr(B)}) = \pi_R(p)\overline{gr(A)}$ for all $p \in M$.
(c) if π_R is open and A is a closed subgroup of G_X, then $\overline{gr(A)}$ is an icer on X with $G_X(\overline{gr(A)}) = A$.

Exercise 12.3 Let:

(i) A and B be closed subgroups of G, and
(ii) $AB = BA$.

Then $\overline{gr(A)} \circ \overline{gr(B)} = \overline{gr(AB)}$.

Exercise 12.4 Let:

(i) A be a closed subgroup of G, and
(i) R an icer on M.

Show that there exists an icer $R_0 \subset R$ such that:

(a) $R_0 \subset R$ is a proximal extension, and
(b) $R_0 \circ \overline{gr(A)}$ is an icer on M.

Hint: Use the quasi-relative product.

PART IV

Subgroups of G and the dynamics of minimal flows

Every minimal flow is determined by an icer R on the universal minimal set M. According to **7.21**, this icer must be of the form

$$R = (R \cap P_0) \circ gr(G(R)),$$

where $G(R) \subset G$ is a τ-closed subgroup of G. This motivates one of the important themes of this book: What does the group $G(R)$ tell us about the minimal flow M/R? We pursue this theme by introducing τ-closed subgroups P, D, and E, of G in sections 13, 14, and 15 respectively. These groups and their relationship to $G(R)$ and to the derived group G' play a key role in the study of the dynamics of the flow $(M/R, T)$.

We show that the proximal relation on M/R is an equivalence relation if and only if $P \subset G(R)$ (see **13.8**). In section 14, we show that $(M/R, T)$ is a distal flow if and only if $R = P_0 \circ gr(G(R))$ with $D \subset G(R)$. Similarly, in section 15, we see that $(M/R, T)$ is an equicontinuous flow if and only if $R = P_0 \circ gr(G(R))$ with $E \subset G(R)$. Showing that $E = DG'$, then implies that a distal flow M/R is equicontinuous if and only if $G' \subset G(R)$ (see **15.22**). Traditionally the group E has been defined as the group of the maximal equicontinuous minimal flow M/S_{eq} where $S_{eq} = P_0 \circ gr(E)$ is the equicontinuous structure relation on M. At the end of section 15 we give an intrinsic description of E.

The *regionally proximal relation* $Q(X)$ for a minimal flow (X, T) is introduced in section 15 in order to study equicontinuity. The relation $Q(X) = \Delta_X$ if and only if the flow (X, T) is equicontinuous (see **15.5**). Moreover if $Q(X)$ is an equivalence relation, then $Q(X) = S_{eq}(X)$, the equicontinuous structure relation on X (**16.2**). Section 16 makes a more detailed study of the regionally proximal relation; including a proof that if $E \subset G(R)G'$, then $Q(M/R)$ is an equivalence relation.

13
The proximal relation and the group P

In this section we wish to examine in more detail the proximal relation

$$P(X) = \{(x, y) \in X \times X \mid \overline{(x, y)T} \cap \Delta_X \neq \emptyset\}$$

for a flow (X, T) (see **4.5**). The relation $P(X)$ is invariant reflexive and symmetric, but is in general neither closed nor transitive. We begin with an investigation of conditions under which $P(X)$ is transitive, i.e. when $P(X)$ is an equivalence relation. Recall that $(x, y) \in P(X)$ if and only if there exists $p \in \beta T$ with $xp = yp$, which is equivalent to saying that $xp = yp$ for all p in some minimal ideal in βT. But this ideal **may depend upon** the pair (x, y). Our first proposition shows that $P(X)$ is an equivalence relation if and only if $P(X) = \{(x, y) \mid xp = yp \text{ for all } p \in M\}$.

Proposition 13.1 Let (X, T) be a flow. Then the following are equivalent:

(a) $E(X)$ contains only one minimal ideal,
(b) $P(X) = \{(x, y) \in X \times X \mid xm = ym \text{ for all } m \in M\}$, and
(c) $P(X)$ is an equivalence relation on X.

PROOF: (a) $\Longrightarrow$ (b)

1. Assume that $I \subset E(X, T)$ is the unique minimal ideal in $E(X, T)$.
2. $\{(x, y) \in X \times X \mid xm = ym \text{ for all } m \in M\} \subset P(X)$. (by **4.4**)
3. Let $\Phi_X : \beta T \to E(X)$ be the canonical map.
4. $\Phi_X(M)$ is a minimal subset of $E(X)$.
5. $\Phi_X(M) = I$. (by 1, 4)
6. Let $(x, y) \in P(X)$.
7. $xm = x\Phi_X(m) = y\Phi_X(m) = ym$. (by 1, 5, 6, and **4.4**)

(b) $\Longrightarrow$ (c)

1. Assume that $P(X) = \{(x, y) \in X \times X \mid xm = ym \text{ for all } m \in M\}$.
2. It suffices to show that $P(X)$ is transitive.

3. Let $(x, y), (y, z) \in P(X)$.
4. $xm = ym = zm$ for all $m \in M$. (by 1, 2)
5. $(x, z) \in P(X)$. (by 1, 4)

(c) $\Longrightarrow$ (a)

1. Assume that $P(X)$ is an equivalence relation.
2. Let I_1 and I_2 be minimal right ideals in $E(X)$.
3. There exists idempotents $u_1 \in I_1$ and $u_2 \in I_2$ with $u_1u_2 = u_1$ and $u_2u_1 = u_2$. (by **3.14**)
4. Let $x \in X$.
5. $(x, xu_1), (x, xu_2) \in P(X)$. (by 1, 3, 4, **4.4**)
6. $(xu_1, xu_2) \in P(X)$. (by 1, 5)
7. There exists a minimal right ideal $I_3 \subset E(X)$ with $xu_1q = xu_2q$ for all $q \in I_3$. (by 6, and **4.4**)
8. There exists an idempotent $u_3 \in I_3$ with $u_3u_1 = u_3$ and $u_1u_3 = u_1$. (by 3, 7, **3.14**)
9. $xu_1 \underset{8}{=} xu_1u_3 \underset{7}{=} xu_2u_3 \underset{3}{=} x(u_2u_1)u_3 = xu_2(u_1u_3) \underset{8}{=} xu_2u_1 \underset{3}{=} xu_2$.
10. $u_1 = u_2$. (by 4, 9)
11. $I_1 \underset{2}{=} u_1I_1 \underset{10}{=} u_2I_1 \underset{2}{=} I_2$.

One consequence of **13.1** is a characterization of those minimal flows for which the proximal relation is an equivalence relation. We identify those icers R on M for which $P(M/R)$ is an equivalence relation using the notion of equivalent minimal idempotents. Recall from **3.13** that a minimal idempotent is an idempotent in some minimal ideal, and $u_1 \sim u_2$ means $u_1u_2 = u_1$ and $u_2u_1 = u_2$.

Proposition 13.2 Let $X = M/R$ with R an icer on M. Then $P(X)$ is an equivalence relation if and only if $(pu_1, pu_2) \in R$ for all $p \in M$ and pairs u_1, u_2 of equivalent minimal idempotents in βT.

PROOF: $\Longrightarrow$

1. Assume that $P(X)$ is an equivalence relation.
2. Let $u_1 \sim u_2$ be minimal idempotents and $p \in M$.
3. $\pi_R(p, pu_1), \pi_R(p, pu_2) \in \pi_R(P(M)) = P(X)$. (by 1, 2, **4.6**, and **4.7**)
4. $\pi_R(pu_1, pu_2) \in P(X)$. (by 1, 3)
5. There exists a minimal ideal $I \subset \beta T$ with $\pi_R(pu_1, pu_2)q \in \Delta$ for all $q \in I$. (by 4, and **4.4**)
6. There exists an idempotent $u \in I$ with $u \sim u_1 \sim u_2$. (by **3.14**)
7. $\pi_R(pu_1) \underset{6}{=} \pi_R(pu_1u) = \pi_R(pu_1)u \underset{5}{=} \pi_R(pu_2)u = \pi_R(pu_2u) \underset{6}{=} \pi_R(pu_2)$.

8. $(pu_1, pu_2) \in R$. (by 7)

$$\Longleftarrow$$

1. Assume that $(pu_1, pu_2) \in R$ whenever $u_1 \sim u_2$ are minimal idempotents in βT.
2. Let I_1 and I_2 be minimal ideals in $E(X)$.
3. There exist minimal idempotents $u_1 \sim u_2 \in \beta T$ with $\Phi_X(u_1) \in I_1$ and $\Phi_X(u_2) \in I_2$. (by **2.9**)
4. Let $x = \pi_R(p) \in X$.
5. $x\Phi_X(u_1) \underset{4}{=} \pi_R(pu_1) \underset{1}{=} \pi_R(pu_2) \underset{4}{=} x\Phi_X(u_2)$.
6. $\Phi_X(u_1) = \Phi_X(u_2)$. (by 4, 5)
7. $I_1 \cap I_2 \neq \emptyset$. (by 2, 3, 6)
8. $I_1 = I_2$. (by 2, 7)
9. $P(X)$ is an equivalence relation. (by 2, 9, **13.1**)

The next two corollaries can be deduced more directly but are easily seen to be immediate consequences of **13.2**.

Corollary 13.3 Let:

(i) $R \subset S$ be icers on M, and
(ii) $P(M/R)$ be an equivalence relation.

Then $P(M/S)$ is an equivalence relation.

Corollary 13.4 Let:

(i) $\{R_i \mid i \in I\}$ be a family of icers on M, and
(ii) $R = \bigcap_{i\in I} R_i$.

Then $P(M/R)$ is an equivalence relation if and only if $P(M/R_i)$ is an equivalence relation for all $i \in I$.

Another important corollary involves the regularizer of an icer S on M. Recall from **8.4** that $reg(S) = \bigcap_{\alpha\in G} \alpha(S)$ and that $G(reg(S)) = \bigcap \alpha^{-1}G(S)\alpha$ is a normal subgroup of G.

Corollary 13.5 Let:

(i) S be an icer on M, and
(ii) $R = \bigcap_{\alpha\in G} \alpha(S)$ be the regularizer of S.

Then $P(M/R)$ is an equivalence relation if and only if $P(M/S)$ is an equivalence relation.

PROOF: This follows immediately from **7.16** and **13.4**.

It is an interesting consequence of **13.2** that for a minimal flow (X, T), if $P(X)$ is closed, then it is also an equivalence relation. This means that in this case $P(X)$ is an icer and the quotient flow $X/P(X)$ is distal. We will examine this situation further in section 13; here we give a proof that if $P(M/R)$ is closed, then it is an equivalence relation.

Proposition 13.6 Let:

(i) R be an icer on M, and
(ii) $S = \bigcap_{\alpha \in G} \alpha(R)$.

Then:

(a) $P(M/R)$ closed implies that $P(M/R)$ is an equivalence relation, and
(b) $P(M/R)$ is closed if and only if $P(M/S)$ is closed.

PROOF: (a) 1. Assume that $P(M/R)$ is closed.
2. Let $p \in M$ and $u_1 \sim u_2$ be minimal idempotents in βT.
3. $(p, pu_2) \in P(M)$.
4. $(pt, pu_2t) \in P(M)$ for all $t \in T$.
5. $\pi_R(pu_1, pu_2) = \pi_R(p, pu_2)u_1 \underset{3,4}{\in} \pi_R(\overline{P(M)}) = \overline{\pi_R(P(M))} \underset{\mathbf{4.7}}{=} \overline{P(M/R)}$
$\underset{1}{=} P(M/R)$.
6. There exists a minimal ideal $K \subset \beta T$ with $\pi_R(pu_1, pu_2)q \in \Delta$ for all $q \in K$. (by 5, **4.4**)
7. There exists $u_3 \in K$ with $u_1 \sim u_3 \sim u_2$. (by 2, 6, and **3.14**)
8. $\pi_R(pu_1, pu_2) = \pi_R(pu_1, pu_2)u_3 \in \Delta$. (by 6, 7)
9. $(pu_1, pu_2) \in R$. (by 8)
10. $P(M/R)$ is an equivalence relation. (by 2, 9, and **13.2**)

(b) $\Longleftarrow$

This follows immediately from **4.7**.

$\Longrightarrow$

1. Assume that $P(M/R)$ is closed.
2. Let $(p, q) \in \overline{P(M)}$.
3. Let $\alpha \in G$.
4. $\alpha(p, q) \in \alpha\left(\overline{P(M)}\right) = \overline{\alpha(P(M))} = \overline{P(M)}$.
5. $\pi_R(\alpha(p), \alpha(q)) \in \pi_R\left(\overline{P(M)}\right) = \overline{P(M/R)} = P(M/R)$.
(by 1, 4, and **4.7**)
6. $P(M/R)$ is an equivalence relation since we are assuming it is closed. (by (a))
7. $\pi_R(\alpha(p), \alpha(q))m \in \Delta$ for all $m \in M$. (by 5, 6, **13.1**)
8. $\alpha(pm, qm) \in R$ for all $\alpha \in G$ and $m \in M$. (by 3, 7)

9. $(pm, qm) \in S$ for all $m \in M$. (by 8, (ii))
10. $\pi_S(p, q) \in P(M/S)$. (by 9, and **4.4**)
11. $\overline{P(M/S)} \underset{\mathbf{4.7}}{=} \overline{\pi_S(P(M))} = \pi_S\left(\overline{P(M)}\right) \underset{2,10}{\subset} P(M/S)$.

It is natural to ask whether the converse of **13.6** (a) is true. A counterexample is given in [Shapiro, *Proximality in minimal transformation groups*, (1970)]. Indeed, in this example not only is $P(X)$ an equivalence relation but every *proximal cell*

$$xP(X) = \{z \in X \mid (x, z) \in P(X)\}$$

is closed, and yet $P(X)$ is not closed. We now investigate this situation from the group-theoretic point of view; our goal is to define closed subgroups $P \subset G$ and $G^J \subset G$ so that $P(M/R)$ is an equivalence relation if and only if $P \subset G(R)$ and $P(M/R)$ is an equivalence relation with closed cells if and only if $PG^J \subset G(R)$.

We first observe that the collection $\{\beta(J) \mid \beta \in G\}$ (where as usual J denotes the set of idempotents in M) forms a partition of M (by **7.4**). On the other hand the subsets $\beta(J) \subset M$ with $\beta \in G$ are **not** the proximal cells in M. In particular for $u \in J$, $uP(M) \neq J$. Indeed $(u, uu_1) \in P(M)$ for any $u \in J$ and any idempotent $u_1 \in \beta T$, so $uu_1 \in uP(M)$ but uu_1 need not be an idempotent. On the other hand uu_1 must be of the form $\alpha(v)$ for some $\alpha \in G$ and $v \in J$. Now if R is any icer for which $P(M/R)$ is an equivalence relation we must have $(\pi_R(u), \pi_R(\alpha(v))) \in P(M/R)$. Using **13.1** this gives $\pi_R(gr(\alpha)) \subset \Delta$ and thus $\alpha \in G(R)$. In this sense the elements $\alpha \in G$ with $(u, \alpha(v)) \in P(M)$ are "obstructions" to $P(M/R)$ being an equivalence relation. This motivates the definition of the subgroup $P \subset G$, which, as the next proposition shows, characterizes those icers on M for which $P(M/R)$ is an equivalence relation.

Definition 13.7 Let $P \subset G$ denote the closed normal subgroup of G generated by the set

$$\{\alpha \mid (u, \alpha(v)) \in P(M) \text{ for some } u, v \in J\}.$$

Proposition 13.8 Let $X = M/R$. Then $P(X)$ is an equivalence relation if and only if $P \subset G(R)$.

PROOF: $\Longrightarrow$

1. Assume that $P(X)$ is an equivalence relation.
2. Let $u, v \in J$ with $(u, \alpha(v)) \in P(M)$ where $\alpha \in G$.
3. Set $S = \bigcap_{\beta \in G} \beta(R)$ and $Y = M/S$.

4. $P(Y)$ is an equivalence relation. (by 1, **13.5**)
5. $\pi_S(u, \alpha(v)), \pi_S(u, v) \in \pi_S(P(M)) = P(Y)$. (by 2, **4.7**)
6. $\pi_S(v, \alpha(v)) \in P(Y)$. (by 4, 5)
7. There exists $q \in \beta T$ with $\pi_S(vq) = \pi_S(\alpha(vq))$. (by 6, and **4.4**)
8. $\alpha \in G(S)$ which is a normal subgroup of G. (by 3, 7, **8.2, 8.4**)
9. $P \subset G(S) \subset G(R)$. (by 2, 8)

$$\Longleftarrow$$

1. Assume that $P \subset G(R)$.
2. Since P is normal in G, $P \subset \bigcap_{\alpha \in G} \alpha G(R)\alpha^{-1} = \bigcap_{\alpha \in G} G(\alpha(R)) = G\left(\bigcap_{\alpha \in G} \alpha(R)\right)$. (by 1, **7.16**, and **7.19**)
3. Let $u_1 \sim u_2$ be minimal idempotents in βT, and $p \in M$.
4. There exists $u^2 = u \in M$ with $u \sim u_1 \sim u_2$. (by **3.14**)
5. There exists $v^2 = v \in M$ and $\beta \in G$ with $\beta(v) = p$. (by **7.4**)
6. There exists $\alpha \in G$ with $\alpha(u) = vu_1$. (by 4, 5, and **7.4**)
7. $(v, \alpha(u)) = (v, vu_1) \in P(M)$. (by 3, 6, and **4.6**)
8. $\alpha \in P$. (by 7)
9. $\beta\alpha\beta^{-1} \in P \subset G(R)$. (by 1, 8)
10. $(pu, pu_1) \underset{5}{=} (\beta(v)u, \beta(v)u_1) \underset{6}{=} (\beta(u), \beta\alpha(u)) - (\beta(u), \beta\alpha\beta^{-1}(\beta(u)) \in gr(\beta\alpha\beta^{-1}) \underset{9}{\subset} R$.
11. $(pu, pu_2) \in R$. (applying 6-10 to u_2)
12. $(pu_1, pu_2) \in R$. (by 10, 11)
13. $P(X)$ is an equivalence relation. (by 3, 12, **13.2**)

As we remarked earlier the partition $\{\beta(J) \mid \beta \in G\}$ is not the collection of proximal cells in M. On the other hand for any icer R for which $P(M/R)$ is an equivalence relation this partition does project under π_R to the collection of proximal cells in M/R, i.e. $\pi_R(\beta(J)) = \pi_R(\beta(v))P(M/R)$. This follows from **13.1**; we leave the proof as an exercise for the reader. In particular, in this case $\pi_R(J)$ is one of the proximal cells in M/R. If the proximal cells are to be closed, then we must have $\pi_R(\overline{J}) = \pi_R(J)$. This means that for $\alpha(u) \in \overline{J}$ we must have $\pi_R(\alpha(u)) \in \pi_R(J)$ which implies that $\alpha \in G(R)$. Once again this motivates the appropriate definition.

Definition 13.9 As usual let $J = \{u \in M \mid u^2 = u\}$ denote the set of idempotents in a fixed minimal ideal $M \subset \beta T$. We denote by:

$$J^* \equiv \{\alpha \in G \mid \alpha(J) \cap \overline{J} \neq \emptyset\},$$

and define G^J to be the closed normal subgroup of G generated by J^*.

The following lemma will be used to prove that the subgroup PG^J characterizes those icers for which M/R is an equivalence relation with closed cells.

Lemma 13.10 Let:

(i) X be a flow,
(ii) $N \subset X \times X$,
(iii) $\alpha \in Aut(X)$, and
(iv) $x \in X$.

Then $\alpha(x)\alpha(N) = \alpha(xN)$.

PROOF: We leave the proof as an exercise for the reader.

Theorem 13.11 Let:

(i) R be an icer on M, and
(ii) $S = \bigcap_{\alpha \in G} \alpha(R)$.

Then the following are equivalent:

(a) $P(M/R)$ is an equivalence relation with closed cells,
(b) $P(M/S)$ is an equivalence relation with closed cells,
(c) $PG^J \subset G(S)$, and
(d) $PG^J \subset G(R)$.

PROOF: (a) $\Rightarrow$ (b)

1. Assume that $P(M/R)$ is an equivalence relation with closed cells.
2. $P(M/S)$ is an equivalence relation. (by 1, **13.5**)
3. Let $x \in \overline{yP(M/S)}$.
4. There exist $p, q \in M$ with $\pi_S(p,q) = (x,y)$.
5. Let $\alpha \in G$.
6. $\pi_R(\alpha(p))$

$$= \pi_R^S(\pi_S(\alpha(p))) \underset{\mathbf{8.11}}{=} \pi_R^S(\chi_S(\alpha)(\pi_S(p))) \underset{3,4}{\in} \pi_R^S\left(\chi_S(\alpha)\left(\overline{\pi_S(q)P(M/S)}\right)\right)$$

$$= \pi_R^S\left(\overline{\chi_S(\alpha)(\pi_S(q))P(M/S))}\right) \underset{\mathbf{13.10}}{=} \pi_R^S\left(\overline{\chi_S(\alpha)(\pi_S(q))\chi_S(\alpha)(P(M/S))}\right)$$

$$\underset{\mathbf{8.11}}{=} \pi_R^S\left(\overline{\pi_S(\alpha(q))\pi_S(\alpha(P(M)))}\right) = \pi_R^S\left(\overline{\pi_S(\alpha(q))\pi_S(P(M))}\right)$$

$$\underset{\mathbf{4.7}}{=} \pi_R^S\left(\overline{\pi_S(\alpha(q)P(M))}\right) = \overline{\pi_R(\alpha(q)P(M))}$$

$$\underset{\mathbf{4.7}}{=} \overline{\pi_R(\alpha(q))P(M/R)} \underset{1}{=} \pi_R(\alpha(q))P(M/R).$$

7. $\pi_R(\alpha(p))m = \pi_R(\alpha(q))m$ for all $m \in M$. (by 1, 6, and **13.1**)
8. $(\alpha(pm), \alpha(qm)) \in R$ for all $m \in M$. (by 7)

9. $(pm, qm) \in S$ for all $m \in M$. (by 5, 8)
10. $xm = \pi_S(pm) = \pi_S(qm) = ym$ for all $m \in M$. (by 9)
11. $(y, x) \in P(M/S)$. (by 10)
12. $x \in yP(M/S)$. (by 11)
13. $yP(M/S)$ is closed. (by 3, 12)

(b) $\Longrightarrow$ (c)

1. Assume that $P(M/S)$ is an equivalence relation with closed cells.
2. $P \subset G(S)$. (by 1, **13.8**)
3. Let $\alpha \in J^*$.
4. There exists $u \in J$ with $\alpha(u) \in \overline{J} \subset \overline{uP(M)}$. (by 3)
5. $\pi_S(\alpha(u)) \underset{4}{\in} \pi_S(\overline{uP(M)}) = \overline{\pi_S(uP(M))} \underset{\mathbf{4.7}}{=} \overline{\pi_S(u)P(M/S)} \underset{1}{=} \pi_S(u)P(M/S)$.
6. $\pi_S(\alpha(m)) = \pi_S(\alpha(u))m = \pi_S(u)m = \pi_S(m)$ for all $m \in M$. (by 1, 5, and **13.1**)
7. $\alpha \in G(S)$. (by 6)
8. $G^J \subset G(S)$. (by 3, 7, and the fact that $G(S)$ is normal in G)

(c) $\Rightarrow$ (d)

$G(S) \subset G(R)$ so this is immediate.

(d) $\Rightarrow$ (a)

1. Assume that $PG^J \subset G(R)$.
2. $P(M/R)$ is an equivalence relation. (by **13.8**)
3. Let $x \in \overline{yP(M/R)}$.
4. There exist $\beta \in G$ and $v \in J$ with $y = \pi_R(\beta(v))$.
5. There exist $\alpha \in G$ and $u \in J$ with $x = \pi_R(\alpha(u))$.
6. $\pi_R(\alpha(u)) \in \overline{\beta(v)P(M/R)} = \overline{\pi_R(\beta(J))} = \pi_R(\beta(\overline{J}))$. (by 2, and **13.1**)
7. There exists $\gamma \in J^*$ and $w \in J$ with $\pi_R(\alpha(u)) = \pi_R(\beta(\gamma(w)))$. (by 6)
8. $\beta\gamma\alpha^{-1} \in G(R)$. (by 7)
9. $\beta\alpha^{-1} = (\beta\gamma^{-1}\beta^{-1})(\beta\gamma\alpha^{-1}) \in G(R)$. (by 1, 8)
10. $(y, x) = \pi_R(\alpha(u), \beta(v)) = \pi_R(\beta(u), \beta(v)) \in P(M/R)$. (by 9)
11. $x \in yP(M/R)$. (by 10)
12. $P(M/R)$ has closed cells. (by 3, 11)

The definition of the subset $J^* \subset G$ though motivated by the fact that when $P(M/R)$ is an equivalence relation $\pi_R(J)$ is a proximal cell in M/R, identifies the elements of G which are obsructions to $\pi_R(J)$ being closed. We make this precise in the following proposition.

Proposition 13.12 Let $X = M/R$ be a minimal flow. Then $J^* \subset G(R)$ if and only if $\pi_R(J)$ is closed.

PROOF: $\Longrightarrow$

1. Assume that $J^* \subset G(X)$.
2. Let $y \in \overline{\pi_R(J)}$.
3. $y = \pi_R(p)$ for some $p \in \overline{J}$. (by 2)
4. There exists an idempotent $u \in J$ and $\alpha \in G$ with $\alpha(u) = p$. (by **7.4**)
5. $\alpha \in J^* \subset G(R)$. (by 1, 3, 4, and **13.9**)
6. $y \underset{3}{=} \pi_R(p) \underset{4}{=} \pi_R(\alpha(u)) \underset{5}{=} \pi_R(u) \in \pi_R(J)$.

$\Longleftarrow$

1. Assume that $\pi_R(J)$ is closed and let $\alpha \in J^*$.
2. There exists an idempotent $u \in J$ with $\alpha(u) \in \overline{J}$.
3. $\pi_R(\alpha(u)) \in \pi_R(\overline{J}) = \pi_R(J)$. (by 1, 2)
4. $(v, \alpha(u)) \in R$ for some $v \in J$. (by 3)
5. $(u, \alpha(u)) \in Ru \subset R$. (by 4)
6. $\alpha \in G(R)$. (by 5)
7. $J^* \subset G(R)$. (by 1, 6)

Let $X = M/R$ be a regular flow and $J_X = \{u \in X \mid u^2 = u\}$. We first recall from **8.6** that the canonical map $\pi_R \colon M \to X$ is a homomorphism both of flows and of semigroups. It follows that

$$\pi_R(J) = J_X \qquad \text{and} \qquad \pi_R(\overline{J}) = \overline{J_X}.$$

Thus the previous proposition shows that J_X is closed if and only if $J^* \subset G(R)$.

EXERCISES FOR CHAPTER 13

Exercise 13.1 Let R be an icer on M. Show that $P(M/R)$ is an equivalence relation if and only if $P(M/R) = \pi_R(P_0)$.

Exercise 13.2 Let:

(i) R be an icer on M,
(ii) $P(M/R)$ be an equivalence relation,
(iii) $u, v \in J$ (the set of idempotents in M), and
(iv) $\beta \in G$.

Then the partition $\{\beta(J) \mid \beta \in G\}$ of M projects onto the proximal cells in M/R, i.e. $\pi_R(\beta(J)) = \pi_R(\beta(v))P(M/R)$.

Exercise 13.3 (See **13.10**) Let:

(i) X be a flow,
(ii) $N \subset X \times X$,

(iii) $\alpha \in Aut(X)$, and
(iv) $x \in X$.

Then $\alpha(x)\alpha(N) = \alpha(xN)$.

Exercise 13.4 Let:

(i) $X = M/R$ be a flow, and
(ii) $N = \{(x, xu) \mid x \in X \text{ and } u \in J\}$.

Then:

(a) N is an invariant equivalence relation on X.
(b) N has closed cells if and only if $G^J \subset G(R)$.
(c) $N = P(X)$ if and only if $P \subset G(R)$.

Exercise 13.5 Let:

(i) $X = M/R$ be a flow,
(ii) $G(R)$ be a normal subgroup of G, and
(iii) $P(X)$ be an equivalence relation with one closed cell.

Then all the cells of $P(X)$ are closed.

Exercise 13.6 Let $X = M/R$ and $Y = M/S$ with $R \subset S$ icers on M. Then the following are equivalent:

(a) X is a proximal extension of Y,
(b) $\left((\pi_S^R)^{-1}(y)\right) p = \{xp\}$ for all $y \in Y$, $x \in (\pi_S^R)^{-1}(y)$, and $p \in M$, and
(c) there exist $y_0 \in Y$ and $p \in M$ such that $\left((\pi_S^R)^{-1}(y_0)\right) p$ is a singleton.

14

Distal flows and the group D

In this section we characterize those icers on M for which the minimal flow M/R is distal. We introduce a τ-closed subgroup $D \subset G$ and show that M/R is distal if and only if $R = P_0 \circ gr(A)$, where $A = G(R)$ is a τ-closed subgroup with $D \subset A$. (See **7.20** for the definition of P_0.) In particular $S_d = P_0 \circ gr(D)$ the so-called distal structure relation is an icer on M, and every minimal distal flow is an image of M/S_d. Indeed a minimal distal flow is completely determined by its group, and the collection of distal minimal flows, $\{M/R \mid S_d \subset R\}$ can be put in one-one correspondence with the collection $\{A = \overline{A} \subset G \mid D \subset A\}$, of τ-closed subgroups of G which contain D.

Let (X, T) be a flow. Recall that as in **4.5**, we say that (X, T) is a *distal flow* if $P(X) = \Delta$ (the diagonal of $X \times X$). Here $P(X)$ is the proximal relation consisting of all the proximal pairs in $X \times X$. We will be interested in minimal distal flows; in this case $X = M/R$ for some icer R on M. We will often use the phrase: R is distal to mean that the flow $(M/R, T)$ is distal. This terminology needs to be used carefully; this **does not mean** that the subflow (R, T) of $(M \times M, T)$ is a distal flow. Note that the homomorphic image of a distal flow is distal (see **4.10**). Hence, using the terminology above, R is distal, if and only if, $\alpha(R)$ is distal for every $\alpha \in G$. (Recall that by **7.6**, α induces an isomorphism of the flows M/R and $M/\alpha(R)$.)

Now assume that R is a distal icer on M; then according to **4.7**, $\pi_R(P(M)) = P(M/R)$, so we must have $P(M) \subset R$. Recalling the notation from **7.20**:

$$P_0 = \{(\alpha(u), \alpha(v)) \mid \alpha, \beta \in G \text{ and } u, v \in J\} \subset P(M).$$

The following proposition shows that M/R is distal if and only if $P_0 \subset R$.

Proposition 14.1 Let:

(i) R be an icer on M,
(ii) $X = M/R$, and
(iii) $J \subset M$ be the set of idempotents in M.

Then the following are equivalent:

(a) X is a distal flow,

(b) $P(M) \subset R$, and

(c) $P_0 \equiv \bigcup_{\alpha \in G} \alpha(J \times J) \subset R$.

PROOF: (a) $\Longleftrightarrow$ (b)

This follows immediately from the fact that $\pi_R(P(M)) = P(X)$ (see **4.7**).

(b) $\Longrightarrow$ (c)

Clear since $P_0 \subset P(M)$.

(c) $\Longrightarrow$ (b)

1. Assume that $P_0 \subset R$ and let $(p, q) \in P(M)$.
2. There exists a minimal ideal $I \subset \beta T$ and an idempotent $w \in I$ with $q = pw$. (by **4.6**)
3. There exists $v \in J$ with $vw = v$ and $wv = w$. (by **3.14**)
4. $(p, pv) \in P_0 \subset R$. (by 1)
5. $(pv, pw) \underset{3}{=} (pvw, pw) = (pv, p)w \underset{4}{\in} Rw \subset R$.
6. $(p, q) \underset{2}{=} (p, pw) \underset{4,5}{\in} R \circ R \subset R$.

We deduce some immediate consequences of **14.1**.

Corollary 14.2 Let $R \subset S$ be distal icers on M. Then the map $\pi_S^R \colon M/R \to M/S$ is open.

PROOF: $R \cap P_0 = P_0 = S \cap P_0$ so this follows immediately from **7.23**.

Corollary 14.3 Let R be a regular icer on M, and $J \subset M$ be the set of idempotents in M. Then R is distal if and only if $J \times J \subset R$.

Corollary 14.4 Let R be an icer on M. Then $(M/R, T)$ is distal if and only if $R = P_0 \circ gr(G(R))$.

PROOF: Recall that by **7.21**, $R = (R \cap P_0) \circ gr(G(R))$.

Suppose that S is a compact Hausdorff topological space with a group structure which is a compactification of T in the sense that there exists a homomorphism of T onto a dense subgroup of S. Assume further that for every $t \in T$ right multiplication by the image of t in S is continuous (this is a much weaker than assuming S is a topological group). Then (S, T) is a minimal flow where T acts on S by right multiplication. In fact in this case $\overline{(s_1, s_2)T} \cap \Delta_S \neq \emptyset \Rightarrow (s_1, s_2)s \in \Delta$ for some $s \in S \Rightarrow s_1 = s_2$. Thus (S, T) is a minimal distal flow.

The following proposition shows that every regular minimal distal flow is of this form.

Proposition 14.5 Let:

(i) $X = M/R$ be a regular distal flow, and
(ii) $u \in J$ be any idempotent in M.

Then:

(a) $X = \{\pi_R(\alpha(u)) \mid \alpha \in G\}$ is a group with identity $\pi_R(u)$,
(b) the map

$$\begin{array}{ccc} T & \to & X \\ t & \to & \pi_R(ut) \end{array}$$

is a homomorphism of T onto a dense subgroup of X, and
(c) the map

$$\begin{array}{ccc} \varphi\colon G & \to & X \\ \alpha & \to & \pi_R(\alpha(u)) \end{array}$$

is an epimorphism which induces an isomorphism

$$G/G(R) \to X.$$

PROOF: (a) 1. Let $x \in X$.
2. There exists $p \in M$ with $\pi_R(p) = x$.
3. There exist $\alpha \in G$ and $v \in J$ with $\alpha(v) = p$. (by **7.4**)
4. $x = \pi_R(p) = \pi_R(\alpha(v)) = \pi_R(\alpha(u))$. (by 2, 3, (i), and **14.1**)
5. $\pi_R\colon M \to X = M/R$ is a semigroup homomorphism. (by (i), **8.6**)
6. $\pi_R(\alpha(u))\pi_R(\beta(u)) = \pi_R(\alpha(u)\beta(u)) = \pi_R(\alpha(\beta(u)))$ for all $\alpha, \beta \in G$. (by 5)
7. $\pi_R(u)\pi_R(\beta(u)) = \pi_R(\beta(u)) = \pi_R(\beta(u))\pi_R(u)$ for all $\beta \in G$. (by 6)
(b) 1. Let $t, s \in T$.
2. $\pi_R(ut)\pi_R(us) = \pi_R(ut)\pi_R(u)s = \pi_R(uts)$. (by part (a))
3. $\overline{\{\pi_R(ut) \mid t \in T\}} = \overline{\{\pi_R(u)t \mid t \in T\}} = X$. ($X$ is minimal)
(c) 1. φ is an epimorphism of the group G onto the group X. (by (a))
2. $\alpha \in \ker\varphi \iff \pi_R(\alpha(u)) = \pi_R(u) \iff (u, \alpha(u)) \in R \iff \alpha \in G(R)$.

It is important to note that **14.5 does not say** that any regular minimal distal flow is a compact topological group. The map φ while it is an isomorphism of groups is not in general continuous. Indeed (by **11.11**) the quotient $G/G(R)$ is a compact Hausdorff topological group if and only if the derived group $G' \subset G(R)$. We will see in the next section that this is the case if and only if (X, T) is an *equicontinuous flow*.

We now wish to define the closed subgroup $D \subset G$ which allows us to identify all of those icers R for which M/R is distal. The group D characterizes those icers R for which $P(M/R)$ is closed in the same sense that the subgroup $P \subset G$ characterizes those for which $P(M/R)$ is an equivalence relation. Since $\pi_R(P(M)) = P(M/R)$, if the latter is closed, then $\pi_R\left(\overline{P(M)}\right) = P(M/R)$ so that $gr(\alpha) \subset \overline{P(M)}$ implies that $\alpha \in G(R)$. This motivates the following definition.

Definition 14.6 Let $K = \{\alpha \in G \mid gr(\alpha) \subset \overline{P(M)}\}$. We define D to be the closed subgroup of G generated by K. More generally, for any regular minimal flow (X, T), we define D_X to be the closed subgroup of G_X generated by $K_X = \{\alpha \in G_X \mid gr(\alpha) \subset \overline{P(X)}\}$. Thus in particular $D = D_M$.

As an immediate consequence of the definitions we have the following lemma.

Lemma 14.7 Let:

(i) $X = M/R$ be a regular minimal flow, and
(ii) $\beta \in G_X$.

Then:

(a) $K_X = \{\alpha \in G \mid gr(\alpha) \cap \overline{P(X)} \neq \emptyset\}$
(b) $\beta^{-1} K_X \beta = K_X$, and
(c) D_X is normal in G_X.

PROOF: (a) This follows immediately from the fact that $\overline{P(X)}$ is closed and invariant.

(b) 1. Let $\alpha \in K_X$.
2. $(\beta(p), \alpha(\beta(p))) \in \overline{P(X)}$. (by 1, (i))
3. $(p, \beta^{-1}\alpha\beta(p)) \in \beta^{-1}\left(\overline{P(X)}\right) = \overline{P(X)}$. (by 2)
4. $\beta^{-1}\alpha\beta \in K_X$. (by 3)
5. $\beta^{-1} K_X \beta \subset K_X$. (by 1, 4)
6. $\beta K_X \beta^{-1} \subset K_X$. (by 5 applied to β^{-1})
7. $\beta^{-1} K_X \beta = K_X$. (by 5, 6)

(c) 1. $K_X \subset \beta^{-1} D_X \beta$ for all $\beta \in G_X$. (by part (b))
2. $D_X \subset \beta^{-1} D_X \beta$ for all $\beta \in G_X$. (by 1)
3. $D_X \subset \beta D_X \beta^{-1}$ for all $\beta \in G_X$. (by 2 applied to β^{-1})
4. D_X is normal in G_X. (by 2, 3)

We use the preceding lemma to verify that the group D does indeed characterize those icers for which $P(M/R)$ is closed.

Proposition 14.8 Let $X = M/R$ be a minimal flow. Then the following are equivalent:

(a) $P(X)$ is closed,
(b) $D \subset G(R)$, and
(c) (X, T) is a proximal extension of the distal flow $(M/(P_0 \circ gr(G(R))), T)$.

PROOF: (a) $\Longrightarrow$ (b)

1. Assume that $P(X)$ is closed.
2. Let $\alpha \in G$ with $gr(\alpha) \subset \overline{P(M)}$ and $u \in J$.
3. $\pi_R(u, \alpha(u)) \in \pi_R(\overline{P(M)}) = \overline{\pi_R(P(M))} = \overline{P(X)} = P(X)$. (by 1, **4.7**)
4. There exists $p \in \beta T$ with $\pi_R(u, \alpha(u))p \in \Delta \subset X \times X$. (by 3, **4.4**)
5. $\pi_R(gr(\alpha)) \cap \Delta \neq \emptyset$. (by 4)
6. $\pi_R(gr(\alpha))$ is a minimal subset of $X \times X$. (by **3.5**, **7.5**)
7. $\pi_R(gr(\alpha)) \subset \Delta$. (by 5, 6)
8. $gr(\alpha) \subset R$. (by 7)
9. $\alpha \in G(R)$. (by 8)
10. $D \subset G(R)$. (by 2, 9)

(b) $\Longrightarrow$ (a)

1. Assume that $D \subset G(R)$.
2. Let $(\alpha(u), \beta(v)) \in \overline{P(M)}$.
3. $(\alpha(u), \beta\alpha^{-1}(\alpha(u))) = (\alpha(u), \beta(u)) = (\alpha(u), \beta(v))u \in \overline{P(M)}u \subset \overline{P(M)}$. (by 2)
4. $gr(\beta\alpha^{-1}) \cap \overline{P(M)} \neq \emptyset$. (by 3)
5. $\beta\alpha^{-1} \in D \subset G(R)$. (by 1, 4, and **14.7**)
6. $\pi_R(\alpha(u), \beta(v)) \underset{5}{=} \pi_R(\alpha(u), \alpha(v)) \in \pi_R(P(M)) \underset{\mathbf{4.7}}{=} P(X)$.
7. $\pi_R(\overline{P(M)}) = P(X)$. (by 1, 6, **4.7**)
8. $P(X)$ is closed. (by 7)

(a) $\Longrightarrow$ (c)

1. Assume that $P(X)$ is closed.
2. $P(X)$ is an icer. (by **13.6**)
3. Let $Y = X/P(X) = M/S$.
4. X is a proximal extension of Y. (by 3)
5. $P(Y) = \pi_S^R(P(X)) = \Delta_Y$. (by 3, and **4.7**)
6. Y is distal with $G(S) = G(R)$. (by 3, 4, 5, **7.11**)
7. $S = P_0 \circ gr(G(S)) = P_0 \circ gr(G(R))$. (by 5, 6, **14.4**)

(c) $\Longrightarrow$ (a)

1. Assume that X is a proximal extension of the distal flow $Y = M/S$.

2. $\overline{P(M)} \subset S$. (by 1)
3. $D \underset{2}{\subset} G(S) \underset{1,7.11}{=} G(R)$.

Recall that the group $P \subset G$ was defined to be the closed normal subgroup of G generated by the set $\{\alpha \mid (u, \alpha(v)) \in P(M)$ for some $u, v \in J\}$. But

$$(u, \alpha(v)) \in P(M) \Rightarrow gr(\alpha) \subset \overline{P(M)},$$

thus since D is closed and normal in G, it follows that $P \subset D$. Since the proof given above that (a) $\Rightarrow$ (b) makes no use of **13.6**, it provides an alternate proof of **13.6**(a). Namely $P(M/R)$ closed $\underset{\mathbf{14.8}}{\Rightarrow} D \subset G(R) \Rightarrow P \subset G(R) \underset{\mathbf{13.8}}{\Rightarrow} P(M/R)$ is an equivalence relation. Thus as an immediate consequence of **14.8**, $P(M/R)$ is an icer if and only if $D \subset G(R)$.

Of course if $P(M/R)$ is closed, then each of its cells are closed. We saw in **13.11** that this is the case if and only if $PG^J \subset G(R)$, so it follows that $G^J \subset D$. To see this directly we observe that

$$\alpha(J) \cap \overline{J} \neq \emptyset \Rightarrow \emptyset \neq gr(\alpha) \cap (J \times \overline{J}) \subset gr(\alpha) \cap \overline{P(M)}$$
$$\Rightarrow gr(\alpha) \subset \overline{P(M)} \Rightarrow \alpha \in D.$$

This shows that $J^* \subset D$ (see **13.9**); since D is closed and normal in G this argument shows that $G^J \subset D$. For emphasis we restate these observations in the following proposition.

Proposition 14.9 Let:

(i) R be an icer on M, and
(ii) P, G^J, and D be the subgroups of G defined in **13.7**, **13.9**, and **14.6** respectively.

Then:

(a) $PG^J \subset D$, and
(b) $P(M/R)$ is an icer if and only if $D \subset G(R)$.

As we remarked earlier, in [Shapiro, L., *Proximality in minimal transformation groups*, (1970)] Shapiro gives an example where $P(X)$ is an equivalence relation and every proximal cell is closed, yet $P(X)$ is not closed. Thus for this example it follows from **13.11** that $PG^J \subset G(X)$. On the other hand it follows from **14.8** that $D \not\subset G(X)$. Thus $PG^J \subset D$, but $PG^J \neq D$.

We saw in **14.1**, that the flow $(M/R, T)$ is distal if and only if $P_0 \subset R$. On the other hand if M/R is distal, then $P(M/R) = \Delta$ is closed, so it follows from **14.8** that $D \subset G(R)$. Thus it follows from **14.4** that any distal icer R is of the form $R = P_0 \circ gr(A)$ for some τ-closed subgroup of G with $D \subset A$.

The main theorem of this section says that $\{P_0 \circ gr(A) \mid D \subset A = \overline{A}\}$ is exactly the collection of all distal icers on M.

Theorem 14.10 Let A be a closed subgroup of G. Then $P_0 \circ gr(A)$ is a distal icer on M if and only if $D \subset A$.

PROOF: $\Longrightarrow$

1. Assume that $R = P_0 \circ gr(A)$ is an icer.
2. R is distal. (by 1, **14.1**)
3. $P(M/R) = \Delta$ is closed. (by 2)
4. $D \subset G(R) = A$. (by 1, 3, **14.8**)

$\Longleftarrow$

1. Assume that $D \subset A$.
2. Let $S = \overline{gr(A)}$.
3. S is an icer on M with $D \subset G(S)$. (by 1, 2, **12.2**)
4. S is a proximal extension of the distal flow $P_0 \circ gr(G(S)) = P_0 \circ gr(A)$. (by 3, **14.8**)

As an immediate consequence of theorem **14.10** we obtain the so-called "Galois theory" of minimal distal flows discussed in [Ellis, R. (1969)] (see proposition 13.23 of that reference).

Theorem 14.11 Let:

(i) $\mathcal{D} = \{R \mid R \text{ is a distal icer on } M\}$,
(ii) $\mathcal{G} = \{H \mid H \text{ is a closed subgroup of } G \text{ with } D \subset H\}$,

(iii) $\varphi\colon \mathcal{D} \to \mathcal{G}$
$\quad N \to G(N)$, and

(iv) $\psi\colon \mathcal{G} \to \mathcal{D}$
$\quad A \to P_0 \circ gr(A)$.

Then:

(a) φ is bijective, its inverse being the map ψ, and
(b) $\psi(A)$ is regular if and only if A is normal.

PROOF: (a) This is basically a restatement of **14.10**.

(b) $\Longrightarrow$

1. Assume that $A \in \mathcal{G}$ is normal, and let $\alpha \in G$.
2. $A = G(\psi(A))$ is normal.
3. $G(\alpha(\psi(A))) = G(\psi(A))$. (by 2, **7.17**)

4. $\alpha(\psi(A)) = \psi(A)$. (by 3, and part (a))
5. $\psi(A)$ is regular. (by 1, 4)

$$\Leftarrow$$

The group of any regular flow is normal by **8.2**.

When R and S are icers on M we know that $R \circ S$ is an icer on M if and only if $R \circ S = S \circ R$. Thus in particular the condition $G(R)G(S) = G(S)G(R)$ (that $G(R)G(S)$ be a group) is necessary in order for $R \circ S$ to be an icer. When one of the icers R or S is distal, this condition is also sufficient.

Proposition 14.12 Let:

(i) R, S be icers on M,
(ii) S be distal, and
(iii) $G(R)G(S)$ be a group.

Then:

(a) $R \circ S = P_0 \circ gr(G(R)G(S))$ is a distal icer on M,
(b) $G(R \circ S) = G(R)G(S)$,
(c) $\pi_R(S)$ is an icer on M/R, and
(d) $(M/R)/\pi_R(S)$ is distal.

PROOF: (a), (b):
1. $S = P_0 \circ gr(G(S))$. (by (ii), and **14.4**)
2. $R \circ S \underset{1,\mathbf{7.21}}{=} (P_0 \cap R) \circ gr(G(R)) \circ P_0 \circ gr(G(S))$

$$= gr(G(R)) \circ (P_0 \cap R) \circ P_0 \circ gr(G(S))$$
$$= gr(G(R)) \circ P_0 \circ gr(G(S)) = gr(G(R)) \circ gr(G(S)) \circ P_0$$
$$= gr(G(S)G(R)) \circ P_0 \underset{\text{(ii)},\text{(iii)},\mathbf{14.10}}{=} P_0 \circ gr(G(R)G(S)).$$

3. $R \circ S$ is a distal icer. (by 2, and **14.10**)
   (c) This follows from part (a) and **6.11**.
   (d) 1. $(M/R)/\pi_R(S) \cong M/(\pi_R^{-1}(\pi_R(S))) = M/(R \circ S)$. (by **7.2**)
2. $(M/R)/\pi_R(S)$ is distal. (by 1, (a))

As one application of the material discussed above, along with the results from sections 4 and 12 on topological transitivity and pointwise almost periodicity, we obtain the following important relationship between the groups G^∞ and D. We will explore these ideas further in section 20 where we show that this theorem is equivalent to the generalized Furstenberg structure theorem for distal extensions.

Theorem 14.13 Let $A \subset G$ be a closed subgroup with $A = A'$. Then $A \subset D$; in particular $G^\infty \subset D$.

PROOF: 1. Let $R = P_0 \circ gr(D)$.
2. $(M/R, T)$ is distal. (by **14.10**)
3. Let $S = \overline{gr(A)}$.
4. S is an icer on M with $G(S) = A$. (by **12.2**)
5. (S, T) is a topologically transitive flow. (by **12.5**)
6. $G(R)G(S) = DA = AD = G(S)G(R)$. (by 1, 2, D is normal in G)
7. $R \circ S$ is an icer on M and $\pi_R(S)$ is an icer on M/R with $\big(M/R\big)/\pi_R(S) \cong M/(R \circ S)$. (by 2, 5, **14.12**)
8. $(\pi_R(S), T)$ is topologically transitive. (by 5, **4.21**)
9. $(\pi_R(S), T)$ is distal. (by 2, 7, **4.13**)
10. $(\pi_R(S), T)$ is minimal. (by 8, 9, **4.24**)
11. $\pi_R(S) = \Delta$ and hence $S \subset R$. (by 7, 10)
12. $A = G(S) \subset G(R) = D$. (by 1, 4, 11)

Using the "classical" terminology, the distal structure relation on M is the smallest icer on M such that the quotient flow is distal. We present the definition from the point of view of theorem **14.10**.

Definition 14.14 We define the *distal structure relation on M* by $S_d = P_0 \circ gr(D)$.

Proposition 14.15 Let R be an icer on M. Then R is distal if and only if $S_d \subset R$.

PROOF: This follows immediately from **14.10**.

More generally for any minimal flow (X, T), we would like to produce $S_d(X)$, the smallest icer on X such that $X/S_d(X)$ is distal. The following corollary shows that the projection of S_d onto X is a distal icer on X which we define to be the distal structure relation on X.

Corollary 14.16 Let $X = M/R$. Then:

(a) $R \circ S_d$ is an icer on M,
(b) $G(R \circ S_d) = G(R)D$,
(c) $\pi_R(S_d)$ is an icer on X, and
(d) $X/\pi_R(S_d)$ is distal.

PROOF: This follows immediately from **14.12** since D is normal in G

Definition 14.17 Let $X = M/R$ be a minimal flow. We define the *distal structure relation on X* by $S_d(X) = \pi_R(S_d)$.

It should be pointed out that $S_d(X)$ depends only on the minimal flow (X, T) not on the icer R, even when X is not regular. This amounts to saying that if

S is an icer on M with $X \cong M/S$, then the isomorphism $\varphi\colon M/R \to M/S$ takes $\pi_R(S_d)$ to $\pi_S(S_d)$. But in this situation $S = \alpha(R)$ for some $\alpha \in G$ (by **7.7**) with $\varphi \circ \pi_R = \pi_S \circ \alpha$. But S_d is regular (since D is normal) and hence $\varphi(\pi_R(S_d)) = \pi_S(\alpha(S_d)) = \pi_S(S_d)$.

We leave it as an exercise for the reader to verify that if S is an icer on X with X/S distal, then $S_d(X) \subset S$. When R is regular, the icer $S_d(X)$ can be constructed directly from the subgroup D_X of the group G_X of automorphisms of X. To this end in analogy with **7.20**, we denote by:

$$P_0(X) = \bigcup_{\alpha \in G_X} \alpha(J_X \times J_X),$$

where as usual J_X is the set of idempotents of X. Note that $P_0(X) \subset P(X)$ is an equivalence relation on X; in general it is neither invariant nor closed. Clearly the analog of the fact that $S_d = P_0 \circ gr(D)$ is the statement that $S_d(X) = P_0(X) \circ gr(D_X)$. We leave it to the reader to verify this in exercise **14.6** below.

NOTES ON SECTION 14

Notation 14.N.1 In [Ellis (1969)] the algebra, $\mathcal{D}$ of distal functions on T was defined. This turned out to be a uniformly closed T-invariant subalgebra of $\mathcal{C}(\beta T)$ such that the associated flow $|\mathcal{D}|$ was minimal, distal, and had the property that every minimal distal flow was a homomorphic image thereof. Hence by **14.15** $|\mathcal{D}|$ is isomorphic to M/S_d and the group D introduced above coincides with $G(\mathcal{D})$ which was also denoted D in the reference above.

EXERCISES FOR CHAPTER 14

Exercise 14.1 (An extension of **Ex. 4.2**.) Let (X, T) be a flow and $\Phi\colon \beta T \to E(X)$ be the canonical map. Then the following are equivalent:

(a) X is distal.
(b) e is the only idempotent in $E(X)$.
(c) $E(X)$ is a group.
(d) $E(X) = \Phi(G(u))$ for all idempotents $u \in J$.
(e) The map $\begin{array}{ccc} G & \to & E(X) \\ \alpha & \to & \Phi(\alpha(u)) \end{array}$ is an epimorphism for all $u \in J$.
(f) $E(X) = \Phi(M)$.
(g) $E(X)$ is minimal.

Exercise 14.2 Let $X = M/R$ be distal and $u \in J$ be any idempotent in M. Then the map

$$\begin{array}{rcc} \varphi\colon G & \to & X \\ \alpha & \to & \pi_R(\alpha(u)) \end{array}$$

induces a bijection

$$G/G(R) \to X.$$

Exercise 14.3 (Local version of **14.13**) Let:

(i) (X, T) be a regular minimal flow,
(ii) $A \subset G_X$ be a closed, and
(iii) $A = A'$.

Then $A \subset D_X$.

Hint: this can be deduced from **14.13**, or from **4.24** directly.

Exercise 14.4 Let:

(i) $R \subset S$ be icers on M,
(ii) $X = M/R$, $Y = M/S$, and
(iii) $f \in Hom(M, X)$.

Then:

(a) $f(S_d) = S_d(X)$.
(b) $\pi_S^R(S_d(X)) = S_d(Y)$.

Exercise 14.5 Let:

(i) $X = M/R$ be a minimal flow, and
(ii) N be an icer on X.

Then X/N is distal if and only if $S_d(X) \subset N$.

Exercise 14.6 Let $X = M/R$ be regular.
Then:

(a) $\chi_R(D) = D_X$.
(b) $S_d(X) = P_0(X) \circ gr(D_X)$.

15

Equicontinuous flows and the group E

In this section we introduce the concept of an equicontinuous flow. Basically a flow (X, T) is equicontinuous when the group T, thought of as a family of continuous elements of X^X, is uniformly equicontinuous [Rudin (1953)]. Every equicontinuous flow is distal (see **15.3**), but the converse is false. The prototype example of an equicontinuous minimal flow is a compact group K with a dense subgroup which is a homomorphic image of T (see **15.9** for a precise statement). In fact every regular equicontinuous minimal flow, (X, T), is of this form (see **15.10**). Indeed in this situation $G/G(X) \cong X \cong E(X, T)$ as compact Hausdorff topological groups. Thus one approach to the study of equicontinuity is via the enveloping semigroup. This approach, which does not require that (X, T) be minimal, will be pursued in the appendix to this section. In the body of this section we study equicontinuity for minimal flows in the context of icers on M and subgroups of G. In particular we introduce both the *regionally proximal* relation $Q(X)$ and a subgroup $E \subset G$ which play roles analogous to those played by the proximal relation $P(X)$ and the subgroup D for minimal distal flows.

Definition 15.1 We say that (X, T) is an *equicontinuous flow* if given any neighborhood U of the diagonal $\Delta \subset X \times X$, there exists a neighborhood V of Δ with $VT \subset U$. Let $x \in X$. Then the flow (X, T) is *equicontinuous at* x if given any neighborhood U of Δ, there exists a neighborhood V of x such that $(xt, yt) \in U$ for all $y \in V$ and $t \in T$. The flow (X, T) is *pointwise equicontinuous* if it is equicontinuous at x for every $x \in X$.

Since for our flows (X, T) the space X is compact, pointwise equicontinuity and equicontinuity are equivalent. For future reference we state this in the following lemma, leaving the proof as an exercise for the reader.

Lemma 15.2 (X, T) is pointwise equicontinuous if and only if (X, T) is equicontinuous.

For the compact Hausdorff space X, the collection of neighborhoods of Δ in $X \times X$ is the unique uniformity on X compatible with its topology (see [James, I.M., (1987)] page 106). Hence **15.1** amounts to requiring that the family of maps $\{\pi^t : X \to X \mid t \in T\}$ is uniformly equicontinuous. (See [Munkres,
$x \to xt$
James R., (1975)] for the definition in the metric case.)

We begin our study of equicontinuity with the elementary observation that any equicontinuous flow is distal.

Proposition 15.3 Let (X, T) be an equicontinuous flow. Then (X, T) is distal.

PROOF: 1. Let $(x, y) \in P(X)$.
2. Let V be a neighborhood of $\Delta \subset X \times X$.
3. There exists a neighborhood U of Δ such that $UT \subset V$.
((X, T) is equicontinuous)
4. $\overline{(x, y)T} \cap \Delta \neq \emptyset$. (by 1)
5. There exists $t \in T$ such that $(x, y)t \in U$. (by 3, 4)
6. $(x, y) \in Ut^{-1} \subset V$. (by 3, 5)
7. $y = x$. (by 2, 6, X is Hausdorff)

The converse of **15.3** is false. Before giving an example of a distal flow which is not equicontinuous we introduce the so-called *regionally proximal relation*, $Q(X)$ which will play an important role in our exposition. $Q(X)$ is an analog of the proximal relation; just as (X, T) is distal if and only if $P(X, T) = \Delta$, we will show that (X, T) is equicontinuous if and only if $Q(X) = \Delta$. By way of motivation note that

$$\begin{aligned}(x, y) \in P(X) &\iff \overline{(x, y)T} \cap \Delta \neq \emptyset \\ &\iff (x, y)T \cap V \neq \emptyset \text{ for all open } V \text{ with } \Delta \subset V.\end{aligned}$$

Thus $P(X) = \bigcap\{VT \mid V \text{ is open and } \Delta \subset V\}$ and it is natural to consider the intersection $\bigcap\{\overline{VT} \mid V \text{ is open and } \Delta \subset V\}$; this is precisely the regionally proximal relation on X.

Definition 15.4 Let (X, T) be a flow. We define the *regionally proximal relation on X* by:

$$Q(X) = \bigcap\{\overline{WT} \mid W \subset X \times X \text{ is open with } \Delta_X \subset W\}.$$

When $X = M$ we will often simply write $Q \equiv Q(M)$.

The following propositions outline a few of the basic properties of $Q(X)$.

Proposition 15.5 Let (X, T) be a flow. Then:

(a) $Q(X)$ is invariant, closed, reflexive, and symmetric,
(b) $\alpha(Q(X)) = Q(X)$ for all $\alpha \in Aut(X)$,
(c) $\overline{P(X)} \subset Q(X)$, and
(d) X is equicontinuous if and only if $Q(X) = \Delta_X$.

PROOF: (a) and (b) are immediate from the definition.

(c) This follows immediately from the fact that $P(X) = \bigcap\{VT \mid V \text{ is open and } \Delta \subset V\}$.

(d) $\Longrightarrow$

1. Assume that (X, T) is equicontinuous and let $V \subset X \times X$ be open with $\Delta \subset V$.
2. There exists $U \subset X \times X$ be open with $\Delta \subset U \subset \overline{U} \subset V$. ($X$ is compact Hausdorff)
3. There exists $W \subset X \times X$ be open with $\Delta \subset W$, and $WT \subset U$. (by 1, 2)
4. $Q(X) \subset \overline{WT} \subset \overline{U} \subset V$. (by 2, 3)
5. $Q(X) = \Delta$. (by 1, 4)

$\Longleftarrow$

1. Assume that $Q(X) = \Delta$ and let $V \subset X \times X$ be open with $\Delta \subset V$.
2. Assume that $\overline{WT} \cap (X \times X \setminus V) \neq \emptyset$ for all $W \subset X \times X$ open with $\Delta \subset W$.
3. $\{\overline{WT} \cap (X \times X \setminus V) \mid \Delta \subset W^{\text{open}} \subset X \times X\}$ are closed sets with the F.I.P. (by 1, 2)
4. $Q(X) \cap (X \times X \setminus V) \neq \emptyset$. (by 3, and compactness)
5. There exists W with $\overline{WT} \cap (X \times X \setminus V) = \emptyset$. (4 contradicts 1)
6. $WT \subset V$. (by 5)
7. (X, T) is equicontinuous. (by 1, 6)

Proposition 15.6 Let X, Y be minimal and $\pi: X \to Y$ be a homomorphism. Then $\pi(Q(X)) = Q(Y)$.

PROOF: Clearly $\pi(Q(X)) \subset Q(Y)$.

1. Let $(x, y) \in Q(Y)$, and W be an open neighborhood of Δ_X.
2. Let $V \subset X$ be closed with nonempty interior and $V \times V \subset W$.
3. There exists $F \subset T$ finite with $X = VF$. (X is minimal)
4. $\pi(V)F = Y$. (by 3, π is surjective since Y is minimal)
5. $U = \text{int}(\pi(V)) \neq \emptyset$. (by 4)
6. $\Delta_Y \subset (U \times U)T \underset{5}{\subset} \text{int}(\pi((V \times V)T)) \underset{2}{\subset} \text{int}(\pi(WT))$.
7. $(x, y) \in \overline{\pi(WT)} \subset \pi(\overline{WT})$. (by 1, 6)
8. $\pi^{-1}(x, y) \cap \overline{WT} \neq \emptyset$. (by 7)

9. $\pi^{-1}(x, y) \cap \bigcap\{\overline{WT} \mid W \text{is an open neighborhood of} \Delta_X\} \neq \emptyset$. (by 1, 8, and compactness)

10. $\pi^{-1}(x, y) \cap Q(X) \neq \emptyset$. (by 9)

11. $(x, y) \in \pi(Q(X))$. (by 10)

Corollary 15.7 Let:

(i) (X, T) be a an equicontinuous minimal flow,
(ii) (Y, T) be a flow, and
(iii) $\pi: X \to Y$ be an epimomorphism.

Then (Y, T) is equicontinuous.

PROOF: $Q(Y) \underset{\mathbf{15.6}}{=} \pi(Q(X)) \underset{\mathbf{15.5}}{=} \pi(\Delta_X) = \Delta_Y$, so (Y, T) is equicontinuous by **15.5**.

In general transitivity is the only property that $Q(X)$ lacks in order that it be an icer. In the next section we will study $Q(X)$ itself in more detail. In this section we will concentrate on the relationship between $Q(X)$ and equicontinuity. Historically $Q(X)$ arose in an attempt to determine the equicontinuous structure relation $S_{eq}(X)$, the smallest icer on X such that $X/S_{eq}(X)$ is equicontinuous.

Note that $(x, y) \in Q(X)$ if and only if there exist nets $\{x_i\}$ and $\{y_i\}$ in X and a net $\{t_i\}$ in T with $x_i \to x$, $y_i \to y$, and $\lim x_i t_i = \lim y_i t_i$. This formulation is often used as the definition of $Q(X)$.

It follows immediately from **15.5** that if $Q(X) = \Delta$, then $P(X) = \Delta$ which gives an alternative proof of the fact that any equicontinuous flow is distal. On the other hand any flow for which $\Delta = P(X) \neq Q(X)$ is distal but not equicontinuous. We take advantage of this observation in the simple example below.

Example 15.8 Let:

(i) $X = \{z \in \mathbf{C} \mid |z| \leq 1\}$ be the unit disk in the complex plane, and
(ii) $T: X \to X$ be the homeomorphism defined by $zT = ze^{i|z|}$.

Then:

(a) (X, T) is a distal flow where $T = \{T^n \mid n \in \mathbf{Z}\}$ is identified with the group of integers, and
(b) (X, T) is not equicontinuous.

PROOF: (a) $P(X, T) = \Delta$ because

$$|z_1 T - z_2 T| = \left|z_1 e^{i|z_1|} - z_2 e^{i|z_2|}\right| = |z_1 - z_2| \qquad \text{if } |z_1| = |z_2|$$

and

$$|z_1 T - z_2 T| \geq \big||z_1| - |z_2|\big| \qquad \text{if } |z_1| \neq |z_2|.$$

(b) This follows from the fact that $Q(X, T) = \{(z_1, z_2) \mid |z_1| = |z_2|\}$; we leave the details to the reader.

We noted in section 14 that a compact Hausdorff topological space S, with a group structure which contains a homomorphic image of T as a dense subgroup whose elements act continuously on S by right multiplication (S might be referred to as a compactification of T) gives rise to a distal flow (S, T). If S is a compact Hausdorff topological group, then the flow (S, T) is equicontinuous. We give the details in the following proposition.

Proposition 15.9 Let:

(i) K be a compact Hausdorff topological group,
(ii) $\psi: T \to K$ a homomorphism with $\overline{\psi(T)} = K$,
(iii) $\pi: K \times T \to K$ be defined by $\pi(k, t) = k\psi(t)$ for all $k \in K$, and $t \in T$,
(iv) f be the continuous extension of ψ to βT, and
(v) $\kappa: \beta T \to E(K, T)$ be the canonical map.

Then:

(a) π defines an action of T on K,
(b) the flow (K, T) is equicontinuous, minimal, and regular,
(c) $f(pq) = f(p)f(q) = f(p)\kappa(q)$ for all $p, q \in \beta T$,
(d) $\begin{array}{ccc} E(K,T) & \to & (K,T) \\ p & \to & ep \end{array}$ is an isomorphism, and
(e) $G(K, T) = \{\alpha \in G \mid f(\alpha(u)) = f(u) \text{ for some } u \in J\}$.

PROOF: (a) and (d) are straightforward and we leave them to the reader.

(b) It is immediate that (K, T) is minimal. The fact that (K, T) is regular follows from **8.9**, and the fact that L_k is an automorphism of (K, T) for every $k \in K$. We give a proof that (K, T) is equicontinuous.

1. Let $\Delta \subset V \subset K \times K$ be open.
2. Let $x, y \in K$.
3. $(y, y)x \in \Delta \subset V$.
4. There exist open sets $(y, y) \in W_{x,y} \subset K \times K$ and $x \in U_{x,y}$, such that $W_{x,y}U_{x,y} \subset V$. (by 3, (i))
5. $\{U_{x,y} \mid x \in K\}$ is an open cover of K.
6. There exist $U_{x_1,y}, \ldots, U_{x_n,y}$ with $K \subset \bigcup_{i=1}^{n} U_{x_i,y}$. (by 5, (i))
7. Set $W_y = \bigcap_{i=1}^{n} W_{x_i,y}$.
8. $W_y K \underset{6}{\subseteq} W_y \left(\bigcup_{i=1}^{n} U_{x_i,y} \right) = \bigcup_{i=1}^{n} W_y U_{x_i,y} \underset{7}{\subseteq} \bigcup_{i=1}^{n} W_{x_i,y} U_{x_i,y} \subset V$.
9. Set $W = \bigcup_{y \in K} W_y$.

10. W is open with $\Delta \subset W$. (by 4, 7, 9)
11. $WT \subset WK \subset V$. (by 8, 9)
12. (K, T) is equicontinuous. (by 1, 10, 11)

(c) 1. $f(pt) \underset{\text{(iv)}}{=} \lim_{s \to p} f(st) \underset{\text{(iv)}}{=} \lim_{s \to p} \psi(st) \underset{\text{(ii)}}{=} \lim_{s \to p} \psi(s)\psi(t) \underset{\text{(iv)}}{=} f(p)\psi(t)$
$\underset{\text{(v)}}{=} f(p)\kappa(t)$.

2. $f(pq) = \lim_{t \to q} f(pt) \underset{1}{=} \lim_{t \to q} f(p)\kappa(t) \underset{\text{(v)}}{=} f(p)\kappa(q)$.

3. $f(pq) = \lim_{t \to q} f(pt) \underset{1}{=} \lim_{t \to q} f(p)\psi(t) \underset{\text{(i)}}{=} f(p) \lim_{t \to q} \psi(t) \underset{\text{(iv)}}{=} f(p)f(q)$.

(e) 1. Let $R = \{(p, q) \in M \times M \mid \kappa(p) = \kappa(q)\}$ so that $E(K, T) = M/R$.
2. $G(K) = G(E(K)) = \{\alpha \in G \mid gr(\alpha) \subset R\}$. (by 1 and part (d))
3. Let $\alpha \in G(K)$, $u \in J$.
4. $\kappa(\alpha(u)) = \kappa(u) = e$. (by 1, 2, 3)
5. $f(\alpha(u)) = f(u\alpha(u)) = f(u)\kappa(\alpha(u)) = f(u)$. (by 4)
6. Now suppose that $f(\beta(u)) = f(u) = e$.
7. $f(\beta(u)) = f(u\beta(u)) = f(u)\kappa(\beta(u)) = \kappa(\beta(u))$.
8. $\kappa(\beta(u)) = e = \kappa(u)$. (by 6, 7)

Our next goal is to prove the converse of **15.9**. Note that when R is regular and distal, we saw in **14.5**, that $X = M/R$ is a group (though multiplication may not be continuous) and that a homomorphic image of T is a dense subgroup. We will show that the equicontinuous regular minimal flows are exactly those for which X is a compact topological group (in fact isomorphic to $G/G(R)$). Again there is a homomorphism of T onto a dense subgroup of X, and the right action of T on X is given by right multiplication in X.

Theorem 15.10 Let:
(i) R be a regular icer on M such that M/R is distal, and
(ii) $u^2 = u \in X = M/R$. (Note that in view of (i) u is unique and acts as the identity in the group X, see **14.5**.)

Then the following are equivalent:

(a) (X, T) is equicontinuous,
(b) $\pi: X \times X \to X$, $(x, y) \to xy$ is continuous and X is a compact topological group,
(c) the map $\varphi: Aut(X) \to X$, $\alpha \to \alpha(u)$ is continuous,
(d) $G/G(R)$ and X are isomorphic as compact topological groups, and
(e) $G' \subset G(R)$.

PROOF: (a) $\Longrightarrow$ (b)
1. Assume that X is equicontinuous and let $(x, y) \in X \times X$.

2. Let $W \subset X$ be open with $x\ y \in W$.
3. There exists $U \subset X \times X$ open with $\Delta \subset U$ and $\{z \mid (xy, z) \in U\} \subset W$. (by 2)
4. Let $U_0 \subset X \times X$ be an open neighborhood of Δ with $U_0 \circ U_0 \subset U$.
5. There exists an open set $V \subset X \times X$ with $\Delta \subset V$, and $\overline{VT} \subset U_0 \cap L_x^{-1}(U_0)$. (by 1, 4, L_x is continuous)
6. Let $(p, q) \in X \times X$ with $(x, p) \in V$ and $(y, q) \in V$.
7. $(xq, pq) = (x, p)q \in Vq \subset \overline{VT} \subset U_0$. (by 5, 6)
8. $(xy, xq) = L_x(y, q) \in L_x(V) \subset U_0$. (by 5, 6)
9. $(xy, pq) \in U_0 \circ U_0 \subset U$. (by 4, 7, 8)
10. $\{(p, q) \mid (x, p) \in V \text{ and } (y, q) \in V\} \subset \pi^{-1}\{z \mid (xy, z) \in U\} \subset \pi^{-1}(W)$. (by 5, 8)
11. $\pi^{-1}(W)$ is a neighborhood of (x, y). (by 10)
12. π is continuous. (by 1, 2, 11)
13. Let W be an open set with $x^{-1} \in W$.
14. Assume that for every closed neighborhood N of x there exists $n \in N$ with $n^{-1} \in X \setminus W$.
15. Let V be any closed neighborhood of the identity $e \in X$.
16. $(N \times (X \setminus W)) \cap \pi^{-1}(V) \neq \emptyset$ for every closed neighborhood N of x.
17. $\{(N \times (X \setminus W)) \cap \pi^{-1}(V) \mid N, V$ closed neighborhoods of x and e resp.$\}$ has the F.I.P.. (by 16)
18. There exists $(x, y) \in \bigcap\{(N \times (X \setminus W)) \cap \pi^{-1}(V)\}$. ($\pi$ is continuous and $X \times X$ is compact)
19. $y = x^{-1} \in X \setminus W$. (by 18)
20. There exists a neighborhood N of x with $N^{-1} \subset W$. (19 contradicts 13)

(b) $\Longrightarrow$ (c)

1. Assume that X is a compact topological group.
2. Let W be open in X.
3. Let $\alpha \in \varphi^{-1}(W) \subset Aut(X)$.
4. $\alpha^{-1}(W)$ is an open neighborhood of u. (by 2, 3)
5. There exists an open neighborhood V of u with $VV^{-1} \subset \alpha^{-1}(W)$. (by 1, 4)
6. Let $\beta \in \langle V, \alpha(V) \rangle$.
7. There exist $p, q \in V$ with $\alpha(p) = \beta(q)$. (by 6)
8. $\beta(u) = \beta(q)q^{-1} \underset{7}{=} \alpha(p)q^{-1} = \alpha(pq^{-1}) \underset{5,7}{\in} \alpha(\alpha^{-1}(W)) = W$.
9. $\alpha \in \langle V, \alpha(V) \rangle \subset \varphi^{-1}(W)$. (by 6, 8)
10. $\varphi^{-1}(W)$ is open in $Aut(X)$. (by 3, 9)

(c) $\Longrightarrow$ (d)

1. Assume that φ is continuous.
2. φ is a homomorphism which is a bijection. (by **14.5**, X is regular and distal)

3. φ is a homeomorphism. (by 1, 2, $Aut(X)$ is compact, X is Hausdorff)
4. $G/G(R)$ is homeomorphic to $G_X = Aut(X)$ and hence is Hausdorff. (by 3, **10.16**)
5. $G' \subset G(R)$. (by 4, **11.10**)
6. $G/G(R)$ and hence $Aut(X)$ is a compact topological group. (by 5, **11.11**)
7. φ is an isomorphism of topological groups. (by 2, 3, 6)
8. $G/G(R) \cong Aut(X) \cong X$.

(d) $\Longrightarrow$ (e)

1. Assume that $G/G(R) \cong Aut(X) \cong X$.
2. $G/G(R)$ is Hausdorff. (by 1)
3. $G' \subset G(R)$. (by 2, **11.10**)

(e) $\Longrightarrow$ (a)

1. Assume that X is not equicontinuous.
2. There exists an open neighborhood U of Δ_X, such that $VT \not\subset U$ for all open neighborhoods V of Δ_X. (by 1)
3. There exist $p_V, q_V \in M$ and $t_V \in T$ with $\pi_R(p_V, q_V) \in V$ and $\pi_R(p_V, q_V)t_V \notin U$. (by 2)
4. There exists $\alpha_V \in G$ with $\pi_R(q_V) = \pi_R(\alpha_V(p_V))$. ($X$ is distal)
5. We may assume that $\pi_R(p_V, \alpha_V(p_V)) \to (z, z) \in \Delta$ and that
$$\pi_R(p_V, \alpha_V(p_V))t_V \to \pi_R(w, \beta(w)) \notin \Delta. \quad \text{(by 3, 4)}$$
6. $\chi_R(\alpha_V) \to 1$ in $Aut(X)$ and $\chi_R(\alpha_V) \to \beta$ in $Aut(X)$. (by 5)
7. $1 \neq \chi_R(\beta) \in G'_X = \chi_R(G')$. (by 5, 6, **11.2** and **Ex. 11.6**)
8. $G' \not\subset \ker \chi_R = G(R)$. (by 7, **8.11**)

Let R be a regular icer on M, $X = M/R$, and $u^2 = u \in M$. In **8.11** we showed that the map $\chi_R\colon G \to Aut(X)$ induces an isomorphism of groups $G/G(R) \cong Aut(X)$; we saw in **10.16** that χ_R is continuous and hence that this isomorphism of groups is also a homeomorphism. Note that χ_R induces a map $\begin{matrix} G & \to & X \\ \alpha & \to & \pi_R(\alpha(u)) = \chi_R(\alpha)(u) \end{matrix}$ and hence a map $\varphi : G/G(R) \to X$. In general $G/G(R)$ is a group while X is only a semigroup and the map φ though it is a homomorphism, depends on the choice of u, and is neither onto nor continuous. In **14.5** we saw that X is distal if and only if φ is an isomorphism of groups. Even when X is distal the map φ need not be a homeomorphism; indeed **15.10** shows that in this case φ is a homeomorphism if and only if X is equicontinuous.

We now proceed to identify those icers R for which M/R is equicontinuous. Just as in the distal case we will define a closed normal subgroup E of G so that M/R is equicontinuous if and only if $R = P_0 \circ gr(A)$ for some closed subgroup $A \subset G$ which contains E. There are various ways to define the subgroup E; our approach will be to use the regionally proximal relation $Q = Q(M)$ defined above.

Definition 15.11 Let $L = \{\alpha \in G \mid gr(\alpha) \subset Q\}$. We define E to be the closed subgroup of G generated by L. More generally when (X, T) is a regular flow we let $L_X = \{\alpha \in G_X \mid gr(\alpha) \subset Q(X)\}$, and define E_X to be the closed subgroup of G_X generated by L_X. Note that since $Q(X)$ is closed and invariant $L_X = \{\alpha \in G_X \mid gr(\alpha) \cap Q(X) \neq \emptyset\}$.

We emphasize a few properties of the group E_X in the following lemma.

Lemma 15.12 Let (X, T) be a regular flow and $\chi_X \colon G \to G_X$ be the canonical map. Then:

(a) E_X is normal in G_X.
(b) $D_X \subset E_X$.
(c) $\chi_X(E) = E_X$.

PROOF: (a) 1. Let $\alpha \in L_X$, and $\beta \in G_X$.
2. $gr(\beta^{-1}\alpha\beta) = \beta^{-1}(gr(\alpha)) \subset \beta^{-1}(Q(X)) = Q(X)$. (by)
3. $\beta^{-1}L_X\beta \subset L_X$ for all $\beta \in G_X$. (by 1, 2)
4. $\beta^{-1}L_X\beta = L_X$ for all $\beta \in G_X$. (by 3)
5. $\beta^{-1}E_X\beta = E_X$ for all $\beta \in G_X$. (by 4)
(b) This follows immediately from the fact that $\overline{P(X)} \subset Q(X)$. (by **15.5**)
(c) We leave this as an exercise for the reader.

Our goal now is to show that $\{P_0 \circ gr(A) \mid E \subset A = \overline{A}\}$ is exactly the collection of equicontinuous icers on M. Our approach is to first show that those icers for which $P(M/R) = Q(M/R)$ are exactly those icers with $E \subset G(R)$. This is the content of the following proposition which can be thought of as saying that E characterizes those flows for which the proximal and regionally proximal relations coincide.

Proposition 15.13 Let $X = M/R$ be a minimal flow. Then the following are equivalent:

(a) $P(X) = Q(X)$, and
(b) $E \subset G(R)$.

PROOF: (a) $\Longrightarrow$ (b)
1. Assume that $P(X) = Q(X)$.
2. Let $\alpha \in G$ with $gr(\alpha) \subset Q$ and $u \in J$.
3. $\pi_R(u, \alpha(u)) \in \pi_R(Q) = Q(X) = P(X)$. (by 1, 2, **15.6**)
4. There exists $p \in \beta T$ with $\pi_R(u, \alpha(u))p \in \Delta \subset X \times X$. (by 3, **4.4**)
5. $\pi_R(gr(\alpha)) \cap \Delta \neq \emptyset$. (by 4)

6. $gr(\alpha) \subset R$. (by 5)
7. $\alpha \in G(R)$. (by 8)
8. $E \subset G(R)$. (by 2, 7)

(b) $\Longrightarrow$ (a)

1. Assume that $E \subset G(R)$.
2. Let $(\alpha(u), \beta(v)) \in Q$.
3. $(\alpha(u), \beta\alpha^{-1}(\alpha(u))) = (\alpha(u), \beta(u)) = (\alpha(u), \beta(v))u \in Qu \subset Q$. (by 2)
4. $gr(\beta\alpha^{-1}) \cap Q \neq \emptyset$. (by 3)
5. $\beta\alpha^{-1} \in E \subset G(R)$. (by 1, 4)
6. $\pi_R(\alpha(u), \beta(v)) \underset{5}{=} \pi_R(\alpha(u), \alpha(v)) \in \pi_R(P(M)) \underset{\mathbf{4.7}}{=} P(X)$.
7. $Q(X) = \pi_R(Q) = P(X)$. (by 2, 6, **15.5**, **15.6**)

Theorem 15.14 Let A be a closed subgroup of G. Then $P_0 \circ gr(A)$ is an equicontinuous icer on M if and only if $E \subset A$.

PROOF: $\Longrightarrow$

1. Assume that $R = P_0 \circ gr(A)$ is an equicontinuous icer.
2. $\pi_R(Q) = Q(M/R) = \Delta$. (by 1, **15.5, and 14.14**)
3. $E \subset G(R) = A$. (by 2, $G(R)$ is a closed subgroup of G)

$\Longleftarrow$

1. Assume that $E \subset A$.
2. $D \subset A$. (by 1, **15.12**)
3. $R = P_0 \circ gr(A)$ is a distal icer. (by 1, 2, **14.10**)
4. $E \subset G(R)$. (by 1, 3)
5. $\Delta = P(M/R) = Q(M/R)$. (by 3, 4, **15.13**)
6. R is equicontinuous. (by 5, **15.5**)

It follows immediately from **15.14** that an icer R is a proximal extension of the equicontinuous icer $P_0 \circ gr(G(R))$ if and only if $E \subset G(R)$, which (by **15.13**) holds if and only if $P(M/R) = Q(M/R)$. Moreover if R is equicontinuous, then $R = P_0 \circ gr(G(R))$ contains the equicontinuous icer $P_0 \circ gr(E)$. In analogy with **14.1**, we call this icer the equicontinuous structure relation on M.

Definition 15.15 The *equicontinuous structure relation on M* is given by $S_{eq} = P_0 \circ gr(E)$.

The group E is a normal subgroup of G by **15.12**, so it follows from **14.11**, that S_{eq} is regular. Since S_{eq} is the smallest equicontinuous icer on M, M/S_{eq} is the maximal equicontinuous flow. In fact as the following proposition shows, S_{eq} is also the smallest icer on M containing Q.

Proposition 15.16 Let $X = M/R$. Then the following are equivalent:

(a) X is equicontinuous,
(b) $Q \subset R$, and
(c) $S_{eq} \subset R$.

PROOF: The fact that (a) $\Longleftrightarrow$ (b) follows immediately from **15.5** and **15.6**. The fact that (a) $\Longleftrightarrow$ (c) follows immediately from **15.14**.

We would like to to define the equicontinuous structure relation $S_{eq}(X)$ on $X = M/R$ to be the projection of S_{eq} (just as $S_d(X) = \pi_R(S_d)$). Thus we need to show that this projection $\pi_R(S_{eq})$ is an icer on X.

Proposition 15.17 Let $X = M/R$ be a minimal flow. Then
(a) $R \circ S_{eq}$ is an icer on M.
(b) $\pi_R(S_{eq})$ is an icer on X.

PROOF: (a) 1. $G(R) \subset G = aut(S_{eq})$. ($S_{eq}$ is regular)
2. $R \cap P_0 \subset P_0 = S_{eq} \cap P_0$. ($S_{eq}$ is distal)
3. $R \circ S_{eq}$ is an icer on M. (1, 2, **7.28**)
(b) This follows from part (a) and **6.11**.

Definition 15.18 Let $X = M/R$ be a minimal flow. We define the it equicontinuous structure relation on X by $S_{eq}(X) = \pi_R(S_{eq})$.

As in the case of the distal structure relation $S_{eq}(X)$ depends only on the minimal flow (X, T) not on the icer R, even when X is not regular. Again an isomorphism $M/R \to M/S$ takes $\pi_R(S_{eq})$ to $\pi_S(S_{eq})$ because S_{eq} is regular (since E is normal). The fact that $S_{eq}(X)$ is the smallest icer on X such that $X/S_{eq}(X)$ is equicontinuous follows easily from the fact that S_{eq} is the smallest icer on M such that M/S_{eq} is equicontinuous.

Proposition 15.19 Let:

(i) $X = M/R$ be a minimal flow, and
(ii) N be an icer on X.

Then X/N is equicontinuous if and only if $S_{eq}(X) \subset N$.

PROOF: $\Longrightarrow$

1. Assume that X/N is equicontinuous and set $S = \pi_R^{-1}(N)$.
2. $M/S \cong X/N$. (by 1, **7.2**)
3. S is an equicontinuous icer on M. (by 1, 2)
4. $S_{eq} \subset S$. (by 3, **15.16**)
5. $S_{eq}(X) = \pi_R(S_{eq}) \subset \pi_R(S) = N$.

$\Longleftarrow$

1. Assume that $S_{eq}(X) \subset N$.

2. $S_{eq} \subset \pi_R^{-1}(\pi_R(S_{eq})) = \pi_R^{-1}(S_{eq}(X)) \subset \pi_R^{-1}(N)$. (by 1)
3. $\pi_R^{-1}(N)$ is an equicontinuous icer on M. (by 2, **15.16**)
4. $X/N \cong M/\pi_R^{-1}(N)$ is equicontinuous. (by 3, **7.2**)

We now show that any homomorphism $\varphi: X \to Y$ of minimal flows takes $S_{eq}(X)$ onto $S_{eq}(Y)$. It suffices to consider the map $\pi_S^R: M/R \to M/S$ with $R \subset S$ icers on M.

Proposition 15.20 Let:

(i) $R \subset S$ be icers on M,
(ii) $X = M/R$, $Y = M/S$, and
(iii) $f \in Hom(M, X)$.

Then:

(a) $f(S_{eq}) = S_{eq}(X)$.
(b) $\pi_S^R(S_{eq}(X)) = S_{eq}(Y)$.

PROOF: (a) 1. There exists $\alpha \in G$ with $f = \pi_R \circ \alpha$. (by **7.6**)
2. $f(S_{eq}) = \pi_R(\alpha(S_{eq})) = \pi_R(S_{eq}) = S_{eq}(X)$. (by 1, S_{eq} is regular)
(b) $\pi_S^R(S_{eq}(X)) = \pi_S^R(\pi_R(S_{eq})) = \pi_S(S_{eq}) = S_{eq}(Y)$.

Given the importance of the groups D and E in understanding distality and equicontinuity, it is not surprising that a better understanding of these groups and the relationship between them leads to interesting results. We devote the remainder of this section to expanding upon this theme.

Proposition 15.21 Let G', D, and E be the subgroups of $G = Aut(M)$ defined in **11.1**, **14.6**, and **15.11** respectively. Then $E = DG'$.

PROOF: Proof that $DG' \subset E$:
1. $R = P_0 \circ gr(E)$ is an equicontinuous regular icer on M. (by **15.14**)
2. $D \subset G(R) = E$. (by 1, **14.10, 15.3**)
3. $G' \subset G(R) = E$. (by 1, **15.10**)

Proof that $E \subset DG'$:
1. $R = P_0 \circ gr(DG')$ is a distal regular icer on M. (by **14.10**, **14.11**)
2. $G' \subset G(R)$. (by 1)
3. R is an equicontinuous icer. (by 1, 2, **15.10**)
4. $E \subset G(R) = DG'$. (by 1, 3, **15.13**)

Corollary 15.22 Let $X = M/R$. Then X is equicontinuous if and only if X is distal and $G' \subset G(R)$.

PROOF: $\Longrightarrow$
1. Assume that X is equicontinuous.

2. $R = P_0 \circ G(R)$ and $E \subset G(R)$. (by **15.14**)
3. $G' \subset G(R)$. (by 2, **15.21**)

$\Longleftarrow$

1. Assume that R is distal and $G' \subset G(R)$.
2. $R = P_0 \circ gr(G(R))$ and $D \subset G(R)$. (by **14.10**)
3. $E = DG' \subset G(R)$. (by 1, 2)
4. R is equicontinuous. (by 2, 3, **15.14**)

Traditionally the group E has been defined as the group of the maximal equicontinuous minimal flow. Proposition **15.21** allows us to give an intrinsic description of the group E.

Proposition 15.23 Let:

(i) J be the set of idempotents in the universal minimal ideal M,
(ii) $\Omega(u, v, t) = L_{ut} \circ L_{vt}^{-1}$ for $u, v \in J$, and $t \in T$,
(iii) Ω be the closed normal subgroup of G generated by $\{\Omega(u, v, t) \mid u, v \in J \text{ and } t \in T\}$,
(iv) $J^* = \{\alpha \in G \mid \alpha(J) \cap \overline{J} \neq \emptyset\}$, and
(v) G^J be the closed normal subgroup of G generated by J^*.

Then:

(a) $S \equiv P_0 \circ gr(\Omega G^J G')$ is an icer on M, and
(b) $\Omega G^J G' = E$.

PROOF: (a) Proof that S an equivalence relation:
1. P_0 is an equivalence relation on M with $aut(P_0) = G$.
2. $\Omega G^J G'$ is a subgroup of G. (by (iii), (iv), (v))
3. $P_0 \circ gr(\Omega G^J G')$ is an equivalence relation. (by **7.24**)

Proof that S is closed:
1. Let $(\alpha(u), \beta(v)) \in \overline{S}$.
2. There exist nets $(\alpha_i(u_i), \beta_i(v_i)) \to (\alpha(u), \beta(v))$ with $\alpha_i\beta_i^{-1} \in \Omega G^J G'$ for all i.
3. We may assume that $u_i \to \gamma(a)$ and $v_i \to \delta(b)$ for some $\gamma, \delta \in G$ and $a, b \in J$. (compactness)
4. $\gamma, \delta \in J^* \subset \Omega G^J G'$. (by 3)
5. $\alpha_i \to \alpha\gamma^{-1}$ and $\beta_i \to \beta\delta^{-1}$. (by 2, 3, **10.14**)
6. $G/\Omega G^J G'$ is a compact Hausdorff topological group. (by **11.11**)
7. $\Omega G^J G' \underset{2}{=} \Omega G^J G'\alpha_i\beta_i^{-1} \underset{5,6}{\to} \Omega G^J G'\alpha\gamma^{-1}\delta\beta^{-1}$
$= \Omega G^J G'\alpha\gamma^{-1}\delta\alpha^{-1}\alpha\beta^{-1} \underset{4}{=} \Omega G^J G'\alpha\beta^{-1}$.
8. $(\alpha(u), \beta(v)) \in S$. (by 7)

Proof that S is invariant:

1. Let $(\alpha(u), \beta(v)) \in S$ and $t \in T$.
2. There exist $a, b \in J$ with $uta = ut$ and $vtb = vt$.
3. $(\alpha(u), \beta(v))t \underset{2}{=} (\alpha(ut)a, \beta(vt)b) = (\alpha L_{ut}(a), \beta L_{vt}(b))$.
4. $(\alpha L_{ut})(\beta L_{vt})^{-1} = \alpha L_{ut} L_{vt}^{-1} \beta^{-1} \in \alpha \Omega G^J G' \beta^{-1} = \Omega G^J G' \alpha\beta^{-1} = \Omega G^J G'$. (by 1)
5. $(\alpha(u), \beta(v))t \in S$. (by 3, 4)

(b) 1. $D \subset \Omega G^J G'$. (by **14.10**, and part (a))
2. $E = DG' \subset \Omega G^J G'$. (by 1, **15.21**)
3. Let $u, v \in J$ and $t \in T$.
4. There exists $b \in J$ with $vtb = vt$.
5. $(vt, \Omega(u, v, t)(vt)) \underset{4}{=} (vt, \Omega(u, v, t)(vtb)) \underset{(ii)}{=} (vt, utb) \underset{4}{=} (vt, ut)b \in P(M)b \subset \overline{P(M)}$.
6. $gr(\Omega(u, v, t)) \cap \overline{P(M)} \neq \emptyset$. (by 5)
7. $\Omega(u, v, t) \in D$. (by 6, **14.6**)
8. $\Omega G^J G' \subset DG' = E$. (by 7, (iii), **14.9**, **15.21**)

In general $E = DG'$ so $G' \subset E$; similarly (see exercise **15.11**) for any regular minimal flow $E_X = D_X G'_X$ and $G'_X \subset E_X$. Of particular interest are those situations in which $E_X = G'_X$; one class of examples are the so-called *Bronstein* flows.

Proposition 15.24 Let:

(i) (X, T) be a regular minimal flow, and
(ii) $\overline{gr(G_X)} = X \times X$. (This says that the set of almost periodic points in $X \times X$ is dense in $X \times X$; this is often referred to as the *Bronstein condition*.)

Then $E_X = G'_X$.

PROOF: 1. Let $\alpha \in G_X$ with $gr(\alpha) \subset \overline{P(X)}$.
2. Let $V, W \subset X$ be any open subsets of X.
3. $(V \times \alpha(V)) \cap P(X) \neq \emptyset$. (by 1, 2)
4. $\Delta_X \subset \overline{(V \times \alpha(V))T}$. (by 3)
5. $\emptyset \neq (V \times \alpha(V))T \cap (W \times W)$ is open. (by 2, 4)
6. There exists $\beta \in G_X$ and $p \in M$ with $(p, \beta(p)) \in (V \times \alpha(V))T \cap (W \times W)$. (by 5, (ii))
7. $\beta(V) \cap \alpha(V) \neq \emptyset$ and $\beta(W) \cap W \neq \emptyset$. (by 6)
8. $\beta \in \, < V, \alpha(V) > \, \cap \, < W, W >$. (by 7)
9. $\alpha \in \overline{< W, W >}$. (by 2, 8)
10. $\alpha \in \bigcap \{\overline{< W, W >} \mid \text{W open in X}\} = G'_X$. (by 2, 9)
11. $D_X \subset G'_X$. (by 1, 10)

12. $E_X = \chi_X(E) = \chi_X(DG') = D_X G'_X = G'_X$.
(by 11, **Ex. 11.6, Ex. 14.6, 15.12, 15.21**)

When the group T is abelian, then the flow (M, T) is Bronstein so we have the following corollary.

Corollary 15.25 Let T be abelian. Then $E = G'$.

PROOF: 1. Let $u \in J$, $p \in M$, and $t \in T$.
2. There exists a net $t_i \to p \in \beta T$.
3. $L_{ut}(p) = \lim utt_i = \lim ut_i t = upt = pt$. (1, 2, T is abelian)
4. $R_t = L_{ut} \in G$. (by 1, 3)
5. $(p, pt) \in gr(G)$. (by 1, 4)
6. $\overline{gr(G)} = M \times M$. (by 1, 5, M is minimal)
5. $E = G'$. (by 6, **15.24**)

We end this section with one more example of the relationship between the groups D and E; its proof relies on the fact that any compact Hausdorff group in which multiplication is unilaterally continuous is a topological group. A proof of the latter result is given in the appendix to this section.

Proposition 15.26 Let $[G, G]$ denote the commutator subgroup of G. Then:

(a) $E \subset D\overline{[G, G]}$.
(b) If T is abelian, then $E = D\overline{[G, G]}$.

PROOF: (a) 1. Let $R = P_0 \circ gr(D\overline{[G, G]})$ and $X = M/R$.
2. R is a regular distal icer with $G(R) = D\overline{[G, G]}$. (by 1, **14.11**)
3. $G/G(R) \cong X$ as groups. (by **14.5**)
4. X is abelian. (by 2, 3)
5. Multiplication in X is unilaterally continuous. (by 4)
6. X is a compact topological group. (by 5, see **15.A.13**)
7. (X, T) is equicontinuous. (by 6, **15.10**)
8. $E \subset G(R) = D\overline{[G, G]}$. (by 1, 2, 7, **15.14**)

(b) 1. Assume that T is abelian.
2. G/E and M/S_{eq} are isomorphic as topological groups. (by **15.10**)
3. $T \to M/S_{eq}$ is a continuous homomorphism of T onto a dense subgroup.
$t \to ut$
(by **14.5, 15.10**)
4. G/E is abelian. (by 1, 2, 3)
5. $\overline{[G, G]} \subset E$. (by 4)
6. $D\overline{[G, G]} \subset E$. (by 5, **15.21**)

APPENDIX TO SECTION 15: EQUICONTINUITY AND THE ENVELOPING SEMIGROUP

In this appendix we examine the enveloping semigroup $E(X)$ more closely in the context of equicontinuous flows (X, T). Intuitively if (X, T) is an equicontinuous flow then T, thought of as a family of maps in $E(X)$ is uniformly equicontinuous. Since every element of $E(X)$ is a limit point of T, they must all act continuously on X. Recall that a regular minimal flow X is isomorphic both as a flow and as a semigroup to $I(X)$, a minimal ideal in $E(X)$ (see **8.5** and **8.6**). When the flow (X, T) is also distal $X \cong I(X) \cong E(X)$ are isomorphic groups, though multiplication is continuous if and only if X is equicontinuous.

The main goal of this appendix is to show that for a minimal flow (X, T) to be equicontinuous it suffices that $E(X)$ be a group and that the map $\begin{array}{l} X \to X \\ x \to xp \end{array}$ be continuous for every $p \in E(X)$. This result has as a consequence that any compact Hausdorff space with a group structure for which multiplication is unilaterally continuous, is a topological group.

Proposition 15.A.1 Let:

(i) (X, T) be a flow, and
(ii) $\pi: X \times E(X) \to X$, $(x, p) \to xp$.

Then (X, T) is equicontinuous if and only if π is continuous.

PROOF: Exercise (similar to **15.10**).

As a consequence we have the following proposition.

Proposition 15.A.2 Let (X, T) be a flow. Then:

(a) if (X, T) is equicontinuous, then its enveloping semigroup $E(X)$ is a topological group, and
(b) if $E(X)$ is topological group and (X, T) is minimal, then (X, T) is equicontinuous.

PROOF: (a) We leave this as an exercise for the reader.

(b) 1. Assume that $E(X)$ is a topological group.
2. Let $x_i \to x \in X$ and $q_i \to q \in E(X, T)$.
3. There exist $p_i \in E(X)$ with $x_i = xp_i$. (using minimality)
4. We may assume that $p_i \to p \in E(X)$. ($E(X)$ is compact)
5. $x_i = xp_i \to xp$. (by 4)
6. $xp = x$. (by 2, 5)

7. $x_i q_i \underset{3}{=} x p_i q_i \underset{1,2,4}{\rightarrow} x p q \underset{6}{=} x q$.

8. $\pi\colon X \times E(X) \rightarrow X$ is continuous. (by 2, 7)
   $(x, p) \rightarrow xp$

9. (X, T) is equicontinuous. (by **15.A.1**)

Note that the proof of **15.A.2**(b) makes essential use of the fact that the flow (X, T) is minimal. Indeed the result is false in general. In fact the flow (X, T) of example **15.8** is distal but not equicontinuous, and yet $E(X, T)$ is a topological group in this case (we leave it to the reader to verify this in exercise **15.2**). Here the elements of $E(X)$ do not act continuously on X; on the other hand their restrictions to each minimal subset Y of X are continuous since they are elements of $E(Y)$.

Our main theorem shows that for any minimal flow (X, T), if every element of $E(X)$ acts continuously on X, then (X, T) is equicontinuous (and hence by **15.A.2**, $E(X)$ is a topological group). We begin with the case where X is metrizable; for this we need the following lemma.

Lemma 15.A.3 Let:

(i) (X, T) be a metrizable flow, and
(ii) p be continuous for all $p \in E(X)$.

Then $E(X)$ is metrizable.

PROOF: 1. Let d be a metric on X, and A a countable dense subset of X.
2. Set $R = \{(p, q) \in E(X) \times E(X) \mid d(ap, aq) = 0 \text{ for all } a \in A\}$.
3. It is clear that R is a closed equivalence relation on $E(X)$.
4. $E(X)/R$ is a metric space. (by 1)
5. $(p, q) \in R$ implies $d(xp, xq) = 0$ for all $x \in X$. (by 1, 2, (ii))
6. $(p, q) \in R$ implies $p = q$. (by 1, 5)
7. $E(X) = E(X)/R$ is a metric space. (by 4, 6)

The metric case of our theorem can now be deduced from the following well known result (see for example [Bourbaki, N., (1949)]).

Proposition 15.A.4 Let X, Y be compact metric spaces, $\mathcal{C}(X, Y)$ the set of continuous functions from X to Y provided with the topology of uniform convergence. Then $\mathcal{C}(X, Y)$ is separable (has a countable dense subset).

Proposition 15.A.5 Let:

(i) $X = M/R$ be minimal.
(ii) X be metrizable.
(iii) p be continuous for every $p \in E(X)$.

Then (X, T) is equicontinuous.

PROOF: 1. Let $\epsilon > 0$.
2. Let $\varphi: X \to C(E(X), X)$ be defined by $\varphi(x): E(X) \to X$.
$x \to \varphi(x)$ $p \to xp$
3. $E(X)$ is metrizable. (by (iii), **15.A.3**)
4. There exists a countable subset B of X such that $\varphi(B)$ is dense in $\varphi(X)$. (by (ii), 2, **15.A.4**)
5. Set $A(b) = \{x \mid d(bt, xt) \leq \epsilon/4 \text{ for all } t \in T\}$.
6. $A(b)$ is closed. ($x \to xt$ is continuous for all $t \in T$)
7. $\bigcup_{b \in B} A(b) = X$. (by 4)
8. There exists $b \in B$ with $int(A(b)) \neq \emptyset$. (by 6, 7, X is a Baire space)
9. Let $V = int(A(b))$.
10. $d(xt, yt) \leq d(xt, bt) + d(bt, yt) \leq \epsilon/2$ for all $x, y \in V$ and $t \in T$. (by 5, 9)
11. For every $z \in X$, there exists $t_z \in T$ with $zt_z \in V$. (by (i))
12. There exists a neighborhood W_z of z with $W_z t_z \subset V$. (by 9, 11)
13. Set $W = \bigcup_{z \in X} (W_z \times W_z) \subset X \times X$.
14. W is a neighborhood of the diagonal in $X \times X$. (by 12, 13)
15. Let $(w_1, w_2) \in W$.
16. There exists $z \in X$ with $(w_1, w_2) \in W_z \times W_z$. (by 13, 15)
17. $(w_1 t_z, w_2 t_z) \in V \times V$. (by 12, 16)
18. $d(w_1 t, w_2 t) = d(w_1 t_z t_z^{-1} t, w_2 t_z t_z^{-1} t) \leq \epsilon/2$ for all $t \in T$. (by 10, 17)

Corollary 15.A.6 Let:

(i) $X = M/R$ be minimal,
(ii) X be metrizable, and
(iii) (X, T) be equicontinuous.

Then $E(X)$ is metrizable.

PROOF: 1. The map $X \times E(X) \to X$ is continuous. (by (iii), **15.A.1**)
$(x, p) \to xp$
2. p is continuous. (by 1)
3. $E(X)$ is metrizable. (by 2, **15.A.3**)

Our next step is to consider the case where the group T is countable. We will use the fact that the topology on a compact space X can be recovered from the *pseudo-metrics* on X (see [Dugundji, J., (1966)]). These are the maps ρ: $X \times X \to \mathbf{R}$ satisfying all the conditions necessary to be a metric except that $\rho(x, y) = 0$ does not imply that $x = y$. We first observe that a continuous pseudo-metric on X gives rise to a quotient space which is metrizable if the group T is countable. The proof will be left to the reader.

Lemma 15.A.7 Let:

(i) (X, T) be a flow,
(ii) $S \subset T$ a subgroup of T,
(iii) d a continuous psuedo-metric on X, and
(iv) $R_{S,d} = \{(x, y) \in X \times X \mid d(xt, yt) = 0 \text{ for all } t \in S\}$.

Then:

(a) $R_{S,d}$ is a closed equivalence relation which is invariant under S, and
(b) if S is countable, then $X/R_{S,d}$ is metrizable.

Now we can prove our result for the countable group case.

Proposition 15.A.8 Let:

(i) (X, T) be minimal.
(ii) p be continuous for all $p \in E(X)$.
(iii) T be countable.

Then (X, T) is equicontinuous.

PROOF: 1. Let V be an open neighborhood of Δ in $X \times X$, and $x \in X$.
2. There exists d, a pseudo-metric on X, and $\epsilon > 0$ such that $\{(x_1, x_2) \mid d(x_1, x_2) < \epsilon\} \subset V$.
3. Set $R = R_{T,d} = \{(v, w) \mid d(vt, wt) = 0 \text{ for all } t \in T\}$.
4. R is an icer on X and X/R is metrizable. (by 1, 3, (iii), and **15.A.7**)
5. If $(v, w) \in R$, then $(vp, wp) \in R$ for all $p \in E(X)$. (by 4)
6. Let $\pi: X \to X/R = Y$ be the canonical map.
7. Let $\theta: E(X) \to E(X/R)$ be an epimorphism such that $\pi(zp) = \pi(z)\theta(p)$ for all $z \in X$, and $p \in E(X)$. (as in **2.10**)
8. Let $\{y_\alpha\}$ be a net in Y with $y_\alpha \to y$ and $r \in E(Y)$.
9. There exist $z_\alpha \in X$ and $p \in E(X)$ with $\pi(z_\alpha) = y_\alpha$ and $\theta(p) = r$.
10. We may assume that $z_\alpha \to z$.
11. $\pi(z) = y$. (by 9, 10)
12. $z_\alpha p \to zp$. (by 10, (ii))
13. $y_\alpha r \underset{9}{=} \pi(z_\alpha)\theta(p) \underset{7}{=} \pi(z_\alpha p) \underset{12}{\to} \pi(zp) \underset{7}{=} \pi(z)\theta(p) \underset{11}{=} yr$.
14. The map $y \to yr$ is continuous for all $r \in E(X/R)$. (by 8, 13)
15. $(X/R, T)$ is equicontinuous. (by 14, **15.A.5**)
16. The map $d: X \times X \to \mathbf{R}$ induces a continuous map $\rho: X/R \times X/R \to \mathbf{R}$ with

$$d(x_1, x_2) = \rho(\pi(x_1), \pi(x_2)). \quad \text{(by 3)}$$

17. $\{(y_1, y_2) \mid \rho(y_1, y_2) < \epsilon\}$ is an open neighborhood of $\Delta_{X/R}$. (by 16)
18. There exists $\pi(x) \in N$ open, such that $\rho(\pi(x)t, zt) < \epsilon$ for all $t \in T$, and $z \in N$. (by 15, 17)

19. Let $U = \pi^{-1}(N)$, $y \in U$ and $t \in T$.
20. $d(xt, yt) = \rho(\pi(xt), \pi(yt)) = \rho(\pi(x)t, \pi(y)t) < \epsilon$. (by 16, 18, 19)
21. $(xt, yt) \in V$ for all $y \in U$ and $t \in T$. (by 2, 20)
22. (X, T) is equicontinuous at x. (by 1, 21)
23. (X, T) is equicontinuous. (by 1, 22, and **15.2**)

In order to deal with the general case we need the following proposition.

Proposition 15.A.9 The flow (X, T) is equicontinuous if and only if the flow (X, H) is equicontinuous for all countable subgroups, H of T.

PROOF: $\Longrightarrow$

This is clear.

$\Longleftarrow$

1. Assume that (X, H) is equicontinuous for any countable subgroup $H \subset T$.
2. Assume that (X, T) is not equicontinuous.
3. There exists an open neighborhood V of Δ, such that $UT \cap \big((X \times X) \setminus V\big) \neq \emptyset$ for all open neighborhoods U of Δ.
4. Let W be an open neighborhood of Δ with $W \circ W \circ W \subset V$.
5. For every $t \in T$ set $B_t = \{s \in T \mid (xt, xs) \in W \cap W^{-1}$ for all $x \in X\}$.
6. Suppose that $T = B_{t_1} \cup \cdots \cup B_{t_n}$.
7. Choose $\Delta \subset U$ open with $Ut_i \subset W$ for $1 \leq i \leq n$.
8. Let $t \in T$ and $(x, y) \in U$.
9. $t \in B_{t_j}$ for some $1 \leq j \leq n$.
10. $(xt, xt_j), (xt_j, yt_j), (yt_j, yt) \in W$. (by 5, 7, 9)
11. $(xt, yt) \in V$. (by 4, 10)
12. $UT \subset V$. (by 8, 11)
13. No finite union of sets of the form B_t contains T. (12 contradicts 3)
14. Choose a sequence $\{t_i\} \subset T$ such that $t_n \notin \bigcup_{i=1}^{n-1} B_{t_i}$.
15. Let H be the subgroup of T generated by the sequence $\{t_n\}$.
16. (X, H) is equicontinuous. (by 1)
17. By Ascoli's theorem (see [Bourbaki, N., (1949)] page 42) $\overline{H}$ is a compact subset of $C(X, X)$ when the latter is provided with the topology of uniform convergence. (by 16)
18. The set $\{t_n \mid n = 1, 2, \ldots\}$ has an accumulation point $p \in C(X, X)$. (by 15, 17)
19. Let W_0 be an open neighborhood of $\Delta \subset X \times X$ with $W_0 \circ W_0 \subset W \cap W^{-1}$.
20. Let $B_0 = \{q \in \overline{H} \mid (xp, xq) \in W_0 \cap W_0^{-1}$ for all $x \in X\}$.
21. There exist $k < m$ with $t_k, t_m \in B_0$. (by 18, 19, 20)
22. $(xt_k, xp), (xp, xt_m) \in W_0$ for all $x \in X$. (by 20, 21)

23. $(xt_k, xt_m) \in W \cap W^{-1}$ for all $x \in X$. (by 19, 22)
24. $t_m \in B_{t_k}$ for all $x \in X$. (by 5, 23)
25. (X, T) is equicontinuous. (24 contradicts 14, so 2 must be false)

Note that given a minimal flow (X, T) and a subgroup $S \subset T$, the flow (X, S) need not be minimal, thus to handle the general case of our theorem we need one last technical lemma.

Lemma 15.A.10 Let:
(i) (X, T) be minimal,
(ii) H a countable subgroup of T, and
(iii) d a continuous pseudo-metric on X.

Then there exists a countable subgroup K of T such that $H \subset K$, and $(X/R_{K,d}, K)$ is a minimal flow.

PROOF: 1. Set $H_0 = H$, $X_0 = X/R_{H_0,d}$, and let $\pi_0 \colon X \to X_0$ be the canonical map.
2. There exists a countable base $\mathcal{U}_0$ for the topology on X_0. (by (i), **15.A.7**)
3. Let $U \in \mathcal{U}_0$.
4. $\pi_0^{-1}(U)T = X$. (by (i))
5. There exists a finite subset F_U of T such that $\pi_0^{-1}(U)F_U = X$. (by 4)
6. Let H_1 be a countable subgroup of T such that $H_0 \cup \bigcup_{U \in \mathcal{U}_0} F_U \subset H_1$.
7. $\pi_0^{-1}(U)H_1 = X$ for all $U \in \mathcal{U}_0$. (by 3, 5, 6)
8. $\pi_0^{-1}(W)H_1 = X$ for all open subsets W of X_0. (by 2, 7)
9. Assume $H_0 \subset H_1 \subset \cdots \subset H_n$ have been chosen so that:

H_i is a countable subgroup of T,

$X_i = X/R_{H_i,d}$,

$\pi_i : X \to X_i$ is the canonical map $0 \leq i \leq n-1$, and

$\pi_i^{-1}(W)H_{i+1} = X$ for every nonempty open set $W \subset X_i$.

10. Set $X_n = X/R_{H_n,d}$, let $\pi_n \colon X \to X_n$ be the canonical map, and $\mathcal{U}_n$ be a countable base for the topology on X_n.
11. $\pi_n^{-1}(U)T = X$ for all $U \in \mathcal{U}_n$. (by (i))
12. As in 5, 6, 7, there exists a countable subgroup H_{n+1} with $H_n \subset H_{n+1}$ and $\pi_n^{-1}(W)H_{n+1} = X$ for all open subsets W of X_n.
13. H_n, X_n, and π_n are defined for all n by induction.
14. Let $K = \bigcup H_n$.
15. $H \subset K$ and K is a countable subgroup of T. (by 14)
16. $X/R_{K,d} = \varprojlim(X_n, \pi_n^{n+1})$ where $\pi_n^{n+1} \colon X_{n+1} \to X_n$ is the canonical map. (by 14)

17. Let W be open in $X/R_{K,d}$ and $\pi\colon X \to X/R_{K,d}$ be the canonical map.
18. There exists n and U open in X_n such that $\phi^{-1}(U) \subset W$ where ϕ: $X/R_{K,d} \to X_n$ is the canonical map. (by 16)
19. $\pi^{-1}(W)K \underset{14}{\supset} \pi^{-1}(W)H_{n+1} \underset{18}{\supset} \pi^{-1}(\phi^{-1}(U))H_{n+1}$
$= \pi_n^{-1}(U)H_{n+1} \underset{12}{=} X.$
20. $\pi^{-1}(W)K = X$ for all open subsets W of $X/R_{K,d}$. (by 17, 19)
21. $WK = X/R_{K,d}$ for every open set $W \subset X/R_{K,d}$. (by 20)
22. $(X/R_{K,d}, K)$ is minimal. (by 21)

We are now in position to deduce the main theorem of this appendix.

Theorem 15.A.11 Let:

(i) (X, T) be minimal, and
(ii) p be continuous for all $p \in E(X)$.

Then (X, T) is equicontinuous.

PROOF: 1. Let H be a countable subgroup of T.
2. Let d be a continuous pseudo-metric on X and $x \in X$.
3. There exists a countable subgroup K of T with $H \subset K$ and $(X/R_{K,d}, K)$ minimal. (by **15.A.10**)
4. As in the proof of **15.A.8** p is continuous for all $p \in E(X/R_{K,d}, K)$. (using (ii))
5. $(X/R_{K,d}, K)$ is equicontinuous. (by 3, 4, **15.A.8**)
6. $(X/R_{K,d}, H)$ is equicontinuous. (by 3, 5)
7. As in the proof of **15.A.8**, (X, H) is equicontinuous.
8. (X, T) is equicontinuous. (by 1, 7, **15.A.9**)

Corollary 15.A.12 Let:

(i) (X, T) be minimal, and
(ii) $E(X, T)$ be abelian.

Then (X, T) is equicontinuous.

PROOF: 1. $L_p = R_p$ for all $p \in E(X, T)$. (by (ii))
2. $p \equiv R_p$ is continuous for all $p \in E(X, T)$. (by 1)
3. (X, T) is equicontinuous. (by **15.A.11**)

Theorem 15.A.13 Let H be a compact Hausdorff space provided with a group structure in which multiplication is unilaterally continuous. Then H is a topological group.

PROOF: 1. Consider the minimal flow (H, H).
2. Let $p \in E(H, H)$ and $\{h_\alpha\} \subset H$ with $h_\alpha \to p$.

3. We may assume that $h_\alpha \to h \in H$.
4. $xp = \lim xh_\alpha = xh$ for all $x \in H$.
(by 3, left multiplication L_x in H is continuous for all x)
5. $p = R_h$. (by 4)
6. p is continuous. (by 5, right multiplication R_h in H is continuous for all h)
7. (H, H) is equicontinuous. (by 6, **15.A.11**)
8. $H = E(H, H)$ is a topological group. (by 2, 5, 7, **15.A.2**)

NOTES ON SECTION 15

Note 15.N.1 Let H be a locally compact Hausdorff space provided with a group structure for which multiplication is unilaterally continuous. Then H is a topological group.

PROOF: This can be proven using methods similar to those of **15.A.13**. For this and related results see [Ellis, R., *Locally compact transformation groups*, (1957)].

Note 15.N.2 Let

(i) $\pi\colon G \to G/G'PG^J$ be the canonical map,
(ii) $\psi\colon T \to G/G'PG^J$, and $t \to ut$
(iii) f the continuous extension of ψ to βT.

Then $f(\alpha(u)) = \pi(\alpha)$ for all $\alpha \in G$ and $u \in J$.

PROOF: 1. Let $\alpha \in G$ and $u \in J$.
2. There exists a net (t_i) in T with $ut_i \to \alpha(u)$. ($M \subset \beta T$ is minimal)
3. There exist $u_i \in J$ with $ut_iu_i = ut_i$ for all i. (by **3.12**)
4. We may assume that $u_i \to \beta(v)$ with $\beta \in G^J$ and $v \in J$.
(by compactness and **3.12**)
5. $L_{ut_i} \to \alpha\beta^{-1}$. (by 2, 3, 4 and **10.14**)
6. $f(t_i) = f(u)f(t_i)f(u) = f(ut_iu)$. (as in **15.9**)
7. $\pi(L_{ut_i}) = f(t_i) \underset{6}{=} f(ut_iu) = f(L_{ut_i}(u))$.
8. $f(\alpha(u)) \underset{2}{=} \lim f(ut_i) \underset{6}{=} \lim f(L_{ut_i}(u)) \underset{7}{=} \lim \pi(L_{ut_i})$
$\underset{5}{=} \pi(\alpha\beta^{-1}) = \pi(\alpha)\pi(\beta)^{-1} \underset{4}{=} \pi(\alpha)$.

Note 15.N.3 Let $\mathcal{E}$ denote the algebra of almost periodic functions on T. Then the Stone space $|\mathcal{E}|$ admits a minimal equicontinuous action of T. Indeed $(|\mathcal{E}|, T) \cong (M/S_{eq}, T)$ thus the E of definition **15.A.6** coincides with the definition of E given in [Ellis, R., 1969].

EXERCISES FOR CHAPTER 15

Exercise 15.1 Show that a flow (X, T) is equicontinuous if and only if it is pointwise equicontinuous.

Exercise 15.2 Let (X, T) be the flow given in example **15.8**. Show that:

(a) $Q(X) = \{(z_1, z_2) \mid |z_1| = |z_2|\}$, and
(b) $E(X)$ is a compact topological group. (Hint: (Y, T) is equicontinuous for every minimal set $Y \subset X$.)

Exercise 15.3 Let:

(i) (X, T) be a minimal flow, and
(ii) R be an icer on X.

Show that $(X/R, T)$ is equicontinuous if and only $Q(X) \subset R$.

Exercise 15.4 Let (X, T) be a weak mixing minimal flow. Show that:

(a) $Q(X) = X \times X$, and
(b) X has no non-trivial equicontinuous factors.

Exercise 15.5 Let:

(i) $\{(X_i, T) \mid i \in I\}$ be a family of equicontinuous flows, and
(ii) $X = \Pi_{i \in I} X_i$.

Then the product flow (X, T) is equicontinuous.

Exercise 15.6 Let (X, T) be an equicontinuous flow. Then:

(a) $(E(X), T)$ is an equicontinuous flow, and
(b) if in addition $X = M/R$, the regularizer $(reg(R), T)$ is an equicontinuous flow.

Exercise 15.7 Let:

(i) (X, T) be a flow, and
(ii) $$\begin{aligned} \pi\colon X \times E(X) &\to X. \\ (x, p) &\to xp \end{aligned}$$

Then (X, T) is equicontinuous if and only if π is continuous.

Exercise 15.8 Let:

(i) (X, T) be a an equicontinuous flow,
(ii) (Y, T) be a flow, and
(iii) $f\colon X \to Y$ be an epimomorphism.

Then (Y, T) is equicontinuous.

Exercise 15.9 Show that a **minimal** flow (X, T) is equicontinuous if and only if its enveloping semigroup $E(X)$ is a compact topological group.

Exercise 15.10 (See **15.23**) Let:

(i) $X = M/R$ be regular,
(ii) $J_X^* = \{\alpha \in G_X \mid \alpha(J_X) \cap \overline{J_X} \neq \emptyset\}$,
(iii) $G_X(J_X)$ be the closed normal subgroup of G_X generated by J_X^*,
(iv) $\Omega_X(u, v, t) = L_{ut}L_{vt}^{-1}$ where $u, v \in J_X$ and $t \in T$, and
(v) Ω_X be the closed normal subgroup of G_X generated by

$$\{\Omega_X(u, v, t) \mid u, v \in J_X \text{ and } t \in T\}.$$

Then:

(a) $E_X = G_X(J_X)\Omega_X G'_X$,
(b) $\chi_R(G^J) = G_X^J$,
(c) $\chi_R(\Omega) = \Omega_X$, and
(d) $\chi_R(E) = E_X$.

Exercise 15.11 (See **15.21**) Let $X = M/R$ be regular. Then:

(a) $E_X = D_X G'_X$.
(b) $S_{eq}(X) = P_0(X) \circ gr(E_X)$.
(c) $G_X(S_{eq}(X)) - E_X$.

Exercise 15.12 (See **15.A.7**) Let:

(i) (X, T) be a flow,
(ii) $S \subset T$ be a countable subgroup of T,
(iii) d be a continuous pseudo-metric on X, and
(iv) $R_{S,d} = \{(v, w) \in X \times X \mid d(vt, wt) = 0 \text{ for all } t \in S\}$.

Show that:

(a) $R_{S,d}$ is a closed equivalence relation on X, and
(b) $X/R_{S,d}$ is metrizable.

16

The regionally proximal relation

In this section we examine in more depth the regionally proximal relation $Q(X)$, which was defined in **15.4**. Here we will be primarily interested in the case where $X = M/R$ is a minimal flow. Note that if $E \subset G(R)$, then by **15.13** $P(M/R) = Q(M/R)$, thus in particular $P(M/R)$ is closed and it follows from **13.6** that $Q(X)$ is an equivalence relation and hence an icer on X. In fact a stronger result holds; the main result of this section is that if $E \subset G(R)G'$, then $Q(M/R)$ is an icer. We begin this section with a few standard results on $Q(X)$ and its relationship to the equicontinuous structure relation $S_{eq}(X)$.

Proposition 16.1 Let N be an icer on the minimal flow (X, T). Then $(X/N, T)$ is equicontinuous if and only if $Q(X) \subset N$.

PROOF: $\Longrightarrow$

1. Assume that $Y = X/N$ is equicontinuous, and let $\pi \colon X \to Y$ be the canonical map.
2. $Q(Y) = \Delta_Y$. (by 1, **15.5**)
3. $\pi(Q(X)) = Q(Y) = \Delta_Y$. (by 2, **15.6**)
4. $Q(X) \subset N$. (by 1, 3)

$\Longleftarrow$

1. Assume that $Q(X) \subset N$.
2. $Q(Y) = \pi(Q(X)) = \Delta_Y$. (by 1, **15.6**)
3. Y is equicontinuous. (by 2, **15.5**)

Proposition 16.2 Let $X = M/R$. Then the following are equivalent:

(a) $S_{eq}(X) = Q(X)$,
(b) $Q(X)$ is an icer, and
(c) $S_{eq}(X) \subset Q(X)$.

PROOF: (a) $\Longrightarrow$ (b) is clear.

(b) $\Longrightarrow$ (c)

1. Assume that $Q(X)$ is an icer.
2. $X/Q(X)$ is equicontinuous. (by **16.1**)
3. Let S be an icer on M with $X/Q(X) = M/S$.
4. $S_{eq} \subset S$. (by 3, **15.16**)
5. $S_{eq}(X) = \pi_R(S_{eq}) \subset \pi_R(S) = Q(X)$. (by 3, 4, **15.6**)

(c) $\Longrightarrow$ (a)

1. Assume that $S_{eq}(X) \subset Q(X)$.
2. $X/S_{eq}(X)$ is equicontinuous. (by **15.19**)
3. $Q(X) \subset S_{eq}(X)$. (by 2, **16.1**)

Corollary 16.3 Let:

(i) R and N be icers on M with $R \subset N$,
(ii) $X = M/R$, $Y = M/N$, and
(iii) $Q(X)$ be an icer on X.

Then $Q(Y)$ is an icer on Y.

PROOF: 1. $S_{eq}(X) \subset Q(X)$. (by **16.2**)
2. $S_{eq}(Y) \underset{\mathbf{15.20}}{=} \pi_N^R(S_{eq}(X)) \underset{1}{\subset} \pi_N^R(Q(X)) \underset{\mathbf{15.6}}{=} Q(Y)$.
3. $Q(Y)$ is an icer on Y. (by 2, **16.2**)

We now prove two technical lemmas which will be used to deduce the main result of this section.

Lemma 16.4 Let:

(i) N be a non-vacuous open subset of M, and
(ii) $A_N = \{\alpha \in G \mid \alpha(J) \cap N \neq \emptyset\}$.

Then:

(a) $\overline{int(\overline{A_N})} \equiv cic(A_N) \neq \emptyset$.
(b) $G' cic(A_N) = cic(A_N)$.

PROOF: (a) 1. The set of almost periodic points of the flow (G, M) is dense in M. ((M, T) is minimal)
2. There exists an almost periodic point $\beta(w)$ of the flow (G, M) with $\beta(w) \in N$. (by 1)
3. $G' \subset cic\{\gamma \mid \gamma(\beta(w)) \in N\}$. (by 2, **11.15**)
4. $\emptyset \neq cic\{\gamma \mid \gamma\beta(w) \in N\} \subset cic\{\gamma \mid \gamma\beta \in A_N\} = cic(A_N\beta^{-1}) = cic(A_N)\beta^{-1}$. (by 3, (ii), **10.6**)

(b) This follows from (a) and **11.3**.

Lemma 16.5 Let:

(i) $v \in J$,
(ii) $A_N = \{\alpha \in G \mid \alpha(J) \cap N \neq \emptyset\}$, and
(iii) $A_v = \bigcap_{N \in \mathcal{N}_v} cic(A_N)$. (Here $\mathcal{N}_v = \{V \mid V$ an open neighborhood of v in $M\}$.)

Then:

(a) $A_v \neq \emptyset$,
(b) $G' A_v = A_v$,
(c) $(\alpha(w), v) \in Q$ for all $\alpha \in A_v$ and $w \in J$,
(d) $A_v \subset E$, and
(e) $E A_v = E$.

PROOF: (a) $\{cic(A(N)) \mid N \in \mathcal{N}_v\}$ is a collection of nonempty closed sets with the finite intersection property so this follows from the fact that G is compact. (by **16.4**, **10.6**)

(b) 1. $A_v \subset G' A_v = G' \bigcap_{N \in \mathcal{N}_v} cic(A_N) \subset \bigcap_{N \in \mathcal{N}_v} G' cic(A_N) = \bigcap_{N \in \mathcal{N}_v} cic(A_N) = A_v$. ($1 \in G'$, **16.4**, **10.12**)

(c) 1. Let $w \in J$ and $\alpha \in \bigcap_{N \in \mathcal{N}_v} \overline{A_N}$.
2. Let U be an open invariant neighborhood of Δ in $M \times M$.
3. Let W be an open neighborhood of $(\alpha(w), v)$ in $M \times M$.
4. Let V be an open neighborhood of v with $\alpha(V) \times \alpha(V) \subset U$ and $\{\alpha(w)\} \times V \subset W$.
5. $\alpha \in \overline{A_V}$. (by 1, 4)
6. There exists $\beta \in A_V \cap < V, \alpha(V) >$. (by 5)
7. There exists $u \in J$ and $p \in V$ with $\beta(u) \in V$ and $\beta(p) \in \alpha(V)$. (by 6, (ii))
8. $(\alpha(w), \beta(u))p = (\alpha(p), \beta(p)) \in \alpha(V) \times \alpha(V) \subset U$. (by 4, 7)
9. $(\alpha(w), \beta(u)) \in U$. (by 2, 8)
10. $\emptyset \neq (\{\alpha(w)\} \times V) \cap U \subset W \cap U$. (by 4, 7, 9)
11. $(\alpha(w), v) \in \overline{U}$. (by 3, 10)
12. $(\alpha(w), v) \in Q$. (by 2, 11)

(d) 1. Let $\alpha \in A_v$.
2. $(\alpha(v), v) \in Q$. (by part (c))
3. $\alpha \in E$. (by 2, **15.11**)

(e) This follows immediately from parts (a) and (d).

Theorem 16.6 Let R be an icer on M with $E \subset G(R)G'$. Then $Q(M/R)$ is an icer.

PROOF: 1. Let $(p, q) \in S_{eq} = P_0 \circ gr(E)$.

2. $(p, q) = (\beta(v), \gamma(w))$ for some $v, w \in J$ and $\beta^{-1}\gamma \in E$. (by 1)
3. $E = \beta^{-1}E\beta \subset \beta^{-1}G(R)G'\beta = \beta^{-1}G(R)\beta G'$. ($E$ and G' are normal)
4. $E = EA_v \subset (\beta^{-1}G(R)\beta G')A_v = \beta^{-1}G(R)\beta A_v$. (by 3, **16.5**)
5. There exist $\rho_R \in G(R)$ and $\alpha \in A_v$ with $\beta^{-1}\gamma = \beta^{-1}\rho_R\beta\alpha$. (by 2, 4)
6. $(v, \alpha(w)) \in Q$. (by 5, **16.5**)
7. $(\alpha(w), \beta^{-1}\gamma(w)) = (\alpha(w), \beta^{-1}\rho_R\beta\alpha(w)) \in gr(\beta^{-1}G(R)\beta) = \beta^{-1}(gr(G(R)))$. (by 5, **7.15**)
8. $(v, \beta^{-1}\gamma(w)) \in Q \circ \beta^{-1}(gr(G(R)))$. (by 6, 7)
9. $(\beta(v), \gamma(w)) = \beta(v, \beta^{-1}\gamma(w)) \in \beta\left(Q \circ \beta^{-1}(gr(G(R)))\right)$
$= \beta(Q) \circ gr(G(R)) = Q \circ gr(G(R))$. (by 8, **15.5**)
10. $S_{eq}(R) = \pi_R(S_{eq}) \subset \pi_R(Q \circ gr(G(R))) = \pi_R(Q) = Q(M/R)$.
(by 1, 2, 9, **15.9**)
11. $Q(M/R)$ is an icer. (by 10, **16.2**)

The converse of **16.6** is false. For a counter-example see [Auslander, J., Ellis, D. B., Ellis, R., *The regionally proximal relation*, (1995)].

Corollary 16.7 Let $X = M/R$ and $P(X)$ be closed. Then $Q(X)$ is an equivalence relation.

PROOF: 1. $D \subset G(R)$. (by **14.8**)
2. $E \underset{\mathbf{15.21}}{=} DG' \underset{1}{\subset} G(X)G'$.
3. $Q(X)$ is an equivalence relation. (by **16.6**)

Corollary 16.8 Let:

(i) $X = M/R$ be a regular flow, and
(ii) X be Bronstein, i. e. $\overline{gr(G_X)} = X \times X$.

Then $E \subset G(X)G'$ and hence $Q(X)$ is an equivalence relation.

PROOF: 1. $E_X = G'_X$. (by (i), (ii), **15.24**)
2. $E \subset \chi_X^{-1}(E_X) = \chi_X^{-1}(G'_X) = G'G(R)$. (by 1, **Ex. 11.6**, **Ex. 15.10**)

The result above holds without the regularity assumption, but the map χ_X: $G \to G_X$ exists only in the regular case. Thus a proof which does not rely on **15.24** is required.

Corollary 16.9 Let:

(i) $X = M/R$ be any minimal flow, and
(ii) X be Bronstein, i. e. $\overline{\pi_R(gr(G))} = X \times X$.

Then $E \subset G(X)G'$ and hence $Q(X)$ is an equivalence relation.

PROOF: 1. Let $u \in J$ and $\alpha \in G$ with $(u, \alpha(u)) \in Q$.

2. $\pi_R(u, \alpha(u)) \in Q(X)$. (by 1, **15.6**)
3. There exist nets $\{p_i\} \subset M$ and $\{\alpha_i\} \subset G$, such that

$$\pi_R(p_i, \alpha_i(p_i)) \to \pi_R(u, \alpha(u)) \text{ and } \pi_R(p_i, \alpha_i(p_i))t_i \to \pi_R(z, z).$$

(by 2, (ii))
4. We may assume that $(p_i, \alpha_i(p_i)) \to (p, q)$ and $(p_i, \alpha_i(p_i))t_i \to (r, m)$. ($M$ is compact)
5. $\pi_R(p, q) = \lim \pi_R(p_i, \alpha_i(p_i)) = \pi_R(u, \alpha(u))$. (by 3, 4)
6. $\pi_R(r, m) = \lim \pi_R(p_i, \alpha_i(p_i))t_i = \pi_R(z, z)$. (by 3, 4)
7. There exist $v_1, v \in J$ and $\beta_1, \beta_2 \in G(R)$ with

$$(p, q) = (\beta_1(v_1), \beta_2\alpha(v)) = (p, \beta_2\alpha\beta_1^{-1}(pv)). \quad \text{(by 5)}$$

8. There exist $w \in J$ and $\beta \in G(R)$ with $(r, m) = (r, \beta(rw))$. (by 6)
9. $\alpha_i \to \beta_2\alpha\beta_1^{-1}$. (by 4, 7, **10.13**)
10. $\alpha_i \to \beta$. (by 4, 8, **10.13**)
11. $\beta_2\alpha\beta_1^{-1}\beta^{-1} \in G'$. (by 9, 10)
12. $\alpha \in G(R)G'G(R) = G'G(R)$. (by 7, 8, 11, G' is normal)

As an immediate consequence we see that $Q(X)$ is also an equivalence relation for any *point distal* minimal flow. A flow is said to be point distal if there exists a point $a \in X$ such that $(a, x) \in P(X) \Rightarrow x = a$.

Proposition 16.10 Let:

(i) $X = M/R$, and
(ii) $a \in X$ with $aP(X) = \{a\}$.

Then the almost periodic points of $(X \times X, T)$ are dense in $X \times X$ and so $Q(X)$ is an equivalence relation.

PROOF: 1. Let $u \in J$ and $t \in T$.
2. $(at, atu) \in P(X)$. (by 1)
3. $at = atu$. (by 2, (ii))
4. (as, at) is an almost periodic point of $X \times X$ for all $s, t \in T$. (by 1, 3)
5. $\overline{\{(as, at) \mid s, t \in T\}} = X \times X$. (by (i), X is minimal)

EXERCISES FOR CHAPTER 16

Exercise 16.1 Show that $S_{eq} = P_0 \circ Q = Q \circ Q$.

Exercise 16.2 Let:

(i) R and S be icers on M, and
(ii) $G(R) = G(S)$.

Then $S \subset R \circ Q$

Exercise 16.3 Let:

(i) R and S be icers on M,
(ii) $G(R) = G(S)$, and
(iii) $\pi_R^{-1}(Q(M/R))$ be an equivalence relation.

Then $\pi_S^{-1}(Q(M/S)) = S \circ Q \circ S \subset R \circ Q \circ R = \pi_R^{-1}(Q(M/R))$.

PART V

Extensions of minimal flows

In this section we study the structure of various types of extensions $R \subset S$ of minimal flows, using the theory of icers and subgroups of G developed in the preceding sections. The results are often analogous to and motivated by the corresponding results on minimal flows, thought of as extensions of the one-point flow. The groups $G(R)$ and $G(S)$ along with the relative product and quasi-relative product play key roles.

In section 17 we use the circle operator of section 5 to characterize open extensions and to define the related notions of RIC and highly proximal extensions. We see in **17.3** that $R \subset S$ is a RIC extension if and only if $S = R \circ \overline{gr(G(S))}$.

Section 18 is devoted to an in-depth study of distal extension of minimal flows, again from the point of view of icers. One key idea here is that an extension $R \subset S$ is distal if and only if $S = R \circ gr(G(S))$ (see **18.9**). From this point of view it is evident that every distal extension is a RIC extension.

Almost periodic extensions, which are the natural generalization of equicontinuous flows, are studied in section 19. This is done in large part with the aid of the regionally proximal relation for equivalence relations, which is the natural generalization of the regionally proximal relation for flows introduced in section 15. The classical result is that $R \subset S$ is an almost periodic extension if and only if it is a distal extension and $G(S)' \subset G(R)$. Finally in section 20 we collect four theorems which are equivalent to the Furstenberg structure theorem which says that any non-trivial distal extension "factors" into a distal extension together with a non-trivial almost periodic extension. The Furstenberg tower is developed as a consequence of this result; its construction being a nice application of the machinery on icers, subgroups of G, and the τ-topology developed in the preceding sections.

17

Open and highly proximal extensions

The notion of highly proximal extensions was introduced in [Auslander, J., Glasner, S., (1997)]. We will see that every almost one-one extension is a highly proximal extension, and every highly proximal extension is a proximal extension. Our account emphasizes the point of view of equivalence relations. The key tool is the action of βT on 2^X (see section 9) given by the circle operator. We include a brief introduction to relatively incontractible (RIC) extensions which are also defined in terms of the circle operator. We begin by using the circle operator to characterize open extensions of minimal flows. The following theorem is a restatement of exercise **5.6**; we give a proof for the sake of completeness and in order to motivate the definition of RIC extensions.

Theorem 17.1 Let:

(i) $R \subset S$ be icers on M, and
(ii) $X = M/R$ and $Y = M/S$.

Then the following are equivalent:

(a) π_S^R is open,
(b) $\pi_R(pS \circ q) = \pi_R((pq)S)$ for all $p, q \in M$, and
(c) $\pi_R(uS \circ p) = \pi_R(pS)$ some $u \in J$ and all $p \in M$.

PROOF: (a) $\Longrightarrow$ (b)

1. Assume that π_S^R is open and let $p, q \in M$.
2. The map $\varphi : Y \to 2^X$ is a homomorphism of flows. (by 1, **5.7**)
$y \to [(\pi_S^R)^{-1}(y)]$
3. $\varphi(\pi_S(pq)) = [(\pi_S^R)^{-1}(\pi_S(pq))] = [\pi_R(\pi_S^{-1}(pq))] = [\pi_R((pq)S)]$.
4. $\varphi(\pi_S(pq)) \underset{2}{=} \varphi(\pi_S(p))q = [(\pi_S^R)^{-1}(\pi_S(p))]q$
$= [\pi_R(pS)]q \underset{\mathbf{5.10}}{=} [\pi_R(pS) \circ q] \underset{\mathbf{5.11}}{=} [\pi_R((pS) \circ q)]$.
5. $\pi_R((pq)S) = \pi_R((pS) \circ q)$. (by 3, 4)

(b) $\Longrightarrow$ (c)

Clear.

(c) $\Longrightarrow$ (a)

1. Assume that $u \in J$ with $\pi_R(uS \circ p) = \pi_R(pS)$ for all $p \in M$.
2. Let $y_i \to y \in Y$ and $\varphi : Y \to 2^X$ be defined as above. $y \to [(\pi_S^R)^{-1}(y)]$
3. There exist $p_i \in M$ with $y_i = \pi_S(p_i)$ for all i. (Y is minimal)
4. We may assume that $p_i \to p \in M$.
5. $y = \lim y_i = \lim \pi_S(p_i) = \pi_S(p)$. (by 2, 3, 4)
6. $\lim \varphi(y_i) = \lim \varphi(\pi_S(p_i)) = \lim[\pi_R(p_i S)] \underset{1}{=} \lim[\pi_R(uS) \circ p_i]$

$$\underset{\mathbf{5.10}}{=} \lim[\pi_R(uS)]p_i = [\pi_R(uS)]p \underset{\mathbf{5.10}}{=} [\pi_R(uS) \circ p]$$

$$\underset{\mathbf{5.11}}{=} [\pi_R(uS \circ p)] \underset{1}{=} [\pi_R(pS)] \underset{2}{=} \varphi(\pi_S(p)) \underset{6}{=} \varphi(y).$$

7. φ is continuous. (by 2, 6)
8. π_S^R is open. (by 8, **5.7**)

Let $R \subset S$ be icers on M, $u \in J$, and $p, q \in M$. It is immediate that $G(S)(p) \subset pS$ and hence that $\pi_R(G(S)(p)) \subset \pi_R(pS)$. On the other hand, by **5.11**,

$$\pi_R(pS \circ q) = \pi_R(pS) \circ q \subset \pi_R((pq)S),$$

from which we obtain

$$\begin{aligned} \pi_R(G(S)(p)) = \pi_R(G(S)(u)p) &\subset \pi_R(G(S)(u) \circ p) \\ &\subset \pi_R((uS) \circ p) \subset \pi_R(pS) \end{aligned} \qquad (*)$$

for **any** extension. Now **17.1** says that the open extensions are those for which the last containment in equation $(*)$ is an equality. In particular any extension for which $\pi_R(G(S)(p)) = \pi_R(pS)$ is an open extension. The latter holds if $R \cap P_0 = S \cap P_0$, or equivalently $R \subset S$ is a *distal extension*; this gives another proof (see **7.23**) that distal extensions are open. This argument also shows that if $\pi_R(G(S)(u) \circ p) = \pi_R(pS)$ for all $p \in M$, then π_S^R is open. This is one motivation for the following definition.

Definition 17.2 Let $R \subset S$ be icers on M. We say that $R \subset S$ is a *relatively incontractible* (RIC) *extension* if there exists $u \in J$ such that

$$\pi_R(G(S)(u) \circ p) = \pi_R(pS)$$

for all $p \in M$. In the case where $S = M \times M$ so that M/R is a RIC extension of the one-point flow we say the M/R is an *incontractible flow*.

It follows from the preceding discussion that every distal extension of minimal flows is RIC, and every RIC extension of minimal flows is open. Intuitively S is a RIC extension of R if it is the smallest icer containing R which also contains $gr(G(S))$. Of course any icer which contains $gr(G(S))$ also contains $\overline{gr(G(S))}$, hence the following characterization of RIC extensions.

Proposition 17.3 Let $R \subset S$ be icers on M. Then the following are equivalent:

(a) $\pi_R(G(S)(v) \circ p) = \pi_R(pS)$ for all $v \in J$ and $p \in M$,
(b) $R \subset S$ is a RIC extension, and
(c) $S = R \circ \overline{gr(G(S))}$.

PROOF: (a) $\Longrightarrow$ (b)
Clear.

(b) $\Longrightarrow$ (c)

1. Assume that $R \subset S$ is a RIC extension and let $(p, q) \in S$.
2. $q \in pS$. (by 1)
3. $\pi_R(q) \in \pi_R(pS) \underset{1}{=} \pi_R(G(S)(u) \circ p) \underset{\mathbf{12.2}}{=} \pi_R\big(p\overline{gr(G(S))}\big)$ for some $u \in J$.
4. There exists $r \in p\overline{gr(G(S))}$ with $\pi_R(q) = \pi_R(r)$. (by 3)
5. $(p, r) \in \overline{gr(G(S))}$ and $(r, q) \in R$. (by 4)
6. $(p, q) \in \overline{gr(G(S))} \circ R$. (by 5)
7. $S \subset \overline{gr(G(S))} \circ R$. (by 1, 6)
8. $S = \overline{gr(G(S))} \circ R = R \circ \overline{gr(G(S))}$.
(by 7, S is an icer with $R \cup \overline{gr(G(S))} \subset S$)

(c) $\Longrightarrow$ (a)

1. Assume that $S = R \circ \overline{gr(G(S))}$ and let $v \in J$ and $p \in M$.
2. $\pi_R(pS) = \pi_R\big(p\overline{gr(G(S))}\big) = \pi_R(G(S)(v) \circ p)$. (by 1, and **12.2**)

We will see in section 18 that S is a distal extension of R if and only if $S = R \circ gr(G(S))$. Here we use **17.3** to show that any extension of minimal flows can be approximated up to proximal extensions by a RIC extension.

Proposition 17.4 Let:

(i) $R \subset S$ be icers on M, and
(ii) $A \subset G$ be closed with $G(R) \subset A \subset G(S)$.

Then there exist icers $R_A \subset R$ and $S_A \subset S$ such that:

(a) $R_A \subset R$ is a proximal extension,
(b) $R_A \subset S_A$ is a RIC extension, and
(c) $G(S_A) = A$.

PROOF: 1. $M \to M/\overline{gr(A)}$ is an open map. (by **12.2**)

2. $S_A \equiv \overline{gr(A)}(R)$ (the quasi-relative product of $\overline{gr(A)}$ with R) is an icer on M. (by 1, **9.8**)
3. $R_A \equiv R \cap \overline{gr(A)}(R)$ is an icer on M. (by 2)
4. $G(R_A) = G(R) \cap G(\overline{gr(A)}(R))$. (by 3, **7.19**)
5. $G(R) \underset{(ii)}{\subset} A \underset{\mathbf{12.2}}{=} G(\overline{gr(A)}) \subset G(\overline{gr(A)}(R))$.
6. $G(R_A) = G(R)$. (by 4, 5)
7. $R_A \subset R$ is a proximal extension. (by 3, 6, **7.11**)
8. $S_A = R_A \circ \overline{gr(A)}$. (by 2, 3, **9.4**)
9. $G(S_A) = G(R_A)G(\overline{gr(A)}) = G(R)A = A$. (by 6, 8, (ii), **7.26**)
10. $R_A \subset S_A$ is a RIC extension. (by 8, 9, **17.3**)

Corollary 17.5 Let $R \subset S$ be icers on M. Then there exist icers $R_0 \subset R$ and $S_0 \subset S$ such that:

(a) $R_0 \subset R$ is a proximal extension,
(b) $R_0 \subset S_0$ is a RIC extension, and
(c) $S_0 \subset S$ is a proximal extension.

PROOF: Take $A = G(S)$ in **17.4**.

We now turn to a discussion of highly proximal extensions; as motivation we recall (see **7.13**) that $R \subset S$ is an almost one-one extension if there exists $y_0 \in Y$ such that $(\pi_S^R)^{-1}(y_0)$ is a singleton. Exercise **13.6** says that $R \subset S$ is a proximal extension if and only if there exists $y_0 \in Y$ and $p \in M$ such that $((\pi_S^R)^{-1}(y_0))\,p$ is a singleton. Since $((\pi_S^R)^{-1}(y_0))\,p \subset ((\pi_S^R)^{-1}(y_0)) \circ p$, requiring that $((\pi_S^R)^{-1}(y_0)) \circ p$ is a singleton leads to a condition which is a priori stronger than the notion of a proximal extension but weaker than that of an almost one-one extension.

Definition 17.6 Let $X = M/R$ and $Y = M/S$ with $R \subset S$ icers on M. We say that X is a *highly proximal extension* of Y (*h. p. extension*) if there exists $y_0 \in Y$ and $p \in M$ such that $((\pi_S^R)^{-1}(y_0)) \circ p$ is a singleton. Here X is a *maximally highly proximal flow* (*m.h.p.*) if it has no non-trivial highly proximal extensions. The flow X is a *maximally highly proximal extension* (*m.h.p.* extension) of Y if it is a highly proximal extension which is maximally highly proximal.

We gather a few elementary properties of highly proximal extensions beginning with a restatement of the definition.

Proposition 17.7 Let $X = M/R$ and $Y = M/S$ with $R \subset S$ icers on M. Then the following are equivalent:

(a) $\left((\pi_S^R)^{-1}(y)\right) \circ p$ is a singleton for all $y \in Y$ and $p \in M$,
(b) $\left((\pi_S^R)^{-1}(y)\right) \circ p = \{xp\}$ for all $y \in Y$, $x \in (\pi_S^R)^{-1}(y)$, and $p \in M$, and
(c) X is a highly proximal extension of Y.

PROOF: (a) $\Longrightarrow$ (b)

1. Assume that (a) holds and let $y \in Y$, $x \in (\pi_S^R)^{-1}(y)$, and $p \in M$.
2. $\{xp\} \subset \left((\pi_S^R)^{-1}(y)\right) p \subset \left((\pi_S^R)^{-1}(y)\right) \circ p$. (by **5.10**)
3. $\{xp\} = \left((\pi_S^R)^{-1}(y)\right) \circ p$. (by 1, 2)

(b) $\Longrightarrow$ (c)

Clear.

(c) $\Longrightarrow$ (a)

1. Assume that $\left((\pi_S^R)^{-1}(y_0)\right) \circ q$ is a singleton, and let $p \in M$, $y \in Y$.
2. There exists $r \in M$ with $yr = y_0$. (Y is minimal)
3. There exists $\hat{q} \in M$ with $p = rq\hat{q}$. (M is minimal)
4. $\left((\pi_S^R)^{-1}(y)\right) \circ p \underset{3,\mathbf{5.10}}{=} \left((\pi_S^R)^{-1}(y) \circ r\right) \circ q\hat{q} \underset{\mathbf{5.11}}{\subset} (\pi_S^R)^{-1}(yr) \circ q\hat{q}$

$\underset{2}{=} \left((\pi_S^R)^{-1}(y_0) \circ q\right) \circ \hat{q}$.

5. $(\pi_S^R)^{-1}(y) \circ p$ is a singleton. (by 1, 4)

Proposition 17.8 Let $\pi : X \to Y$ be an extension of minimal flows. Then π is highly proximal if and only if every open subset of X contains a fiber of π.

PROOF: $\Longrightarrow$

1. Assume that π is highly proximal and let U be an open subset of X.
2. Let $x \in U$ and $u \in J$ with $xu = x$.
3. Set $y = \pi(x) \in Y$.
4. $\pi^{-1}(y) \circ u = \{xu\} = \{x\} \subset U$. (by 2, 3, **17.7**)
5. There exists $t \in T$ with $\pi^{-1}(yt) = \left(\pi^{-1}(y)\right) t \subset U$. (by 5, **5.10**)

$\Longleftarrow$

1. Assume that every open subset of X contains a fiber of π.
2. Let $y_0 \in Y$, $x_0 \in \pi^{-1}(y_0)$ and U be an open neighborhood of x_0.
3. Let $V = \{y \in Y \mid \pi^{-1}(y) \subset U\}$.
4. V is a nonempty open subset of Y. (by 1, 2, **5.8**)
5. There exists $t_U \in T$ with $y_0 t_U \in V$. (by 4, Y is minimal)
6. $\pi^{-1}(y_0 t_U) = \pi^{-1}(y_0) t_U \subset U$.
7. There exists $q \in \beta T$ with $\pi^{-1}(y_0) \circ q = \{x_0\}$. (by 2, 6)
8. Let $u \in J$ with $y_0 u = y_0$, and set $p = uq$.
9. $p \in M$. (by 8)
10. $\pi^{-1}(y_0) \circ p \underset{\mathbf{5.10}}{=} \left(\pi^{-1}(y_0) \circ u\right) \circ q \underset{\mathbf{5.11}}{\subset} \pi^{-1}(y_0 u) \circ q = \pi^{-1}(y_0) \circ q \underset{7}{=} \{x_0\}$.
11. π is highly proximal. (by 9, 10)

Corollary 17.9 Let:

(i) $X = M/R$ and $Y = M/S$ with $R \subset S$ icers on M, and
(ii) X be an almost one-one extension of Y.

Then X is a highly proximal extension of Y.

PROOF: This follows immediately from the definitions or from **17.8**.

As we remarked above, the following proposition follows immediately from **13.6**; we provide a different proof here for the sake of completeness.

Proposition 17.10 Let:

(i) $X = M/R$ and $Y = M/S$ with $R \subset S$ icers on M, and
(ii) X be a highly proximal extension of Y.

Then X is a proximal extension of Y.

PROOF: 1. Let $x_1, x_2 \in X$ with $\pi_S^R(x_1) = y = \pi_S^R(x_2)$.
2. Let $p \in M$.
3. $x_2 p \in \left((\pi_S^R)^{-1}(y)\right) \circ p = \{x_1 p\}$. (by 1, 2, (ii), **17.7**)
4. $x_1 p = x_2 p$. (by 3)
5. $(x_1, x_2) \in P(X)$. (by 4)

Corollary 17.11 Let A be a closed subgroup of G. Then $Y = M/\overline{gr(A)}$ is a maximally highly proximal flow.

PROOF: 1. Assume that $X = M/R$ is a highly proximal extension of Y with $R \subset \overline{gr(A)}$.
2. X is a proximal extension of Y. (by 1, **17.10**)
3. $G(R) = G(\overline{gr(A)}) = A$. (by 2, **7.11**)
4. $\overline{gr(A)} \subset R$. (by 3)
5. $R = \overline{gr(A)}$ and $X = Y$. (by 1, 4)
6. Y is maximally highly proximal. (by 1, 5)

In view of **17.10** it is natural to ask for an example of a proximal extension which is not highly proximal. Note that if $\pi : X \to \{pt\}$, then π is a highly proximal extension only if $X = \{pt\}$ (by **17.8** any open subset of X would have to contain X, the only fiber of π). Thus in this case any non-trivial proximal flow X will be a proximal but not a highly proximal extension of the one point flow. Of course not every group T admits non-trivial proximal flows. Indeed T admits such flows if and only if $\overline{gr(G)} \neq M \times M$. Indeed when the group T is abelian, the map $x \to xt$ is an automorphism of M for every $x \in M$, so $\overline{gr(G)} = M \times M$. On the other hand there are plenty of examples where $\overline{gr(G)} \neq M \times M$ (see for example [Glasner, S., *Compressibility properties in topological dynamics*, (1975)])

In view of **17.9** it is natural to ask for an example of a highly proximal extension which is not an almost one-one extension. One such example is provided as follows.

Example 17.12 (two circle) Let:

(i) $Y = S^1 = \{e^{i\alpha} \mid 0 \leq \alpha < 2\pi\}$ be the circle,
(ii) $T = \mathbf{Z}$ the integers act on Y by an irrational rotation θ, and
(iii) $X = \{1\} \times Y \cup \{2\} \times Y$ be provided with a topology for which a neighborhood base at $(1, e^{i\alpha})$ is given by $\{U_\epsilon \mid \epsilon > 0\}$ where

$$U_\epsilon = \{(1, e^{i\beta}) \mid \alpha \leq \beta < \alpha + \epsilon\} \cup \{(2, e^{i\beta}) \mid \alpha < \beta < \alpha + \epsilon\},$$

and a neighborhood base at $(2, e^{i\alpha})$ is given by $\{V_\epsilon \mid \epsilon > 0\}$ where

$$V_\epsilon = \{(2, e^{i\beta}) \mid \alpha - \epsilon < \beta \leq \alpha\} \cup \{(1, e^{i\beta}) \mid \alpha - \epsilon < \beta < \alpha\}.$$

(iv) T act on X by θ on both circles,
(v) $\pi : X \to Y$ be defined by $\pi(j, e^{i\alpha}) = e^{i\alpha}$ for $j \in \{1, 2\}$.

Then:

(a) X is a compact Hausdorff space and the action of T on X is minimal,
(b) X is a highly proximal extension of Y, and
(c) X is <u>not</u> an almost one-one extension of Y.

PROOF: (a) We leave this as an exercise for the reader.
(b) This follows from **17.8** since every open subset of X contains a fiber of π.
(c) This is clear since every fiber of π consists of exactly two points.

Given an icer R on M, an extension $R_0 \subset R$ is a proximal extension if and only if $G(R_0) = G(R)$; we now investigate the collection of highly proximal extensions of R.

Proposition 17.13 Let:

(i) $X = M/R$ and $Y = M/S$ with $R \subset S$ icers on M, and
(ii) $u \in J$.

Then the following are equivalent:

(a) X is a highly proximal extension of Y.
(b) $qS \circ p \subset (qp)R$ for all $p, q \in M$.
(c) $uS \circ u \subset uR$.

PROOF: (a) $\Longrightarrow$ (b)

1. Assume that X is a highly proximal extension of Y and let $p, q \in M$.
2. $\pi_R(qS \circ p) \underset{\mathbf{5.11}}{=} \pi_R(qS) \circ p = (\pi_S^R)^{-1}(\pi_S(q)) \circ p \underset{\mathbf{17.7}}{=} \{\pi_R(q)p\} = \{\pi_R(qp)\}$.

3. $qS \circ p \subset (qp)R$. (by 2)

(b) $\Longrightarrow$ (c)

Setting $q = u = p$ in (b) yields (c).

(c) $\Longrightarrow$ (a)

1. Assume that $uS \circ u \subset uR$.
2. $(\pi_S^R)^{-1}(\pi_S(u)) \circ u = \pi_R(uS) \circ u \underset{\mathbf{5.11}}{=} \pi_R(uS \circ u) \underset{1}{=} \{\pi_R(u)\}$.
4. X is a highly proximal extension of Y. (by 2, **17.7**)

Proposition 17.14 Let:

(i) $R \subset S \subset N$ be icers on M, and
(ii) $X = M/R$, $Y = M/S$ and $Z = M/N$.

Then X is a highly proximal extension of Z if and only if X is a highly proximal extension of Y and Y is a highly proximal extension of Z.

PROOF: $\Longrightarrow$

1. Assume that X is a highly proximal extension of Z and let $u \in J$.
2. $uS \circ u \subset uN \circ u \subset uR \subset uS$. (by 1, 2, (i), **17.13**)
3. X is a h. p. extension of Y and Y is a h. p. extension of Z. (by 2, **17.13**)

$\Longleftarrow$

1. Assume that X is a h. p. extension of Y, Y is a h. p. extension of Z, and let $u \in J$.
2. $uN \circ u = (uN \circ u) \circ u \subset uS \circ u \subset uR$. (by 1, **5.10**, **17.13**)
3. X is a highly proximal extension of Z. (by 2, **17.13**)

Corollary 17.15 Let:

(i) $X = M/R$ and $Y = M/S$ with $R \subset S$ icers on M,
(ii) X be a highly proximal extension of Y, and
(iii) π_S^R be open.

Then $R = S$.

PROOF: 1. $\pi_R(pS) = \pi_R(uS \circ p) \subset \pi_R(pR) = \{\pi_R(p)\}$ for all $p \in M$ and $u \in J$. (by (ii), (iii), **17.1**, **17.13**)
2. $S \subset R$. (by 1)
3. $S = R$. (by 2, (i))

We will now construct for an icer R, the maximal highly proximal extension $R_{hp} \subset R$; this is a highly proximal extension of R which has the property that $R_{hp} \subset R_0$ for any R_0 where $R_0 \subset R$ is a highly proximal extension. The construction of R_{hp} requires a lemma which analyzes the action of βT on the cells $\{pR \mid p \in M\}$. Its proof applies results of section 5, in particular **5.3** and **5.10**, to do calculations in 2^M.

Lemma 17.16 Let:

(i) R be an icer on M,
(ii) $p, q \in M$, and
(iii) $u, v \in J$.

Then:

(a) $pR \circ q \subset (pq)R$.
(b) $p \in uR \circ q$ implies $uR \circ p \subset uR \circ q$.
(c) $p \in uR \circ u$ implies $uR \circ p$ is a subsemigroup of M.
(d) $uR \circ p = vR \circ p$.

PROOF: (a) 1. Let $r \in pR \circ q$.
2. There exist $r_i \in pR$ and $t_i \to q$ with $\lim r_i t_i = r$.
3. $(r, pq) = \lim(r_i t_i, pt_i) \in \overline{R} = R$. (by 2, (i))
4. $r \in (pq)R$. (by 3)
(b) 1. Assume that $p \in uR \circ q$.
2. There exist $p_i \in uR$ and $t_i \to q$ with $\lim p_i t_i = p$.
3. $[uR \circ p] = [uR]p = \lim[uR]p_i t_i = \lim[(uR \circ p_i)t_i]$. (by 2, **5.10**)
4. $(uR \circ p_i)t_i \subset (p_i R)t_i = (uR)t_i$. (by 2, and part (a))
5. $\lim[(uR)t_i] = \lim[uR]t_i = [uR]q = [uR \circ q]$. (by 2, **5.10**)
6. $uR \circ p \subset uR \circ q$. (by 3, 4, 5, **5.3**)
(c) 1. Assume that $p \in uR \circ u$.
2. Let $r_1, r_2 \in uR \circ p$.
3. $uR \circ r_1 r_2 \underset{\mathbf{5.10}}{=} (uR \circ r_1) \circ r_2 \underset{2,(b)}{\subset} (uR \circ p) \circ r_2 \underset{1,(b)}{\subset} (uR \circ u) \circ r_2 \underset{\mathbf{5.10}}{=} uR \circ r_2 \underset{2,(b)}{\subset} uR \circ p$.
4. $r_1 r_2 = u r_1 r_2 \in uR \circ r_1 r_2 \subset uR \circ p$. (by 3)
(d) 1. $uR \circ p \underset{\mathbf{5.10}}{=} (uR \circ v) \circ p \underset{(a)}{\subset} vR \circ p$.
2. $vR \circ p \underset{\mathbf{5.10}}{=} (vR \circ u) \circ p \underset{(a)}{\subset} uR \circ p$.
3. $uR \circ p = vR \circ p$. (by 1, 2)

Definition 17.17 Let R be an icer on M and $u \in J$. We define

$$R_{hp} = \{(p, q) \in M \mid uR \circ p = uR \circ q\}.$$

It follows from **17.16** that R_{hp} is independent of the choice of u. Note also that by **5.10**

$$R_{hp} = \{(p, q) \in M \mid [uR]p = [uR]q\},$$

so R_{hp} is an icer on M and M/R_{hp} is a quasi-factor of M.

The study of highly proximal extensions of M/R is facilitated by the study of R_{hp}. In particular we will use the following lemma to show that $R_{hp} \subset R$ is a maximal highly proximal extension of R in the sense mentioned earlier.

Lemma 17.18 Let:

(i) R be an icer on M, and
(ii) $u \in J$.

Then:

(a) $R_{hp} \subset R$.
(b) there exists $m \in uR \circ u$ such that $uR \circ m \subset mR_{hp}$.
(c) $pR_{hp} = uR \circ p$ for all $p \in M$.

PROOF: (a) 1. Let $(p, q) \in R_{hp}$.
2. $\emptyset \neq uR \circ p = uR \circ q \subset pR \cap qR$. (by 1, **17.16**)
3. $pR = qR$. (by 2, since R is an equivalence relation)
(b) 1. Set $\mathcal{C} = \{uR \circ p \mid p \in uR \circ u\}$.
2. Let $\{uR \circ p_i \mid i \in I\} \subset \mathcal{C}$ be a decreasing chain with respect to inclusion.
3. Set $B = \bigcap_{i \in I}(uR \circ p_i)$.
4. We may assume by passing to a subnet if necessary that $p_i \to p \in \overline{uR \circ u} = uR \circ u$.
5. $[uR \circ p] = [uR]p = \lim[uR]p_i = \lim[uR \circ p_i]$. (by 4, **5.10**)
6. $B \subset uR \circ p$. (by 5, **5.3**)
7. Let $i \in I$.
8. $uR \circ p_j \subset uR \circ p_i$ for all $j > i$.
9. $uR \circ p \subset uR \circ p_i$. (by 5, 8, **5.3**)
10. $uR \circ p \subset B$. (by 7, 9)
11. $B = uR \circ p = \inf\{uR \circ p_i \mid i \in I\} \in \mathcal{C}$. (by 6, 10)
12. $\mathcal{C}$ is inductive when ordered by inclusion.
13. There exists a minimal element $uR \circ m \in \mathcal{C}$. (by 12 and Zorn's Lemma)
14. Let $p \in uR \circ m$.
15. $p = up \in (uR)p \subset uR \circ p \subset uR \circ m \subset uR \circ u$. (by 1, 13, 14, **17.16**)
16. $uR \circ p = uR \circ m$. (by 13, 14, 15)
17. $p \in mR_{hp}$. (by 16)
18. $uR \circ m \subset mR_{hp}$. (by 14, 17)
(c) 1. Let $p \in M$.

2. There exists $\hat{p} \in M$ with $m\hat{p} = p$. (M is minimal)
3. $uR \circ p = (uR \circ m) \circ \hat{p} \subset (mR_{hp}) \circ \hat{p} \subset (m\hat{p})R_{hp} = pR_{hp}$.
(by 2, (b), **17.16**)
4. $pR_{hp} = \{q \in M \mid uR \circ p = uR \circ q\} \subset uR \circ p$. (since $q \in uR \circ q$)
5. $pR_{hp} = uR \circ p$. (by 3, 4)

Proposition 17.19 Let R, N be icers on M with $N \subset R$. Then:

(a) $\pi_{R_{hp}}$ is open.
(b) M/N is a highly proximal extension of M/R if and only if $R_{hp} \subset N$.
(c) $(R_{hp})_{hp} = R_{hp}$.

PROOF: (a) 1. $pR_{hp} \circ q = (uR \circ p) \circ q = uR \circ (pq) = (pq)R_{hp}$.
(by **17.18**)
2. $\pi_{R_{hp}}$ is open. (by 1, **17.1**)
(b) M/N is a h. p. extension of M/R if and only if $pR_{hp} = uR \circ p \subset pN$ for all $p \in M$. (by **17.13**, **17.18**)
(c) $p(R_{hp})_{hp} = (uR_{hp}) \circ p = (uR \circ u) \circ p = uR \circ p = pR_{hp}$ for all $p \in M$.
(by **17.18**)

We now use the preceding results to characterize maximally highly proximal flows; we show in the appendix that $X = M/R$ is maximally highly proximal if and only if X is *extremely disconnected.*

Proposition 17.20 Let R be an icer on M. Then the following are equivalent:

(a) M/R is maximally highly proximal.
(b) $R = R_{hp}$.
(c) π_R is open.

PROOF: (a) $\Longrightarrow$ (b)
1. Assume that M/R is maximally highly proximal.
2. M/R_{hp} is a highly proximal extension of M/R. (by **17.19**(b))
3. $R_{hp} = R$. (by 1, 2)

(b) $\Longrightarrow$ (c)

This follows immediately from **17.19**(a).

(c) $\Longrightarrow$ (a)

1. Assume that π_R is open and let M/S be a h.p. extension of M/R with $S \subset R$.
2. $pR = uR \circ p \subset pS$ for all $u \in J$ and $p \in M$. (by 1, **17.1**, **17.13**)
3. $S = R$. (by 1, 2)
4. M/R is maximally highly proximal. (by 1, 3)

APPENDIX TO SECTION 17: EXTREMELY DISCONNECTED FLOWS

We saw in **17.20** that M/R is maximally highly proximal (m.h.p.) if and only if π_R is open. This leads to an interesting topological characterization of maximally highly proximal flows: a flow (X, T) is m.h.p. if and only if X is extremely disconnected. We begin by recalling what it means for a space to be extremely disconnected.

Definition 17.A.1 The topological space X is *extremely disconnected* if the closure of any open subset of X is open.

Note that it follows from **1.2** that the Stone-Cech compactification βT, of T, is extremely disconnected. The same is true of any minimal ideal in βT.

Proposition 17.A.2 The space M is extremely disconnected.

PROOF: 1. Let $V \subset M$ be open and $\alpha(u) \in \overline{V}$.
2. Set $A = \{t \in T \mid ut \in \alpha^{-1}(V)\}$.
3. $\overline{A} \cap M \subset \overline{\alpha^{-1}(V)}$. (by 2)
4. $u \in \alpha^{-1}(\overline{V}) = \overline{\alpha^{-1}(V)}$. (by 1)
5. Let $W \subset T$ with $u \in \overline{W} \subset \beta T$.
6. $\overline{W}$ is open in βT. (by 5, **1.2**)
7. There exists $q \in \alpha^{-1}(V) \cap \overline{W} \cap L_u^{-1}(\overline{W})$. (by 4, 5, 6)
8. There exists $t_W \in W$ with $ut_W \in \alpha^{-1}(V)$. (by 7)
9. There exists a net $\{t_W\} \subset A$ with $t_W \to u$. (by 5, 8, **1.2**)
10. $u \in \overline{A} \cap M \subset \alpha^{-1}(\overline{V})$. (by 3, 9)
11. $\alpha(u) \in \alpha(\overline{A} \cap M) \subset \overline{V}$. (by 10)
12. $\alpha(u) \in int(\overline{V})$. (by 11, **1.2**)
13. $\overline{V}$ is open. (by 1, 12)

Proposition 17.A.3 Let:

(i) X, Y be compact Hausdorff spaces,
(ii) X be extremely disconnected, and
(iii) $\pi : X \to Y$ be continuous, open, and onto.

Then Y is extremely disconnected.

PROOF: 1. Let $V \subset Y$ be open.
2. $\overline{V} = \overline{\pi(\pi^{-1}(V))} \subset \pi(\overline{\pi^{-1}(V)}) \subset \pi(\pi^{-1}(\overline{V})) = \overline{V}$. (by (iii))
3. $\overline{V} = \pi(\overline{\pi^{-1}(V)})$. (by 2)
4. $\overline{\pi^{-1}(V)}$ is open. (by 1, (ii), (iii))
5. $\overline{V}$ is open. (by 3, 4, (iii))

Proposition 17.A.4 Let:

(i) $X = M/R$, and $Y = M/S$ be minimal flows with $R \subset S$, and
(ii) X be extremely disconnected.

Then Y is extremely disconnected if and only if π_S^R is open.

PROOF:

This follows immediately from **17.A.3**, since π_S^R is continuous and onto.

$\Longrightarrow$

1. Assume that Y is extremely disconnected.
2. Let $U \subset X$ be open and $y \in \overline{\pi_S^R(U)}$.
3. Let $W \subset Y$ be open with $y \in W$.
4. $\emptyset \neq U \cap \left(\pi_S^R\right)^{-1}(W)$ is open in X. (by 2, 3, π_S^R is continuous)
5. There exists $V \subset X$ open with $V = \overline{V} \subset U \cap \left(\pi_S^R\right)^{-1}(W)$. (by 4, (ii))
6. There exists a finite set $F \subset T$ such that $VF = X$. (5, $X = M/R$ is minimal)
7. $\pi_S^R(\overline{V})F = \pi_S^R(V)F = Y$. (by 6)
8. $\emptyset \underset{7}{\neq} int\left(\pi_S^R(V)\right) \underset{5}{\subset} int\left(\pi_S^R(U)\right) \cap W$.
9. $y \in \overline{int\left(\pi_S^R(U)\right)}$. (by 3, 8)
10. $\overline{int\left(\pi_S^R(U)\right)} = \overline{\pi_S^R(U)}$. (by 2, 9)
11. For every $x \in U$, let $U_x \subset X$ be open with $x \in U_x \subset \overline{U}_x \subset U$.
12. $U = \bigcup_{x \in U} \overline{U}_x$. (by 11)
13. $\pi_S^R(U) \underset{12}{=} \bigcup_{x \in X} \pi_S^R(\overline{U}_x) = \bigcup_{x \in X} \overline{\pi_S^R(U_x)} \underset{10}{=} \bigcup_{x \in X} \overline{int(\pi_S^R(U_x))}$.
14. $\pi_S^R(U)$ is open. (by 1, 12)

Corollary 17.A.5 Let R be an icer on M. Then the following are equivalent:

(a) M/R is extremely disconnected,
(b) π_R is open, and
(c) M/R is maximally highly proximal.

PROOF: The equivalence of (a) and (b) follows from **17.A.4** since M is extremely disconnected by **17.A.2**. We showed that (b) and (c) are equivalent in **17.20**.

We end this appendix with two lemmas concerning extremely disconnected spaces in general.

Lemma 17.A.6 Let:

(i) X and Y be compact Hausdorff spaces,
(ii) $\pi : X \to Y$ be continuous and onto,

(iii) Y be extremely disconnected, and
(iv) $\pi(A) \neq Y$ for any closed proper subset of X.

Then:

(a) $\pi(U) \subset \overline{Y \setminus \pi(X \setminus U)}$ for all open sets $U \subset X$.
(b) π is one-one.

PROOF: (a) 1. Let $U \subset X$ be open.
2. Let $x \in U$ and $V \subset Y$ be open with $\pi(x) \in V$.
3. Assume that $V \cap \big(Y \setminus \pi(X \setminus U)\big) = \emptyset$.
4. $V \subset \pi(X \setminus U)$. (by 3)
5. There exists an open set $W \subset X$ with $x \in W \subset U \cap \pi^{-1}(V)$.
6. $\pi(W) \subset V \subset \pi(X \setminus U) \subset \pi(X \setminus W)$. (by 4, 5)
7. $\pi(X \setminus W) = \pi(X) = Y$. (by 6)
8. $V \cap \big(Y \setminus \pi(X \setminus U)\big) \neq \emptyset$. (7 contradicts (iv))
9. $\pi(U) \subset \overline{Y \setminus \pi(X \setminus U)}$. (by 2, 8)
(b) 1. Let $x_1 \neq x_2 \in X$.
2. There exists open sets U_1, and U_2 with $x_i \in U_i$ and $U_1 \cap U_2 = \emptyset$.
3. $X \setminus U_1 \cup X \setminus U_2 = X$. (by 2)
4. $\pi(X \setminus U_1) \cup \pi(X \setminus U_2) = Y$. (by 3, (ii))
5. $Y \setminus \pi(X \setminus U_1) \subset \pi(X \setminus U_2)$. (by 4)
6. $\overline{Y \setminus \pi(X \setminus U_1)} \subset \pi(X \setminus U_2)$. (by 5)
7. $Y \setminus \pi(X \setminus U_2) \subset Y \setminus \overline{Y \setminus \pi(X \setminus U_1)}$. (by 6)
8. $\overline{Y \setminus \pi(X \setminus U_1)}$ is open. (by (i), (ii), (iii))
9. $\overline{Y \setminus \pi(X \setminus U_2)} \subset Y \setminus \overline{Y \setminus \pi(X \setminus U_1)}$. (by 7, 8)
10. $\pi(U_1) \cap \pi(U_2) \underset{(a)}{\subset} \overline{Y \setminus \pi(X \setminus U_1)} \cap \overline{Y \setminus \pi(X \setminus U_2)} \underset{9}{=} \emptyset$.
11. $\pi(x_1) \neq \pi(x_2)$. (by 2, 10)
12. π is one-one. (by 1, 11)

The following result, (see [Gleason, A., *Projective topological spaces*, (1958)]) is important for the study of extremely disconnected spaces.

Proposition 17.A.7 Let:

(i) X and Y be compact Hausdorff spaces,
(ii) $\pi : X \to Y$ be continuous and onto, and
(iii) Y be extremely disconnected.

Then there exists a continuous function $g : Y \to X$ such that $\pi(g(y)) = y$ for all $y \in Y$.

PROOF: 1. Let $\mathcal{C} = \{C \mid C = \overline{C} \subset X \text{ and } \pi(C) = Y\}$.

2. Applying Zorn's lemma to $\mathcal{C}$ yields a closed subset $X_0 \subset X$ such that $\pi(X_0) = Y$ and $\pi(A) \neq Y$ for any proper closed subset of X_0.
3. The restriction π_0 of π to X_0 is one-one. (by 2, **17.A.6**)
4. $g = \pi_0^{-1} : Y \to X$ is continuous. (by 3, (i), (ii))
5. $\pi(g(y)) = y$ for all $y \in Y$. (by 3, 4)

NOTES ON SECTION 17

Note 17.N.1 Let X be compact Hausdorff space, and $\mathcal{A}$ be a uniformly closed subalgebra of the algebra $\mathcal{C}(X)$, of continuous functions $f : X \to \mathbf{R}$. We say that $\mathcal{A}$ is *complete* if given any $\mathcal{F} \subset \mathcal{A}$ with $\|f\| \leq 1$ for all $f \in \mathcal{F}$, there exists $g \in \mathcal{A}$ such that:

(i) $f \leq g$ for all $f \in \mathcal{F}$.
(ii) if $f \leq h \in \mathcal{A}$ for all $f \in \mathcal{F}$, the $g \leq h$.

Clearly such a g, if it exists, is unique. In this case we write $g = \bigvee \mathcal{A}$ or $g = \bigvee_{f \in \mathcal{A}} f$.

Note 17.N.2 Let X be a set. Then $\mathcal{C}(\beta(X))$ is complete.

PROOF: 1. Let $\mathcal{F} \subset \mathcal{C}(\beta(X))$ with $\|f\| \leq 1$ for all $f \in \mathcal{F}$.
2. Let $h(x) = \sup_{f \in \mathcal{F}} f(x)$ for all $x \in X \subset \beta X$.
3. There exists a continuous function $\hat{h} : \beta X \to [-1, 1]$ such that $\hat{h}(x) = h(x)$ for all $x \in X$.
4. $f \leq \hat{h}$ for all $f \in \mathcal{F}$. (by 2, 3)
5. Let $g \in \mathcal{C}(\beta X)$ with $f \leq g$ for all $f \in \mathcal{F}$.
6. $\hat{h}(x) \leq g(x)$ for all $x \in X$. (by 2, 5)
7. $\hat{h} \leq g$. (by 6 since $\overline{X} = \beta X$)

Note 17.N.3 Let X be a compact Hausdorff space. Then X is extremely disconnected if and only if $\mathcal{C}(X)$ is complete.

PROOF: $\Longrightarrow$
1. Let X be extremely disconnected and $\mathcal{F} \subset \mathcal{C}(X)$ with $\|f\| \leq 1$ for all $f \in \mathcal{F}$.
2. Let $\mathcal{T}$ be the topology on X and $\mathcal{D}$ be the discrete topology on X.
3. The map $(X, \mathcal{D}) \to (X, \mathcal{T})$ has a continuous extension $\pi : \beta X \to (X, \mathcal{T})$.
$x \to x$
4. There exists a continuous map $\sigma : (X, \mathcal{T}) \to \beta X$ with $\pi \circ \sigma = 1_X$. (by **17.A.7**)

5. Let $\mathcal{G} = \{f \circ \pi \mid f \in \mathcal{F}\} \subset \mathcal{C}(\beta X)$.
6. $g = \bigvee \mathcal{G} \in \mathcal{G}$ exists. (by **17.N.2**)
7. $g \circ \sigma = \bigvee \mathcal{F}$. (exercise)

$\Longleftarrow$

1. Assume that $\mathcal{C}(X)$ is complete.
2. Let $\emptyset \neq U \subset X$.
3. Let $\mathcal{F} = \{f \in \mathcal{C}(X) \mid f(X) \subset [0, 1]$ and $f(x) = 0$ for all $x \notin U\}$.
4. Let $g = \bigvee \mathcal{F}$.
5. $g : X \to [0, 1]$ is continuous and $g(U) \subset (0, 1]$.
6. $\overline{U} = g^{-1}((0, 1])$. (exercise)
7. $\overline{U}$ is open. (by 6)

EXERCISES FOR CHAPTER 17

Exercise 17.1 Show that if the group T is abelian, then every minimal flow is incontractible.

Exercise 17.2 Let $R \subset S$ be icers on M. Show that the following are equivalent:

(a) $R \subset S$ is a RIC extension, and
(b) $R \subset S$ is an open extension and $\pi_R(G(S)(u) \circ u) = \pi_R(uS)$ for some $u \in J$.

Exercise 17.3 In this exercise we describe a general method for constructing almost one-one extensions. Let:

(i) $Y = M/S$ with S an icer on M,
(ii) $y_0 \in Y$, and
(iii) $u \in J$ and $\alpha \in G$ with $\pi_S(\alpha(u)) = y_0$,
(iv) $f : y_0 T \to K$ be a continuous function where K is a compact Hausdorff space, (e.g. $K = [0, 1]$)
(v) $g : \beta T \to K$ be the continuous extension to βT of the map $\begin{array}{l} T \to K \\ t \to f(y_0 t), \end{array}$
(vi) $N = \{(p, q) \in M \times M \mid g(pt) = g(qt)$ for all $t \in T\}$, and
(vii) $R = \alpha(N) \cap S$.

Then:

(a) N is an icer on M.
(b) R is an icer on M.
(c) $X = M/R$ is an almost one-one extension of Y such that $(\pi_S^R)^{-1}(y_0)$ consists of a single point.

Note that the extension X depends on the function f. When f is constant, $N = M \times M$ and $X = Y$. The next exercise provides a non-trivial example.

Exercise 17.4 Let:

(i) X be the unit circle,
(ii) $\varphi : X \to X$ be an "irrational rotation",
(iii) T the group generated by φ,
(iv) I an open arc of X with $\emptyset \neq I \neq X$,
(v) $x_0 \in I$ be such that $x_0 T \cap \{a, b\} = \emptyset$ where $\{a, b\}$ is the set of endpoints of I,
(vi) $f(x_0 t) = \begin{cases} 1 & \text{if } x_0 t \in I \\ 0 & \text{if } x_0 t \notin I, \end{cases}$ and
(vii) $\pi : Z \to X$ the almost one-one extension induced by f as in **17.3**.

Then $\pi^{-1}(x_0)$ is a singleton and both $\pi^{-1}(a)$ and $\pi^{-1}(b)$ have exactly two points.

Exercise 17.5 (two circle): (see **17.12**) Let:

(i) $Y = S^1 = \{e^{i\alpha} \mid 0 \leq \alpha < 2\pi\}$ be the circle,
(ii) $T = \mathbf{Z}$ the integers act on Y by an irrational rotation θ,
(iii) $X = \{1\} \times Y \cup \{2\} \times Y$ be provided with a topology for which a neighborhood base at $(1, e^{i\alpha})$ is given by $\{U_\epsilon \mid \epsilon > 0\}$ where

$$U_\epsilon = \{(1, e^{i\beta}) \mid \alpha \leq \beta < \alpha + \epsilon\} \cup \{(2, e^{i\beta}) \mid \alpha < \beta < \alpha + \epsilon\},$$

and a neighborhood base at $(2, e^{i\alpha})$ is given by $\{V_\epsilon \mid \epsilon > 0\}$ where

$$V_\epsilon = \{(2, e^{i\beta}) \mid \alpha - \epsilon < \beta \leq \alpha\} \cup \{(1, e^{i\beta}) \mid \alpha - \epsilon < \beta < \alpha\},$$

(iv) T act on X by θ on both circles, and
(v) $\pi : X \to Y$ be defined by $\pi(j, e^{i\alpha}) = e^{i\alpha}$ for $j \in \{1, 2\}$.

Then:

(a) X is a compact Hausdorff space and the action of T on X is minimal,
(b) $\pi : X \to Y$ is a homomorphism of minimal flows.

Exercise 17.6 Let $R \subset S$ be a proximal extension and assume that $(\pi_S^R)^{-1}(y)$ is finite for some $y \in M/S$. Then $R \subset S$ is a highly proximal extension.

Exercise 17.7 Let R be an icer on M. Then M/R is maximally highly proximal if and only if $[pR]$ is an almost periodic point of $(2^M, T)$ for all $p \in M$.

Exercise 17.8 Let:

(i) $R \subset S$ be icers on M,

(ii) M/R be a highly proximal extension of M/S, and
(iii) $\alpha \in G$.

Then:

(a) $M/\alpha(R)$ is a highly proximal extension of $M/\alpha(S)$.
(b) $(\alpha(S))_{hp} = \alpha(S_{hp})$.

Exercise 17.9 Let $R \subset S$ icers on M. Then π_S^R is open if and only if $S = S_{hp} \circ R$.

18

Distal extensions of minimal flows

In section 4 we defined the notion of a distal homomorphism of flows (see **4.14**). In this section we will focus on distal homomorphisms of minimal flows (distal extensions), studying them from the point of view of icers on M. Note that a flow is distal if and only if it is a distal extension of a point; thus the results in section 14 on minimal distal flows can be thought of as results on distal extensions. The results in this section are to a large extent motivated by, and analogous to those of section 14. We begin by restating **4.14** in the context of icers on M.

Definition 18.1 Let $X = M/R$, and $Y = M/S$ be flows with $R \subset S$ icers. Recall that

$$\pi_S^R : X \to Y$$

denotes the canonical homomorphism. We say that X is a *distal extension of* Y if π_S^R is a distal homomorphism in the sense of **4.14**, that is:

$$(x_1, x_2) \in P(X) \text{ with } \pi_S^R(x_1) = \pi_S^R(x_2) \text{ implies that } x_1 = x_2.$$

Note that since $Y = M/S \cong X/\pi_R(S)$, this is equivalent to saying that

$$P(X) \cap \pi_R(S) = \Delta_X.$$

When X is a distal extension of Y, we also write π_S^R is distal, R is a distal extension of S, or simply $R \subset S$ is distal.

The following proposition follows immediately from the definition.

Proposition 18.2 Let $R \subset S$ be icers on M. Then the following are equivalent:

(a) R is a distal extension of S,
(b) $P(M) \cap S \subset R$, and
(c) $P(M) \cap S = P(M) \cap R$.

PROOF: (a) $\Longrightarrow$ (b)

1. Let $R \subset S$ be distal and $(p, q) \in P(M) \cap S$.
2. $(\pi_R(p), \pi_R(q)) \in P(M/R)$. (by 1, **4.7**)
3. $\pi_S^R(\pi_R(p)) = \pi_S(p) \underset{1}{=} \pi_S(q) = \pi_S^R(\pi_R(q))$.
4. $\pi_R(p) = \pi_R(q)$. (by 1, 2, 3)
5. $(p, q) \in R$. (by 4)
6. $P(M) \cap S \subset R$. (by 1, 5)

(b) $\Longrightarrow$ (c)

This is immediate since $R \subset S$.

(c) $\Longrightarrow$ (a)

1. Assume that $P(M) \cap S = P(M) \cap R$ and let $(x_1, x_2) \in P(M/R)$ with $\pi_S^R(x_1) = \pi_S^R(x_2)$.
2. There exist $p_1, p_2 \in M$ with $(p_1, p_2) \in P(M), \pi_R(p_1) = x_1$, and $\pi_R(p_2) = x_2$. (by 1, **4.7**)
3. $\pi_S(p_1) = \pi_S^R(\pi_R(p_1)) \underset{2}{=} \pi_S^R(x_1) \underset{1}{=} \pi_S^R(x_2) \underset{2}{=} \pi_S(p_2)$.
4. $(p_1, p_2) \in P(M) \cap S = P(M) \cap R \subset R$. (by 1, 2, 3)
5. $x_1 \underset{2}{=} \pi_R(p_1) \underset{4}{=} \pi_R(p_2) \underset{2}{=} x_2$.
6. $R \subset S$ is distal. (by 1, 5)

Corollary 18.3 Let:

(i) R, S, N be icers on M, and
(ii) $R \subset S$ be distal.

Then $R \cap N \subset S \cap N$ is distal.

PROOF: 1. $P(M) \cap S \subset R$. (by **18.2**)
2. $P(M) \cap S \cap N \subset R \cap N$.
3. $R \cap N$ is a distal extension of $S \cap N$. (by 2, **18.2**)

One of the keys to our study of distal extensions is the equivalence relation

$$P_0 = \{(\alpha(u), \alpha(v) \mid \alpha \in G \text{ and } u, v \in J\} \subset P(M) \subset M \times M$$

which was defined in **7.20**. We have seen that $M \times M = P_0 \circ gr(G)$, and more generally (in **7.21**) that any icer R on M is of the form

$$R = (R \cap P_0) \circ gr(G(R)).$$

Moreover $(M/R, T)$ is a distal flow (a distal extension of the one-point flow) if and only if $P_0 \subset R$. Our first goal is to prove the analogous result for distal extensions. We showed in **7.23** that if $R \subset S$ is distal, then $R \cap P_0 = S \cap P_0$ and the map π_S^R is open. Now we show that $R \subset S$ is a distal extension if and only if $R \cap P_0 = S \cap P_0$. We deduce this from the fact that for any icer R on M, $R \cap P(M)$ is completely determined by $R \cap P_0$.

Lemma 18.4 Let R, S be icers on M, and $J \subset M$ be the set of idempotents in M. Then the following are equivalent:

(a) $R \cap P(M) \subset S \cap P(M)$,
(b) $\alpha(R) \cap (J \times J) \subset \alpha(S) \cap (J \times J)$ for all $\alpha \in G$, and
(c) $R \cap P_0 \subset S \cap P_0$.

PROOF: (a) $\Longrightarrow$ (b)
1. Assume $R \cap P(M) \subset S \cap P(M)$ and let $(u, v) \in \alpha(R) \cap (J \times J)$.
2. $(\alpha^{-1}(u), \alpha^{-1}(v)) \in R \cap P(M) \subset S \cap P(M)$.
3. $(u, v) \in \alpha(S) \cap (J \times J)$.

(b) $\Longrightarrow$ (c)
1. Assume that $\alpha(R) \cap (J \times J) \subset \alpha(S) \cap (J \times J)$ for all $\alpha \in G$.
2. $R \cap \alpha(J \times J) \subset S \cap \alpha(J \times J)$ for all $\alpha \in G$. (by 1)
3. $R \cap P_0 = \bigcup_{\alpha \in G} R \cap \alpha(J \times J) \subset \bigcup_{\alpha \in G} S \cap \alpha(J \times J) = S \cap P_0$. (by 2)

(c) $\Longrightarrow$ (a)
1. Assume that $R \cap P_0 \subset S \cap P_0$, and let $(p, q) \in R \cap P(M)$.
2. There exists a minimal idempotent $w \in \beta T$ such that $(p, q) = (p, pw)$. (by **4.6**)
3. There exists an idempotent $v \in M$ with $vw = v$ and $wv = w$. (by **3.14**)
4. $(pv, pw) \underset{3}{=} (pv, pwv) = (p, pw)v \underset{2}{=} (p, q)v \in Rv \subset R$.
5. $(p, pv) \in R$. (by 1, 2, 4)
6. There exists $\alpha \in G$ and $u \in J$ with $\alpha(u) = p$.
7. $(p, pv) = (\alpha(u), \alpha(v)) \in R \cap P_0 \subset S \cap P_0$. (by 1, 5, 6)
8. $(pv, pw) \underset{3}{=} (pvw, pw) = (pv, p)w \underset{7}{\in} Sw \subset S$.
9. $(p, q) = (p, pw) \in S$. (by 7, 8)
10. $R \cap P(M) \subset S \cap P(M)$. (by 1, 9)

Theorem 18.5 Let $R \subset S$ be icers on M. Then the following are equivalent:

(a) R is a distal extension of S,
(b) $R \cap \alpha(J \times J) = S \cap \alpha(J \times J)$ for all $\alpha \in G$,
(c) $\alpha(R) \cap (J \times J) = \alpha(S) \cap (J \times J)$ for all $\alpha \in G$, and
(d) $R \cap P_0 = S \cap P_0$.

PROOF: This follows immediately from **18.2** and **18.4**.

As immediate consequences we have the following.

Corollary 18.6 Let $R \subset S$ be regular icers on M. Then R is a distal extension of S if and only if $R \cap (J \times J) = S \cap (J \times J)$.

Corollary 18.7 Let R, S be icers on M with $R \subset S$ distal. Then:

(a) $\alpha(R) \subset \alpha(S)$ is distal for all $\alpha \in G$, and
(b) $reg(R) \subset reg(S)$ is distal.

PROOF: (a) 1. Let $\alpha \in G$.
2. $R \cap P_0 = S \cap P_0$. (by **18.5**)
3. $\alpha(R) \cap P_0 = \alpha(R) \cap \alpha(P_0) = \alpha(R \cap P_0) = \alpha(S \cap P_0) = \alpha(S) \cap P_0$. (by 2)
4. $\alpha(R)$ is a distal extension of $\alpha(S)$. (by 3, **18.5**)

(b) 1. $$reg(R) \cap P_0 = \left(\bigcap_{\alpha \in G} \alpha(R)\right) \cap P_0 = \bigcap_{\alpha \in G} (\alpha(R) \cap P_0) \underset{(a)}{=} \bigcap_{\alpha \in G} (\alpha(S) \cap P_0) = reg(S) \cap P_0.$$

2. $reg(R)$ is a distal extension of $reg(S)$. (by 1, **18.5**)

Corollary 18.8 Let:

(i) R, N, S be icers on M, and
(ii) $R \subset N \subset S$.

Then $R \subset S$ is distal if and only if $R \subset N$ and $N \subset S$ are both distal.

PROOF: $\Longrightarrow$

1. Assume that $R \subset S$ is distal.
2. $P_0 \cap R = P_0 \cap S$. (by **18.5**)
3. $P_0 \cap R = P_0 \cap N = P_0 \cap S$. (by 1, 2, (ii))
4. $R \subset N$ and $N \subset S$ are both distal. (by 3, **18.5**)

$\Longleftarrow$

1. Assume that $R \subset N$ and $N \subset S$ are both distal.
2. $R \cap P_0 = N \cap P_0 = S \cap P_0$. (by 1, **18.5**)
3. $R \subset S$ is distal. (by 2, **18.5**)

One key theme, which is embodied in **18.5** is that distal extensions are completely determined by their groups. We emphasize this by combining **18.5** with results from sections 4 and 7 to characterize distal extensions.

Theorem 18.9 Let $R \subset S$ be icers on M. Then the following are equivalent:

(a) $R \subset S$ is a distal extension,
(b) $\pi_R(S \cap P_0) = \Delta_{M/R}$,
(c) $S = R \circ gr(G(S))$,
(d) $\pi_R(S) = \pi_R(gr(G(S)))$,
(e) $(\pi_R(S), T)$ is a pointwise almost periodic icer.

PROOF: (a) $\Longleftrightarrow$ (b)

$$R \subset S \text{ is distal} \underset{\mathbf{18.5}}{\Longleftrightarrow} R \cap P_0 = S \cap P_0 \Longleftrightarrow \pi_R(S \cap P_0) = \Delta.$$

(a) $\Longrightarrow$ (c)

1. Assume that $R \subset S$ is distal so that $R \cap P_0 = S \cap P_0$.
2. $S \underset{\mathbf{7.21}}{=} (S \cap P_0) \circ gr(G(S)) \underset{1}{=} (R \cap P_0) \circ gr(G(S))$
   $= (R \cap P_0) \circ gr(G(R)) \circ gr(G(S)) \underset{\mathbf{7.21}}{=} R \circ gr(G(S))$.

(c) $\Longrightarrow$ (d)

This is immediate.

(d) $\Longrightarrow$ (e)

1. Assume that $\pi_R(S) = \pi_R(gr(G(S)))$.
2. $\pi_R(S)$ is an icer on $X = M/R$ with $X/\pi_R(S) = M/S$. (by **6.10**)
3. $(gr(G(S)), T)$ is pointwise almost periodic.
4. $(\pi_R(S), T)$ is pointwise almost periodic. (by 1, 3)

(e) $\Longrightarrow$ (a)

This follows immediately from **4.15**.

Proposition 18.10 Let:

(i) R, N, S be icers on M,
(ii) $S, N \subset R$,
(iii) N be a distal extension of R, and
(iv) $G(S) \subset G(N)$.

Then $S \subset N$.

PROOF: $S = (S \cap P_0) \circ gr(G(S)) \underset{\text{(ii)}}{\subset} (R \cap P_0) \circ gr(G(S))$
$\underset{\text{(iii)}}{=} (N \cap P_0) \circ gr(G(S)) \underset{\text{(iv)}}{\subset} (N \cap P_0) \circ gr(G(N)) = N.$

Corollary 18.11 Let $N \subset R$, and $S \subset R$ be distal extensions of R with $G(S) = G(N)$. Then $S = N$.

PROOF: This follows immediately from **18.10**.

We now consider, for any icer R on M, the collection $\mathcal{R} = \{S \mid S \subset R$ is distal$\}$ of all distal extensions of R. When $R = M \times M$ so that M/R is the one-point flow, the collection $\mathcal{R}$ is simply the collection of all minimal distal flows. In this case as we saw in **14.15**, the collection contains a minimal element S_d, the distal structure relation on M. The following proposition, which is an immediate consequence of **18.5**, allows us to define an analog in the general case.

Proposition 18.12 Let $\mathcal{N}$ be a collection of distal extensions of R. Then $S = \bigcap \mathcal{N}$ is a distal extension of R.

PROOF: 1. $S \cap P_0 = \bigcap_{N \in \mathcal{N}} (N \cap P_0) \underset{\mathbf{18.5}}{=} \bigcap_{N \in \mathcal{N}} (R \cap P_0) = R \cap P_0$.
2. S is a distal extension of R. (by 1, **18.5**)

Corollary 18.13 Let R, S be icers on M with $R \subset S$ distal.

Then $\left(\bigcap_{\alpha \in aut(S)} \alpha(R) \right) \subset S$ is distal.

PROOF: 1. Let $\alpha \in aut(S)$.
2. $\alpha(R) \subset \alpha(S) = S$ is distal. (by **18.7**)
3. $\left(\bigcap_{\alpha \in aut(S)} \alpha(R) \right) \subset S$ is distal. (by 1, **18.12**)

Definition 18.14 Let R be an icer on M, and $\mathcal{R}$ the collection of distal extensions of R. Then we define $R^* = \bigcap \mathcal{R}$. Note that by **18.12**, R^* is a distal extension of R. When $R = M \times M$, $R^* = S_d$ the distal structure relation (see **14.14**). For this reason we also refer to R^* as the *R-distal structure relation on M* and sometimes write $S_d(R)$ for R^*.

Clearly the icer $R^* = S_d(R)$ is defined so that the collection $\{S \subset R \mid R^* \subset S\}$ of extensions of R which contain R^* is exactly the collection of distal extensions of R.

Proposition 18.15 Let N, R be icers on M with $N \subset R$. Then N is a distal extension of R if and only if $R^* \subset N$.

PROOF: $\Longrightarrow$

1. Assume that N is a distal extension of R.
2. $N \in \mathcal{R}$ (the collection of distal extensions of R).
3. $R^* = \bigcap \mathcal{R} \subset N$. (by 2)

$\Longleftarrow$

1. Assume that $R^* \subset N$.
2. $R^* \subset R$ is distal. (by **18.12**)
3. $R^* \subset N$, and $N \subset R$ are both distal. (by **18.8**)

Proposition 18.16 Let $R \subset S$ be icers on M. Then:

(a) $R^* \subset S^*$,

(b) $(\alpha(R))^* = \alpha(R^*)$ for all $\alpha \in G$ (in particular $aut(R) \subset aut(R^*)$), and
(c) $R^{**} = R^*$.

PROOF: (a) 1. $S^* \subset S$ is distal. (by **18.12**)
2. $S^* \cap R \subset S \cap R = R$ is distal. (by 1, **18.3**)
3. $R^* \subset S^* \cap R \subset S^*$. (by **18.15**)
(b) 1. $\alpha(R^*) \subset \alpha(R)$ is distal. (by **18.7** and **18.12**)
2. $(\alpha(R))^* \subset \alpha(R^*)$. (by 1, **18.15**)
3. $(\alpha^{-1}(R))^* \subset \alpha^{-1}(R^*)$. (by 2 applied to α^{-1})
4. $R^* = (\alpha^{-1}(\alpha(R)))^* \subset \alpha^{-1}((\alpha(R))^*)$. (by 3 applied to $\alpha(R)$)
5. $\alpha(R^*) \subset (\alpha(R))^*$. (by 4)
6. $\alpha(R^*) = (\alpha(R))^*$. (by 2, 5)
(c) 1. $R^{**} \subset R^*$ is distal. (by **18.12**)
2. $R^* \subset R$ is distal. (by **18.12**)
3. $R^{**} \subset R$ is distal. (by 1, 2, and **18.8**)
4. $R^* \subset R^{**}$. (by 3, **18.15**)
5. $R^{**} = R^*$. (by 1, 4)

The distal structure relation on M is of the form $S_d = P_0 \circ gr(D)$ where D is the subgroup of G defined in **14.6**. In fact by **14.11**, the minimal distal flows are all of the form $P_0 \circ gr(A)$ where A is a closed subgroup of G which contains D. We now define a subgroup $D_R \subset G$ so that the R-distal structure relation $R^* = S_d(R) = (R \cap P_0) \circ gr(D_R)$; then we prove an analog of **14.11** for distal extensions of R.

Definition 18.17 Let R be an icer on M. We define D_R to be the closed subgroup of G generated by $\{\alpha \in G \mid gr(\alpha) \subset \overline{P(M) \cap R}\}$.

Note that the generators of D_R are contained in $G(R)$ so that $D_R \subset G(R)$. The fact that $\gamma\left(\overline{P(M) \cap R}\right) = \overline{P(M) \cap R}$ for all $\gamma \in aut(R)$ implies that D_R is a normal subgroup of $aut(R)$ (and hence of $G(R)$).

Lemma 18.18 Let R be an icer on M. Then:
(a) $\overline{P(M) \cap R} \subset (R \cap P_0) \circ gr(D_R)$, and
(b) if N is an icer on M with $R \cap P_0 = N \cap P_0$, then $D_R \subset G(N) \subset aut(R \cap P_0)$.

PROOF: (a) 1. Let $(\alpha(u), \beta(v)) \in \overline{P(M) \cap R}$.
2. $(\alpha(u), \beta\alpha^{-1}(\alpha(u))) = (\alpha(u), \beta(u)) \in \left(\overline{P(M) \cap R}\right) u \subset \overline{P(M) \cap R}$.
(by 1, $P(M) \cap R$ is invariant)
3. $\beta\alpha^{-1} \in D_R$. (by 2)
4. $(\alpha(u), \alpha(v)) \in P_0 \cap R$. (by 1)

5. $(\alpha(u), \beta(v)) \in (P_0 \cap R) \circ gr(D_R)$. (by 3, 4)

(b) 1. Assume that N is an icer on M with $R \cap P_0 = N \cap P_0$.

2. Let $\alpha \in G$ with $gr(\alpha) \subset \overline{P(M) \cap R}$.

3. $gr(\alpha) \subset \overline{P(M) \cap R} = \overline{P(M) \cap N} \subset N$. (by 2, **18.4**)

4. $\alpha \in G(N)$. (by 2)

5. Let $\beta \in G(N)$.

6. $\beta(R \cap P_0) = \beta(N \cap P_0) = \beta(N) \cap \beta(P_0) = N \cap P_0 = R \cap P_0$. (by 1, 5)

7. $G(N) \subset aut(R \cap P_0)$. (by 5, 6)

Let R be an icer on M and suppose that N is an icer on M with $N \cap P_0 = R \cap P_0$. Then by **7.21**, we must have

$$N = (R \cap P_0) \circ gr(H)$$

for some closed subgroup $H \subset G$. Lemma **18.18** shows that $D_R \subset H \subset aut(R \cap P_0)$. Note also that $\overline{gr(H)} \cap P_0 \subset N \cap P_0 = R \cap P_0$. The following proposition shows that these necessary conditions are also sufficient and hence allows us to identify all of those icers N for which $N \cap P_0 = R \cap P_0$.

Proposition 18.19 Let:

(i) R be an icer in M,

(ii) $\mathcal{N} = \{N \mid N \text{ is an icer with } R \cap P_0 = N \cap P_0\}$, and

(iii) $\mathcal{H} = \{H \mid H = \overline{H} \text{ is a subgroup of } G, D_R \subset H \subset aut(R \cap P_0), P_0 \cap \overline{gr(H)} \subset R\}$.

Then

$$\varphi : \mathcal{N} \to \mathcal{H}$$
$$N \to G(N)$$

is bijective, its inverse being the map ψ defined by $\psi(H) = (R \cap P_0) \circ gr(H)$ for all $H \in \mathcal{H}$. In particular $R^* = (R \cap P_0) \circ gr(D_R)$.

PROOF: 1. Let $H \in \mathcal{H}$.

2. $\psi(H)$ is an equivalence relation on M since $H \subset aut(R \cap P_0)$. (by **7.24**)

3. $\overline{\psi(H)T} \underset{\text{(ii)}}{=} \overline{((P_0 \cap R) \circ gr(H))T} \subset \overline{(P_0 \cap R)T \circ gr(H)}$

$\subset \overline{P(M) \cap R} \circ \overline{gr(H)} \underset{\mathbf{18.18, 7.21}}{\subset} (P_0 \cap R) \circ gr(D_R) \circ \left(P_0 \cap \overline{gr(H)}\right) \circ$

$gr(H) \underset{\text{(iii)}}{\subset} \psi(H) \circ (P_0 \cap R) \circ gr(H) = \psi(H) \circ \psi(H) \underset{2}{\subset} \psi(H)$.

4. $\psi(H)$ is an icer. (by 2, 3)

5. $\psi(H) \in \mathcal{N}$. (by 1, 4)

6. $\varphi(\psi(H)) = H$.

7. Let $N \in \mathcal{N}$.

8. $D_R \subset G(N) \subset aut(R \cap P_0)$. (by 7, **18.18**)
9. $P_0 \cap \overline{gr(G(N))} \subset N \cap P_0 = R \cap P_0 \subset R$. (by 7, (ii))
10. $\varphi(N) \in \mathcal{H}$. (by 8, 9)
11. $\psi(\varphi(N)) = \psi(G(N)) = (R \cap P_0) \circ gr(G(N)) = (N \cap P_0) \circ gr(G(N)) = N$. (by 10, **7.21**)

Note that **18.19** is a generalization of **14.11**; we see this by taking $R = M \times M$. In this case

$$\mathcal{N} = \{N \mid N \text{ is an icer with } P_0 \subset N\} \quad \text{and} \quad \mathcal{H} = \{H \mid D \subset H = \overline{H} \subset G\};$$

the bijections $N \to G(N)$ and $H \to P_0 \circ gr(H)$ give the one-one correspondence (referred to in theorem **14.11**) between the distal icers on M and the closed subgroups of G which contain D.

It is worth restating the previous remark from the point of view of the spaces $X = M/R$ and $Z = M/S$. In this context the assignment

$$A \to X/\pi_R(gr(A))$$

gives a one-one correspondence:

$$\{A \mid G(R) \subset A = \overline{A} \subset G(S)\} \to \{Y \mid X \to Y \to Z\},$$

between the closed subgroups of G which sit between $G(R)$ and $G(S)$, and the distal extensions of Z which sit between X and Z.

We will have occasion to use the previous result in a slightly different form which we articulate in the following corollary.

Corollary 18.20 Let:

(i) $R \subset S$ be distal,
(ii) $\mathcal{N} = \{N \mid N \text{ is an icer with } R \subset N \subset S\}$, and
(iii) $\mathcal{H} = \{H \mid H \text{ is a subgroup of } G \text{ with } G(R) \subset H = \overline{H} \subset G(S)\}$.

Then

$$\varphi : \mathcal{N} \to \mathcal{H}$$
$$N \to G(N)$$

is bijective, its inverse being the map ψ defined by $\psi(H) = (R \cap P_0) \circ gr(H) = R \circ gr(H)$.

In proposition **14.5** we showed that any regular distal flow M/R is a group. We now show that if $R \subset S$ is a distal extension which is regular in the sense of **8.16**, then M/S is a quotient of M/R by a subgroup of the group of automorphisms of M/R. Moreover this subgroup is isomorphic to the group $G(S)/G(R)$. (Note that for a regular extension $G(S) \subset aut(R)$ and hence $G(R)$ is normal in $G(S)$.)

Proposition 18.21 Let:

(i) $X = M/R$ be a distal extension of $Y = M/S$,
(ii) $G(S) \subset aut(R)$ (so that $R \subset S$ is a regular extension, see **8.16**),
(iii) $\chi_R \colon aut(R) \to Aut(X)$ be the canonical epimorphism (see **7.10**), and
(iv) $L = G(S)/G(R) \cong \chi_R(G(S)) \subset Aut(X)$.

Then:

(a) $\pi_S^R(x_1) = \pi_S^R(x_2)$ if and only if $x_2 = \chi_R(\alpha)(x_1)$ for some $\alpha \in G(S)$.
(b) π_S^R induces an isomorphism $(L\backslash X, T) \cong (Y, T)$.

PROOF: (a) $\Longrightarrow$

1. Assume that $\pi_S^R(x_1) = \pi_S^R(x_2)$ for some $x_1, x_2 \in X$.
2. There exists $u \in J$ with $x_1 u = x_1$.
3. $\pi_S^R(x_2 u) \underset{1}{=} \pi_S^R(x_1 u) \underset{2}{=} \pi_S^R(x_1) \underset{1}{=} \pi_S^R(x_2)$.
4. $x_2 u = x_2$. (by 3, (i))
5. There exist $\alpha_1, \alpha_2 \in G$ with $\pi_R(\alpha_1(u)) = x_1$ and $\pi_R(\alpha_2(u)) = x_2$. (by 2, 4)
6. $(\alpha_1(u), \alpha_2(u)) \in S$. (by 1, 5)
7. $\alpha_2\alpha_1^{-1} \in G(S) \subset aut(R)$. (by 6, (ii))
8. $x_2 \underset{5}{=} \pi_R(\alpha_2(u)) = \pi_R(\alpha_2\alpha_1^{-1}(\alpha_1(u))) \underset{\mathbf{7.10}}{=} \chi_R(\alpha_2\alpha_1^{-1})(\pi_R(\alpha_1(u)) \underset{5}{=} \chi_R(\alpha_2\alpha_1^{-1})(x_1)$.

$\Longleftarrow$

1. Let $x_2 = \chi_R(\alpha)(x_1)$ with $\alpha \in G(S)$ and $\pi_R(p) = x_1$.
2. $x_2 = \pi_R(\alpha(p))$. (by 1, **7.10**)
3. $\pi_S^R(x_2) \underset{2}{=} \pi_S(\alpha(p)) \underset{1}{=} \pi_S(p) \underset{1}{=} \pi_S^R(x_1)$.

(b) This follows immediately from part (a).

In the previous proposition, $L = G(S)/G(R)$ is a group since $G(R) = \ker \chi_R$ is normal in $aut(R)$. In general $G(S)$ is a closed subgroup of G (in the τ-topology), but L will be Hausdorff only if $G(S)' \subset G(R)$ (by **11.10**). In addition it is important to note that the action of L on X is not continuous in general. Looking in particular at the case where $S = M \times M$ we saw in **15.10** that $G' \subset G(R)$ if and only if M/R is an equicontinuous regular flow and $X \cong G/G(R)$ is a compact Hausdorff topological group (so $L \cong X$ acts continuously on itself by left multiplication). We will prove in the next section that for a regular distal extension $R \subset S$, the derived group $G(S)' \subset G(R)$ if and only if $R \subset S$ is an *almost periodic extension*, and L is a compact Hausdorff topological group whose action on X is continuous. In this case $M/R \to M/S$ is a *compact group extension* with group L.

Given an icer R on M, the icer $S_d(R)$ is the smallest icer on M such that $S_d(R) \subset R$ is a distal extension. We now define for any minimal flow X/N, and icer R on M with $N \subset R$, an icer $S_d(X; R)$ on X which is the smallest icer on X such that $X/S_d(X; R) \to M/R$ is a distal extension.

Definition 18.22 Let N, R be icers on M with $N \subset R$, and set $X = M/N$. Then the *R-distal structure relation, $S_d(X; R)$ on X*, is defined by

$$S_d(X; R) = \bigcap \{S \mid X/S \to M/R \text{ is distal}\}.$$

Clearly $S_d(X; R)$ is the smallest icer S on X such that X/S is a distal extension of M/R. In particular when $N = \Delta$, $S_d(M; R) = R^*$ is the R-distal structure relation on M.

We now show that $S_d(X; R) = \pi_N(R^*)$.

Proposition 18.23 Let:

(i) R, N be icers on M,
(ii) $N \subset R$, and
(iii) $X = M/N$.

Then:

(a) $N \circ R^* = R^* \circ N = R^* \circ gr(G(N))$,
(b) $R^* \circ N$ is an icer on M,
(c) $\pi_N(R^*)$ is an icer on X, and
(d) $\pi_N(R^*) = S_d(X; R)$.

PROOF: (a), (b), (c) 1. $G(N) \underset{\text{(ii)}}{\subset} G(R) \subset aut(R) \underset{\mathbf{18.16}}{\subset} aut(R^*)$.
2. $N \cap P_0 \subset R \cap P_0 = R^* \cap P_0$. (by **18.5** since R^* is a distal extension of R)
3. $N \circ R^* = R^* \circ N$ is an icer on M and $\pi_N(R^*) = \pi_N(N \circ R^*)$ is an icer on X. (by 1, 2, **7.28**)
4. $R^* \circ N \underset{\mathbf{7.21}}{=} R^* \circ (N \cap P_0) \circ gr(G(N)) \underset{\mathbf{1,7.24}}{=} gr(G(N)) \circ R^* \circ (N \cap P_0)$
$\underset{\mathbf{7.21}}{=} gr(G(N)) \circ gr(G(R^*)) \circ (R^* \cap P_0) \circ (N \cap P_0)$
$\underset{2}{=} gr(G(N)) \circ gr(G(R^*)) \circ (R^* \cap P_0) \underset{\mathbf{7.21,7.24}}{=} R^* \circ gr(G(N))$.

(d) 1. $R^* \subset R^* \circ N \subset R$. (by (ii))
2. $X/\pi_N(R^*) = X/\pi_N(R^* \circ N) \cong M/(R^* \circ N)$ is a distal extension of M/R. (by 1, **18.8**, **18.9**, and the previous parts)
3. Let S be an icer on X such that X/S is a distal extension of M/R.
4. $X/S \cong M/\pi_N^{-1}(S)$.
5. $R^* \subset \pi_N^{-1}(S) \subset R$. (by 3, 4, **18.15**)
6. $\pi_N(R^*) \subset S$. (by 5)
7. $\pi_N(R^*) = S_d(X; R)$. (by 2, 3, 6)

One immediate consequence of **18.23** is that the construction of the R-distal structure relation is natural in the sense that a homomorphism $X \to Y$ of minimal flows maps $S_d(X; R)$ onto $S_d(Y; R)$.

Proposition 18.24 Let:

(i) $N \subset S \subset R$ be icers on M, and
(ii) $X = M/N$, and $Y = M/S$.

Then:

(a) $\pi_S^N(S_d(X; R)) = S_d(Y; R)$, and
(b) In the commutative diagram:

$$\begin{array}{ccc} X & \longrightarrow & Y \\ \downarrow & & \downarrow \\ X/S_d(X; R) & \longrightarrow & Y/S_d(Y; R) \end{array} \qquad (*)$$

where all the maps are the canonical ones, fibers are mapped onto fibers.

PROOF: (a) 1. $\pi_S^N(S_d(X; R)) = \pi_S^N(\pi_N(R^*)) = \pi_S(R^*) = S_d(Y; R)$. (by **18.23**)

(b) 1. $N \circ R^*$ and $S \circ R^*$ are both icers. (by **18.23**)
2. $X/S_d(X; R) \cong X/\pi_N(R^*) \cong M/(N \circ R^*)$. (by **18.23**)
3. $Y/S_d(Y; R) \cong Y/\pi_S(R^*) \cong M/(S \circ R^*)$. (by **18.23**)
4. $S \circ R^* = (N \circ R^*) \circ S$. (by 1, (i))
5. Fibers are mapped to fibers in diagram $(*)$. (by 4, **6.12**)

EXERCISES FOR CHAPTER 18

Exercise 18.1 Let:

(i) $R \subset S$ be a distal extension,
(ii) $A \subset G(S)$ be a τ-closed subgroup, and
(iii) $AG(R) = G(R)A$.

Then $\pi_R(gr(A))$ is an icer on M/R.

Exercise 18.2 Let:

(i) $R \subset S$ be a RIC extension, and
(ii) $G(R)G(S)' = G(S)$.

Then:

(a) $\pi_R(S)$ is topologically transitive (so $R \subset S$ is a weak mixing extension), and
(b) $R \subset S$ is a distal extension if and only if $R = S$.

Exercise 18.3 Let:

(i) $N \subset R$ be icers on M,
(ii) $S \subset R$ be distal,
(iii) $G(S) \subset D_R G(N)$, and
(iii) $X = M/N$.

Then:

(a) $S \circ N = R^* \circ N$ (and hence $S \circ N$ is an icer by **18.23**), and
(b) $\pi_N(S) = S_d(X; R)$.

Exercise 18.4 Let:

(i) R, S be an icers on M with $R \subset S$ distal,
(ii) H be a closed subgroup of G with $G(R) \subset H \subset G(S)$,
(iii) $R^H = \bigcup_{\alpha \in H} (1 \times \alpha)(R)$, and
(iv) $R_H = \{(\alpha(u), \beta(v)) \mid \alpha\beta^{-1} \in H, (\alpha(u), \alpha(v)) \in R\}$.

Then $R_H = R^H = R \circ gr(H) \subset S$.

Exercise 18.5 Let:

(i) R be an icer on M,
(ii) $N \subset R$ be distal, and
(iii) $\gamma \in aut(R)$ with $G(N)\gamma = \gamma G(N)$.

Then $\gamma \in aut(N)$.

Exercise 18.6 Let:

(i) R be an icer on M, and
(ii) $N \subset R$ be distal.

Then:

(a) if $G(N)$ is a normal subgroup of $aut(R)$, then $aut(R) \subset aut(N)$, and
(b) if $G(N)$ be a normal subgroup of $G(R)$, then $G(R) \subset aut(N)$.

Exercise 18.7 Let $A \subset G$ be a closed subgroup with $A = A'$. Then the following are equivalent:

(a) $gr(A) = \overline{gr(A)}$,
(b) $M \to M/\overline{gr(A)}$ is distal,
(c) $M \to M/\overline{gr(A)}$ is trivial,
(d) $A = \{id\}$, and
(e) $P_0 \cap \overline{gr(A)} = \Delta$.

Exercise 18.8 Let R be an icer on M with $M \to M/R$ distal. Then $G(R)^\infty = \{id\}$.

19

Almost periodic extensions of minimal flows

The notion of an equicontinuous minimal flow generalizes in a natural way to that of an almost periodic extension of minimal flows. Recall that proximal and distal extensions are defined so that proximal and distal extensions of the point flow are proximal and distal flows respectively. Similarly almost periodic extensions are defined so that an almost periodic extension of the point flow is an equicontinuous flow. We will use a generalization of the regionally proximal relation to obtain analogs for almost periodic extensions of many of the results on equicontinuous flows and distal extensions obtained earlier. This section explores two main themes; the first involves a generalization of **15.10** where we showed that a regular equicontinuous minimal flow is a compact topological group containing a homomorphic image of T as a dense subset. The second theme involves defining an analog of the equicontinuous structure relation for an arbitrary icer R on M.

Recall that (X, T) is an equicontinuous flow if for every open set W with $\Delta \subset W \subset X \times X$, there exists an open set V with $\Delta \subset V \subset X \times X$ and $VT \subset W$. Now if X is minimal and N is an icer on X, so that $\pi_N : X \to X/N$ is an extension of minimal flows there is a natural weakening of this notion to fibers of the map π_N. Namely we can require that

$$(x, y)T \subset W \text{ for any } (x, y) \in V \text{ with } \pi_N(x) = \pi_N(y),$$

or equivalently that $(V \cap N)T \subset W$. This is the definition of an almost periodic extension of minimal flows.

Definition 19.1 Let $X = M/R$, and $Y = M/S$ be flows with $R \subset S$ icers. We say that X is an *almost periodic extension of* Y ($R \subset S$ is *almost periodic*) if for every open set W with $\Delta_{M/R} \subset W \subset M/R \times M/R$, there exists an open set V with $\Delta \subset V$ and $(V \cap \pi_R(S))T \subset W$.

Clearly taking $S = M \times M$ in the definition above, we see that X is an almost periodic extension of a point if and only if (X, T) is an equicontinuous flow.

Next we define for a minimal flow (X, T) and an icer N on X, the relation $Q(N) \subset X \times X$. This is done so that $Q(N)$ is the analog of the regionally proximal relation on X.

Definition 19.2 Let (X, T) be a minimal flow and N be an icer on X. We define the *generalized (relativized) regionally proximal relation*, $Q(N)$ by:

$$Q(N) = \bigcap \{\overline{WT} \mid W \subset N \text{ is open with } \Delta_X \subset W\}.$$

Note that the fact that $Q(N) \subset N$ follows immediately from the fact that N is closed and invariant. Note also that when $N = X \times X$, $Q(X \times X)$ (which we sometimes write $Q(X)$ by abuse of notation), is the regionally proximal relation on X. We will be particularly interested in the case where $R \subset S$ are icers on M, $X = M/R$ and $N = \pi_R(S)$, so that $X/N = M/S$. In this case we will sometimes use the notation:

$$Q(R, S) \equiv Q(\pi_R(S)).$$

Let $\pi : X \to X/N$ be the canonical map. The relation $Q(N)$ is discussed in [Auslander, J., (1988)], and [McMahon, D., Wu, T.S., *Distal homomorphisms of nonmetric minimal flows* (1980)] where it is referred to as $Q(\pi)$. It is also discussed in [Ellis, R., (1969)] (see page 134). Though different definitions of an almost periodic extension are given in these references they both show (as we do in the following proposition) that $X \to X/N$ is an almost periodic extension if and only if $Q(N) = \Delta_X$. This of course generalizes the fact (proven in **15.5**) that a flow (X, T) is equicontinuous if and only if $Q(X) = \Delta$.

Proposition 19.3 Let $R \subset S$ be icers on M. Then $R \subset S$ is an almost periodic extension if and only if $Q(\pi_R(S)) = \Delta_{M/R}$.

PROOF: $\Longrightarrow$

1. Assume that $R \subset S$ is almost periodic and let W be a closed neighborhood of $\Delta_{M/R}$.
2. There exists V open with $\overline{(V \cap \pi_R(S))T} \subset W$. (by 1)
3. $Q(\pi_R(S)) \subset W$. (by 2)
4. $Q(\pi_R(S)) = \Delta_{M/R}$. (by 1, 3)

$\Longleftarrow$

1. Assume $Q(\pi_R(S)) = \Delta$ and let W be open with $\Delta_{M/R} \subset W \subset M/R \times M/R$.
2. $W^c \equiv (M/R \times M/R) \setminus W$ is closed.

3. Assume for contradiction that $W^c \cap \overline{(V \cap \pi_R(S))T} \neq \emptyset$ for all open V with $\Delta \subset V$.
4. $\emptyset \neq W^c \cap Q(\pi_R(S))$. (by 3, compactness)
5. There exists V open with $\Delta \subset V$ and $\overline{(V \cap \pi_R(S))T} \subset W$.
(4 contradicts 1)

For future reference we include a few properties of the generalized regionally proximal relation which are direct consequences of its definition.

Lemma 19.4 Let (X, T) be a minimal flow and N be an icer on X. Then:

(a) $Q(N)$ is closed, invariant, reflexive and symmetric,
(b) if (N, T) is a topologically transitive flow, then $Q(N) = N$,
(c) $(x, y) \in Q(N)$ if and only if there exist nets $\{(x_i, y_i)\} \subset N$ and $\{t_i\} \subset T$ with $x_i \to x$, $y_i \to y$, and $\lim x_i t_i = \lim y_i t_i$, and
(d) $N \cap P_0 \subset N \cap P(M) \subset Q(N)$.

PROOF: (a) This is immediate from the definition.

(b) If (N, T) is a topologically transitive flow, then $\overline{WT} = N$ for all open sets $W \subset N$ and hence $Q(N) = N$.

(c) We leave this for the reader to check.

(d) 1. Let $(x, y) \in N$ be a proximal pair.
2. $(x, y)T \cap W \neq \emptyset$ for all W open with $\Delta \subset W \subset X \times X$. (by 1)
3. $(x, y) \in N \cap \bigcap\{WT \mid \Delta \subset W$, and W is open in $X \times X\}$
$= \bigcap\{WT \mid \Delta \subset W$, and W is open in $N\} \subset Q(N)$. (by 1, 2)
4. $N \cap P_0 \subset N \cap P(M) \subset Q(N)$. (by 1, 3)

Proposition 19.5 Let:

(i) $\pi : (X, T) \to (Y, T)$ be a homomorphism of minimal flows,
(ii) N_X be an icer on X,
(iii) N_Y be an icer on Y, and
(iv) $\pi(N_X) = N_Y$.

Then $\pi(Q(N_X)) \subset Q(N_Y)$.

PROOF: 1. Let W be open with $\Delta_Y \subset W \subset Y \times Y$.
2. There exists V open with $\Delta_X \subset V \subset X \times X$ and $\pi(V) \subset W$. (by 1, (i))
3. $\pi(Q(N_X)) \subset \pi\left(\overline{(V \cap N_X)T}\right) = \overline{\pi(V \cap N_X)T} \subset \overline{(W \cap N_Y)T}$.
(by 1, 2, (iv))
4. $\pi(Q(N_X)) \subset Q(N_Y)$. (by 1,3)

Proposition 19.6 Let:

(i) $\pi : (X, T) \to (Y, T)$ be a homomorphism of minimal flows,

(ii) N_X be an icer on X,
(iii) N_Y be an icer on Y, and
(iv) $N_X = \pi^{-1}(N_Y)$.

Then $\pi(Q(N_X)) = Q(N_Y)$.

PROOF: It follows from **19.5** that $\pi(Q(N_X)) \subset Q(N_Y)$.
1. Let $(y_1, y_2) \in Q(N_Y)$ and W be open with $\Delta_X \subset W \subset X \times X$.
2. Let $V \subset X$ be closed with nonempty interior and $V \times V \subset W$.
3. There exists a finite set $F \subset T$ with $VF = X$. ((i), 2, X is compact)
4. $\pi(V)F = Y$. (by (i), 3)
5. $\pi(V)$ has nonempty interior. (by 2, 4)
6. $\Delta_Y \subset \text{int}(\pi((V \times V)T)) \subset \text{int}(\pi(WT))$. (by 2, 5)
7. $(y_1, y_2) \underset{1,6}{\in} \overline{(N_Y \cap \text{int}(\pi(WT)))T} \subset \overline{(N_Y \cap \pi(WT))T}$

$$= \overline{(N_Y \cap \pi(W))T} \underset{(iv)}{=} \overline{(\pi(N_X \cap W))T} = \pi\left(\overline{(N_X \cap W)T}\right).$$

8. $\pi^{-1}(y_1, y_2) \cap \overline{(N_X \cap W)T} \neq \emptyset$. (by 7)
9. $\pi^{-1}(y_1, y_2) \cap Q(N_X) \neq \emptyset$. (by 1, 8, compactness)
10. $(y_1, y_2) \in \pi(Q(N_X))$. (by 9)

Corollary 19.7 Let $R \subset S$ be icers on M. Then $\pi_R(Q(S)) = Q(\pi_R(S))$.

PROOF: In this case $\pi_R^{-1}(\pi_R(S)) = R \circ S \circ R = S$ so this follows immediately from **19.6**.

We now wish to prove an analog of **15.10** which says that a distal flow is equicontinuous if and only if its group contains G'. We will show that a distal extension $R \subset S$ is almost periodic if and only if $G(S)' \subset G(R)$. The key is to show that if $R \subset S$ is distal, then $Q(\pi_R(S)) = \pi_R(gr(G(S)'))$. We begin with a lemma.

Lemma 19.8 Let S be an icer on M. Then $gr(G(S)') \subset Q(S)$.

PROOF: 1. Let V be an open neighborhood of Δ in $M \times M$.
2. $gr(1_M) \subset VT$. (by 1)
3. $gr(G(S)') \subset \overline{VT \cap gr(G(S))} \subset \overline{VT \cap S} = \overline{(V \cap S)T}$. (by 1, 2, **12.4**)
4. $gr(G(S)') \subset \bigcap \overline{(V \cap S)T} = Q(S)$. (by 1, 3)

Proposition 19.9 Let:

(i) $R \subset S$ be a distal extension, and
(ii) $A \subset G(S)$ be a closed subgroup with $AG(R) = G(S)$.

Then $\pi_R(gr(A')) = \pi_R(gr(G(S)')) = Q(\pi_R(S))$.

PROOF: 1. $\pi_R(gr(A')) \subset \pi_R(gr(G(S)')) \subset \pi_R(Q(S)) = Q(\pi_R(S))$. (by (ii), **19.8**, **19.7**)

2. Let $(x, y) \in Q(\pi_R(S))$.
3. There exist nets $\{(x_i, y_i)\} \subset \pi_R(S)$ and $\{t_i\} \subset T$ with

$$x_i \to x, \quad y_i \to y, \quad \text{and} \quad \lim x_i t_i = \lim y_i t_i. \qquad \text{(by 2, 19.4)}$$

4. There exist $p_i \in M$ and $\alpha_i \in A$ with

$$\pi_R(p_i, \alpha_i(p_i)) = (x_i, y_i). \qquad \text{(by (i), (ii), 3, 18.9)}$$

5. By passing to subnets we may assume that

$$(p_i, \alpha_i(p_i)) \to (p, \alpha(pv)) \in \overline{gr(A)} \subset S$$

where $\alpha \in A$ and $v \in J$. (by (ii))
6. Again we may assume that $(p_i t_i, \alpha_i(p_i t_i)) \to (q, \beta(qu)) \in \overline{gr(A)}$.
7. $\pi_R(q, \beta(qu)) \underset{6}{=} \lim \pi_R(p_i t_i, \alpha(p_i t_i)) \underset{4}{=} \lim(x_i t_i, y_i t_i) \underset{3}{\in} \Delta_{M/R}$.
8. $\beta \in A \cap G(R)$. (by 6, 7)
9. Let V, W be nonempty open subsets of M.
10. Since $(p, \alpha(pv))v = (pv, \alpha(pv))$ and $(q, \beta(qu))u = (qu, \beta(qu))$ there exist $s, t \in T$ such that

$$(ps, \alpha(pvs)) \in V \times \alpha(V), \quad \text{and} \quad (qt, \beta(qut)) \in W \times \beta(W).$$

11. There exists i such that

$$(p_i, \alpha_i(p_i))s \in V \times \alpha(V) \quad \text{and} \quad (p_i t_i t, \alpha_i(p_i) t_i t) \in W \times \beta(W).$$

(by 5, 6, 9, 10)

12. There exists i such that $\alpha_i \in A \cap < V, \alpha(V) > \cap < W, \beta(W) >$. (by 11)
13. $\beta^{-1}\alpha \in A'$. (by 8, 9, 12, **11.2**)
14. $(x, y) \underset{3}{=} \lim(x_i, y_i) \underset{4}{=} \lim \pi_R(p_i, \alpha_i(p_i))$
$\underset{5}{=} \pi_R(p, \alpha(pv)) \underset{(i),5}{=} \pi_R(p, \alpha(p)) \underset{8}{=} \pi_R(p, \beta^{-1}\alpha(p)) \underset{13}{\in} \pi_R(gr(A'))$.

Corollary 19.10 Let $R \subset S$ be a distal extension. Then $Q(\pi_R(S))$ is an equivalence relation.

PROOF: 1. $G(S)'$ is a closed normal subgroup of $G(S)$. (by **11.7**)
2. $R \circ gr(G(R)G(S)')$ is an icer on M. (by 1, **18.20**)
3. $Q(\pi_R(S)) = \pi_R(gr(G(S)')) = \pi_R(R \circ gr(G(R)G(S)'))$ is an icer on M/R. (by **19.9**, **6.11**)

Theorem 19.11 Let $R \subset S$ be icers on M. Then $R \subset S$ is an almost periodic extension if and only if $R \subset S$ is distal and $G(S)' \subset G(R)$.

PROOF: $\Longrightarrow$

1. Assume that $R \subset S$ is almost periodic.
2. $Q(\pi_R(S)) = \Delta_{M/R}$. (by **19.3**)
3. $\pi_R(S \cap P_0) \underset{\mathbf{19.4}}{\subset} \pi_R(Q(S)) \underset{\mathbf{19.7}}{=} Q(\pi_R(S)) \underset{2}{=} \Delta_{M/R}$.
4. $R \cap P_0 = S \cap P_0$ so $R \subset S$ is distal. (by 3)
5. $\pi_R(gr(G(S)')) \underset{4,\mathbf{19.9}}{=} Q(\pi_R(S)) \underset{2}{=} \Delta_{M/R}$.
6. $R \subset S$ is distal and $G(S)' \subset G(R)$. (by 4, 5)

$\Longleftarrow$

1. Assume that $R \subset S$ is distal and $G(S)' \subset G(R)$.
2. $Q(\pi_R(S)) = \pi_R(gr(G(S)')) = \Delta_{M/R}$. (by 1, **19.9**)
3. $R \subset S$ is almost periodic. (by **19.3**)

The characterization of almost periodic extensions given in **19.11** has many interesting consequences. We begin with two corollaries which will be used to further study almost periodic extensions.

Corollary 19.12 Let:

(i) R be an icer on M,
(ii) $\mathcal{S}$ be a collection of icers on M,
(iii) $S \subset R$ be almost periodic for all $S \in \mathcal{S}$, and
(iv) $N = \bigcap \mathcal{S}$.

Then M/N is an almost periodic extension of M/R.

PROOF: 1. M/N is a distal extension of M/R. (by **18.12**)
2. $G(R)' \subset G(S)$ for all $S \in \mathcal{S}$. (by (iii))
3. $G(R)' \underset{2}{\subset} \bigcap_{S \in \mathcal{S}} G(S) \underset{\mathbf{7.19}}{=} G(N)$.

Corollary 19.13 Let:

(i) R, S, N be icers on M,
(ii) $S \subset N$ be an almost periodic extension, and
(iii) $\alpha \in G$.

Then:

(a) $R \cap S \subset R \cap N$ is an almost periodic extension, and
(b) $\alpha(S) \subset \alpha(N)$ is an almost periodic extension.

PROOF: (a) 1. $S \subset N$ is a distal extension. (by (ii), **19.11**)
2. $R \cap S \subset R \cap N$ is a distal extension. (by 1, **18.3**)
3. $G(N)' \subset G(S)$. (by (ii), **19.11**)

4. $G(R \cap N)' \subset G(R) \cap G(N)' \subset G(R) \cap G(S) = G(R \cap S)$. (by 3, **7.19**)

5. $R \cap S \subset R \cap N$ is an almost periodic extension. (by 2, 4, **19.11**)

(b) 1. $\alpha(S) \subset \alpha(N)$ is distal. (by (ii), **18.7**, **19.11**)

2. $G(\alpha(N))' \underset{\mathbf{7.16}}{=} (\alpha G(N)\alpha^{-1})' \underset{\mathbf{11.6}}{=} \alpha G(N)'\alpha^{-1} \underset{\text{(ii)}}{\subset} \alpha G(S)\alpha^{-1} \underset{\mathbf{7.16}}{=} G(\alpha(S))$.

3. $\alpha(S) \subset \alpha(N)$ is an almost periodic extension. (by 1, 2, **19.11**)

Note that **19.11** together with **11.10** imply that if $R \subset S$ is almost periodic, then $G(S)/G(R)$ is compact Hausdorff. We will show that if in addition $R \subset S$ is a regular extension, then $M/R \to M/S$ is a *compact group extension* (see the definition below) with group $G(S)/G(R)$. More generally any almost periodic extension is "group-like", a notion we make precise in **19.16**.

Definition 19.14 Let $\pi : X \to Y$ be a homomorphism of minimal flows. We say that π is a *compact group extension* if there exists a compact Hausdorff topological group H and an action $\varphi : H \times X \to X$ such that $\pi : X \to Y$
$$(h, x) \to hx$$
is a *principal bundle with group* H, that is:

(i) φ is continuous,

(ii) the action of H commutes with that of T, that is:

$$(hx)t = h(xt) \text{ for all } x \in X,\ h \in H \text{ and } t \in T, \text{ and}$$

(iii) π induces an isomorphism $(H\backslash X, T) \cong (Y, T)$; i. e.

$$\pi(x_1) = \pi(x_2) \text{ if and only if } x_2 = hx_1 \text{ for some } h \in H.$$

Note that since (X, T) is minimal, (i) and (ii) imply that $hx_0 = kx_0$ for some $x_0 \in X$, if and only if $hx = kx$ for all $x \in X$.

Proposition 19.15 Let:

(i) R, N be icers on M,

(ii) $N \subset R$ be an almost periodic extension, and

(iii) $G(R) \subset aut(N)$ (so that $N \subset R$ is a regular extension, see **8.16**).

Then

(a) $G(R)/G(N)$ is a compact topological group.

(b) the map

$$\begin{aligned} \varphi : G(R) \times M &\to M/N \\ (\alpha, p) &\to \alpha(p)N \end{aligned}$$

induces a continuous action of $G(R)/G(N)$ on M/N such that:

$$\pi_R^N(pN) = \pi_R^N(qN) \Longleftrightarrow (G(R)/G(N))\,(pN) = (G(R)/G(N))\,(qN).$$

In other words π_R^N is a compact group extension with group $G(R)/G(N)$.

PROOF: (a) This follows immediately from **7.10** and **11.11**.

(b) 1. We first show that the map φ is continuous.

2. Let $\{\alpha_i\} \subset G(R)$ and $\{p_i\} \subset M$ be nets with $\alpha_i \to \alpha$ and $p_i \to p$.

3. Using compactness we may assume that $\alpha_i(p_i) \to \beta(pv)$ for some $\beta \in G$ and $v \in J$.

4. $\alpha_i \to \beta$. (by 2, 3, **10.13**)

5. $\alpha, \beta \in G(R)$. (by 2, 4, **10.8**)

6. $G(R)'\alpha_i \to G(R)'\alpha$ and $G(R)'\alpha_i \to G(R)'\beta$. (by 2, 4)

7. $G(R)'\alpha = G(R)'\beta$. ($G(R)/G(R)'$ is Hausdorff by **11.9**)

8. $\beta\alpha^{-1} \underset{7}{\in} G(R)' \underset{\text{(ii)},\mathbf{19.11}}{\subset} G(N)$.

9. $(\alpha(p), \beta(p)) = (\alpha(p), \beta\alpha^{-1}(\alpha(p))) \in N$. (by 8)

10. $(p, \beta(pv)) = \lim(p_i, \alpha_i(p_i)) \in \overline{R} = R$. (by 2, 3)

11. $(\beta(p), \beta(pv)) \underset{5,10}{\in} R \cap P(M) = N \cap P(M)$.

($N \subset R$ is distal by (ii), **19.11**)

12. $\varphi(\alpha_i, p_i) = \alpha_i(p_i)N \underset{3}{\to} \beta(pv)N \underset{11}{=} \beta(p)N \underset{9}{=} \alpha(p)N = \varphi(\alpha, p)$.

We leave it to the reader to verify the remaining details.

Proposition 19.16 Let:

(i) R, N be icers on M,

(ii) $N \subset R$ be an almost periodic extension, and

(iii) $S = \bigcap_{\alpha \in G(R)} \alpha(N)$.

Then

(a) π_R^S is a group extension with group $H_1 = G(R)/G(S)$.

(b) π_N^S is a group extension with group $G(N)/G(S) = H_2 \subset H_1$.

(c) $M/N \to M/R$ is isomorphic to the fiber bundle associated to the principal H_2-bundle $M/S \to M/R$ with fiber H_1/H_2.

PROOF: (a) 1. $S \subset R$ is a regular almost periodic extension.

(by **19.12**, **19.13**, (ii) and (iii), **8.17**)

2. $S \subset R$ is a group extension with group $G(R)/G(S)$. (by 1, **19.15**)

(b) 1. $S = S \cap N \subset R \cap N = N$ is almost periodic . (by part (a) and **19.13**)

2. $G(N) \subset G(R) \subset aut(S)$. (by (ii), (iii))

3. $S \subset N$ is a compact group extension with group $G(N)/G(S)$.

(by **19.15**)

(c) 1. Let $Z = H_1/H_2 \times M/S$.

2. The map $(h, (H_2h_1, z)) \to (H_2h_1h^{-1}, hz)$ defines an action of H_1 on Z.

3. The map $(H_2h_1, z) \to \pi_N^S(h_1z)$ defines an isomorphism of the fiber bundle $Z/H_1 \to M/R$ with $M/N \to M/R$.

We now define for any icer R on M, an icer $S_{eq}(R)$ which is the natural generalization of the equicontinuous structure relation S_{eq}.

Definition 19.17 Let R be an icer on M. We define the *R-equicontinuous structure relation* by

$$S_{eq}(R) = \bigcap\{S \mid S \subset R \text{ is an almost periodic extension of } R\}.$$

We define the subgroup E_R by $E_R = G(S_{eq}(R))$.

It follows immediately from **19.12** that $M/S_{eq}(R)$ is an almost periodic extension of M/R. When $R = M \times M$, $S_{eq}(R) = S_{eq}$, and we are back to the "absolute" case of equicontinuous flows. In this case the results of this section coincide with those of section 13. This also motivates the use of the notation E_R since $E = G(S_{eq})$ (see **15.15**).

Proposition 19.18 Let R and S be icers on M with $R \subset S$. Then:

(a) $R \subset S$ is almost periodic if and only if $S_{eq}(S) \subset R$, and
(b) $S_{eq}(R) \subset S_{eq}(S)$.

PROOF: (a) The first implication is clear, and the converse follows from **18.15** and **19.11**. The details are left to the reader.

(b) 1. $S_{eq}(S) \subset S$ is an almost periodic extension.
2. $R \cap S_{eq}(S) \subset R \cap S = R$ is an almost periodic extension. (by 1, **19.13**)
3. $S_{eq}(R) \subset R \cap S_{eq}(S) \subset S_{eq}(S)$. (by 2, and part (a))

Proposition 19.19 Let R be an icer on M and $\alpha \in G$. Then:

(a) $\alpha(S_{eq}(R)) = S_{eq}(\alpha(R))$,
(b) $aut(R) \subset aut(S_{eq}(R))$,
(c) E_R is a normal subgroup of $aut(R)$, and
(d) E_R is a normal subgroup of $G(R)$.

PROOF: (a) 1. $S_{eq}(R) \subset R$ is almost periodic. (by **19.12**, **19.17**)
2. $\alpha(S_{eq}(R)) \subset \alpha(R)$ is an almost periodic extension. (by 1, **19.13**)
3. $S_{eq}(\alpha(R)) \subset \alpha(S_{eq}(R))$. (by 2, **19.17**)
4. $S_{eq}(R) \subset \alpha(S_{eq}(\alpha^{-1}(R)))$. (by 3, replace R by $\alpha^{-1}(R)$)
5. $\alpha(S_{eq}(R)) \subset S_{eq}(\alpha(R))$. (by 4, replace α by α^{-1})
6. $\alpha(S_{eq}(R)) = S_{eq}(\alpha(R))$. (by 3, 5)

(b) This follows immediately from part (a).
(c) This follows immediately from part (b) and **7.10**.
(d) This follows immediately from part (c).

We saw in Proposition **15.21** that $E = DG'$. Viewing this as a result for the case $R = M \times M$, the next proposition gives the analogous result when R is any icer on M.

Proposition 19.20 Let R be an icer on M. Then $E_R = D_R G(R)'$.

PROOF: Proof that $D_R G(R)' \subset E_R$:
1. $S_{eq}(R) \subset R$ is almost periodic.
2. $G(R)' \subset G(S_{eq}(R)) = E_R$. (by 1, **19.11**, **19.17**)
3. $S_{eq}(R) \subset R$ is distal. (by 1, **19.11**)
4. $R^* = S_d(R) \subset S_{eq}(R)$. (by 3, **18.15**)
5. $D_R \underset{\mathbf{18.19}}{=} G(R^*) \underset{4}{\subset} G(S_{eq}(R)) \underset{\mathbf{19.17}}{=} E_R$.
6. $D_R G(R)' \subset E_R$. (by 2, 5)

Proof that $E_R \subset D_R G(R)'$:
1. D_R is a closed normal subgroup of $G(R)$. (by **18.17**)
2. $G(R)'$ is a closed normal subgroup of $G(R)$. (by **11.7**)
3. $D_R G(R)'$ is a closed normal subgroup of $G(R)$. (by 1, 2, and **10.10**)
4. There exists a distal extension $N \subset R$ with $G(N) = D_R G(R)'$. (by **18.19**)
5. $N \subset R$ is almost periodic. (by 4, **19.11**)
6. $S_{eq}(R) \subset N$. (by 5, **19.17**)
7. $E_R = G(S_{eq}(R)) \underset{6}{\subset} G(N) \underset{4}{=} D_R G(R)'$.

We now prove an analog of **15.24**.

Proposition 19.21 Let:

(i) R be an icer on M, and
(ii) the almost periodic points of the flow (R, T) be dense in R.

(by **7.5** this is equivalent to saying $R = \overline{gr(G(R))}$)

Then:

(a) $D_R \subset G(R)'$.
(b) $E_R = G(R)'$.

PROOF: (a) 1. Let $\alpha \in G$ with $gr(\alpha) \subset \overline{P(M) \cap R}$ and let V be any open subset of M.
2. Let U, W be open subsets of M with $\alpha \in \langle U, W \rangle$.
3. There exists $p \in M$ with $(p, \alpha(p)) \in U \times W$. (by 2)
4. There exists $(x, y) \in P(M) \cap R \cap (U \times W)$. (by 1, 3)
5. There exists $t \in T$ with $(xt, yt) \in V \times V$. (by 1, 4)
6. $\emptyset \neq (V \times V) \cap (U \times W)t \cap R$. (by 4, 5)
7. There exists $\beta \in G(R)$ and $m \in M$ with $(m, \beta(m)) \in V \times V \cap (U \times W)t$. (by 1, 2, 6, and (ii))

8. $\beta(mt^{-1}) \in \beta(U) \cap W$. (by 7)
9. $\beta \in < V, V > \cap \overline{G(R) \cap < U, W >}$. (by 7, 8)
10. $\alpha \in \overline{G(R) \cap < V, V >}$. (by 2, 9)
11. $\alpha \in \bigcap \overline{G(R) \cap < V, V >} = G(R)'$. (by 1, 10, see **11.1**)

(b) This follows from part (a) and **19.20**.

In contrast to **18.16**, which states that $S_d(S_d(R)) = S_d(R)$ for all icers R, $S_{eq}(S_{eq}(R))$ need not equal $S_{eq}(R)$ even for the "absolute" case $R = M \times M$. To see this we need only consider a group T for which $D \neq E$. (For example $T = \mathbf{Z}$ will do.) Then $S_d = P_0 \circ gr(D) \neq P_0 \circ gr(E) = S_{eq}$, so $S_d \subset S_{eq}$ is a non-trivial distal extension. By the generalized Furstenberg structure theorem (see the following section) there exists an icer $S_d \subset R \subset S_{eq}$, such that $R \underset{\neq}{\subset} S_{eq}$ is a non-trivial almost periodic extension. Hence $S_{eq}(S_{eq}) \underset{\neq}{\subset} S_{eq}$.

Proposition **18.8** which says that a distal extension of a distal extension is distal, does not hold for almost periodic extensions. Indeed in the example considered above, $R \subset S_{eq}$ is an almost periodic extension, and $S_{eq} \subset M \times M$ is an almost periodic extension, but $R \subset M \times M$ is not almost periodic.

Given an icer R on M, the factorization

$$M \to M/S_{eq}(R) \to M/R$$

gives the maximal almost periodic extension of M/R. It is natural to generalize this to any extension $X \to M/R$; we now define $S_{eq}(X; R)$ so that the factorization

$$X \to M/S_{eq}(X; R) \to M/R$$

gives the maximal almost periodic extension of M/R which is a factor of X.

Definition 19.22 Let N and R be icers on M and $X = M/N$. We define the *R-almost periodic structure relation on* X, $S_{eq}(X; R)$ to be the smallest icer S on X such that X/S is an almost periodic extension of M/R. Note that $S_{eq}(M; R) = S_{eq}(R)$. Thus when $R = M \times M$, $S_{eq}(M; R) = S_{eq}(R) = S_{eq}$, the equicontinuous structure relation defined in section 13. Therefore the R-almost periodic structure relation on X is also called the *R-equicontinuous structure relation on* X.

Proposition 19.23 Let:

(i) R, N be icers on M,
(ii) $N \subset R$, and
(iii) $X = M/N$.

Then:

(a) $N \circ S_{eq}(R) = S_{eq}(R) \circ N$,

(b) $S_{eq}(R) \circ N$ is an icer on M,
(c) $\pi_N(S_{eq}(R) \circ N) = \pi_N(S_{eq}(R))$ is an icer on X, and
(d) $\pi_N(S_{eq}(R)) = S_{eq}(X; R)$.

PROOF: (a), (b), (c) 1. $G(N) \underset{\text{(ii)}}{\subset} G(R) \underset{\mathbf{7.10}}{\subset} aut(R) \underset{\mathbf{19.19}}{\subset} aut(S_{eq}(R))$.
2. $P_0 \cap N \subset P_0 \cap R = P_0 \cap S_{eq}(R)$. (by **18.5**, since $S_{eq}(R) \subset R$ is distal)
3. $N \circ S_{eq} = S_{eq} \circ N$ is an icer on M and $\pi_N(S_{eq}(R)) = \pi_N(N \circ S_{eq}(R))$ is an icer on X. (by 1, 2, **7.28**)
(d) 1. $S_{eq}(R) \subset S_{eq}(R) \circ N \subset R$. (by (ii))
2. $X/\pi_N(S_{eq}(R)) = X/\pi_N(S_{eq}(R) \circ N) \cong M/(S_{eq}(R) \circ N)$ is an almost periodic extension of M/R. (by 1, previous parts)
3. $S_{eq}(X; R) \subset \pi_N(S_{eq}(R))$. (by 2, **19.22**)
4. Let S be an icer on X such that X/S is an almost periodic extension of M/R.
5. $X/S \cong M/\pi_N^{-1}(S)$.
6. $S_{eq}(R) \subset \pi_N^{-1}(S) \subset R$. (by 4, 5, **19.17**)
7. $\pi_N(S_{eq}(R)) \subset S$. (by 6)
8. $\pi_N(S_{eq}(R)) = S_{eq}(X; R)$. (by 3, 7)

Corollary 19.24 Let:

(i) $R \subset N \subset S$ be icers on M, and
(ii) $X = M/N$, and $Y = M/S$.

Then $\pi_S^N(S_{eq}(X; R)) = S_{eq}(Y; R)$.

PROOF: $\pi_S^N(S_{eq}(X; R)) = \pi_S^N(\pi_N(S_{eq}(R))) = \pi_S(S_{eq}(R)) = S_{eq}(Y; R)$. (by **19.23**)

NOTES ON SECTION 19

Note 19.N.1 *PI-flows.* Following [Ellis, R., Glasner, S., Shapiro, L., *Proximal-isometric (PI) flows*, (1975)] a flow X is defined to be *proximal-isometric* (*PI* for short) if there exists a family $(Y_\gamma \mid \gamma \leq \nu)$ of flows such that:

(i) $Y_0 =$ a point,
(ii) $Y_{\gamma+1}$ is either an almost periodic or a proximal extension of Y_γ for all $\gamma < \nu$,
(iii) $Y_\beta = \lim(Y_\alpha \mid \alpha < \beta)$ if $\beta \leq \nu$ is a limit ordinal, and
(iv) Y_ν is a proximal extension of X.

As in **11.17** we define groups:

(i) $G^0 = G$,
(ii) $G^{\gamma+1} = (G^\gamma)'$ for all $\gamma < \nu$,

(iii) $G^\beta = \bigcap(G^\alpha \mid \alpha < \beta)$ if β is a limit ordinal, and
(iv) $G^\infty = G^{\nu(G)}$ where $\nu(G)$ is the smallest ordinal α with $G^\alpha = G^{\alpha+1}$.

These definitions lead to the following (Proposition 7.3 of the reference mentioned earlier).

Proposition 19.N.2 Let R be an icer on M. Then M/R is a *PI*-flow if and only if $G^\infty \subset G(R)$.

We prove a more general "relative version".

Theorem 19.N.3 (The Relative (generalized) *PI-tower*) Let:

(i) $R \subset S$ be an icers on M, and
(ii) $G(S)^\infty \subset G(R)$.

Then there exists a family $(S_\alpha \mid \alpha \le \nu)$ such that:

(a) $S_0 = S$ and $G(S_\nu) = G(R)$,
(b) $S_{\alpha+1} \subset S_\alpha$ is either an almost periodic extension or a proximal extension, and
(c) $S_\beta = \bigcap_{\alpha<\beta} S_\alpha$ for all limit ordinals $\beta \le \nu$.

PROOF: We proceed by induction.
1. Assume that for every ordinal $\beta < \gamma$, a family $(S_\alpha^\beta \mid \alpha \le \nu(\beta))$ has been constructed such that:

(a) $S_0^\beta = S$ and $G(S_{\nu(\beta)}^\beta) = G(R)G(S)^\beta$,
(b) $S_{\alpha+1}^\beta \subset S_\alpha^\beta$ is either an almost periodic extension or a proximal extension,
(c) $S_\rho^\beta = \bigcap_{\alpha<\rho} S_\alpha^\beta$ for all limit ordinals $\rho \le \beta$, and
(d) if $\beta_1 \le \beta_2$, then $\nu(\beta_1) \le \nu(\beta_2)$ and $S_\alpha^{\beta_1} = S_\alpha^{\beta_2}$ for all ordinals $\alpha \le \nu(\beta_1)$.

2. If γ is a limit ordinal, set $S_{\nu(\gamma)}^\gamma = \bigcap_{\beta<\gamma} S_{\nu(\beta)}^\beta$.

3. $G(S_{\nu(\gamma)}^\gamma) = G\Big(\bigcap_{\beta<\gamma} S_{\nu(\beta)}^\beta\Big) \underset{\mathbf{7.19}}{=} \bigcap_{\beta<\gamma} G(S_{\nu(\beta)}^\beta)$
$\underset{1}{=} \bigcap_{\beta<\gamma} G(R)G(S)^\beta \underset{\mathbf{10.12}}{=} G(R) \bigcap_{\beta<\gamma} G(S)^\beta \underset{\mathbf{11.17}}{=} G(R)G(S)^\gamma$.

4. If $\gamma = \beta + 1$, and $S_{\nu(\beta)}^\beta = \overline{gr(G(R)G(S)^\beta)}$, set

$$\nu(\gamma) = \nu(\beta) + 1 \quad \text{and} \quad S_{\nu(\gamma)}^\gamma = \overline{gr(G(R))} \circ S_{eq}\left(S_{\nu(\beta)}^\beta\right).$$

5. $S_{\nu(\gamma)}^\gamma$ is an icer. (by 4, **19.23**)
6. $S_{\nu(\gamma)}^\gamma$ is an almost periodic extension of $S_{\nu(\beta)}^\beta$. (by 4, and **19.18**)

7. $G(S^{\gamma}_{\nu(\gamma)}) \underset{\mathbf{19.17}}{=} G(R)E_{S^{\beta}_{\nu(\beta)}} \underset{4,\mathbf{19.21}}{=} G(R)\big(G(S^{\beta}_{\nu(\beta)})\big)'$
$\underset{4}{=} G(R)\big(G(R)G(S)^{\beta}\big)' \underset{\mathbf{11.14}}{=} \overline{G(R)G(S)^{\beta+1}} \underset{8}{=} G(R)G(S)^{\gamma}.$

8. If $\gamma = \beta + 1$, and $S^{\beta}_{\nu(\beta)} \neq \overline{gr(G(R)G(S)^{\beta})}$, set $\nu(\gamma) = \nu(\beta) + 2$ and

$$S^{\gamma}_{\nu(\beta)+1} = \overline{gr(G(R)G(S)^{\beta})}, \quad S^{\gamma}_{\nu(\gamma)} = \overline{gr(G(R))} \circ S_{eq}(\overline{gr(G(R)G(S)^{\beta})}).$$

9. $S^{\gamma}_{\nu(\beta)+1}$ is a proximal extension of $S^{\gamma}_{\nu(\beta)} = S^{\beta}_{\nu(\beta)}$. (by 1(a), 8, and **7.11**)
10. $S^{\gamma}_{\nu(\gamma)}$ is an almost periodic extension of $S^{\gamma}_{\nu(\beta)+1}$. (by 8, **19.18**, **19.23**)
11. $G(S^{\gamma}_{\nu(\gamma)}) \underset{\mathbf{19.17}}{=} G(R)E_{S^{\gamma}_{\nu(\beta)+1}} \underset{8,\mathbf{19.21}}{=} G(R)G(S^{\gamma}_{\nu(\beta)+1})'$
$\underset{\mathbf{8}}{=} G(R)\big(G(R)G(S)^{\beta}\big)' \underset{\mathbf{11.14}}{=} G(R)G(S)^{\beta+1} \underset{8}{=} G(R)G(S)^{\gamma}.$

EXERCISES FOR CHAPTER 19

Exercise 19.1 Let $R \subset S$ be icers on M. Show that $R \subset S$ is an almost periodic extension if and only if $Q(S) \subset R$.

Exercise 19.2 Let:

(i) $R \subset S$ be distal, and
(ii) $X = M/R$ and $Y = M/S$.

Then $X/Q(\pi_R(S))$ is an almost periodic extension of Y. Moreover this extension is trivial only if $X = Y$.

Exercise 19.3 Let:

(i) $R \subset S$ be a RIC extension, and
(ii) $A \subset G(S)$ be a closed subgroup with $AG(R) = G(S)$.

Then $\pi_R(\overline{gr(A')}) = \pi_R(\overline{gr(G(S)')}) = Q(\pi_R(S))$.

Exercise 19.4 (See **19.18**) Let $R \subset S$ be icers on M. Show that $R \subset S$ is an almost periodic extension if and only if $S_{eq}(S) \subset R$.

Exercise 19.5 In **19.15** it was shown that if $N \subset R$ is an almost periodic extension and $G(R) \subset aut(N)$, then π^N_R is a group extension with group $G(R)/G(N)$. This exercise provides the converse of that result.

Let $\pi : X \to Y$ be a homomorphism of minimal flows which is a compact group extension with group H. Then there exist icers N and R on M such that:

(a) $N \subset R$ is a distal extension,
(b) $H \cong G(R)/G(N)$ as topological groups,
(c) $G(R)' \subset G(N)$,

(d) $G(R) \subset aut(N)$, and
(e) There is an isomorphism of group extensions:

$$\begin{array}{ccccc} & M/N & \to & X & \\ \pi_R^N & \downarrow & & \downarrow & \pi. \\ & M/R & \to & Y & \end{array}$$

20

A tale of four theorems

In this section we examine four theorems, the Furstenberg Structure Theorem for distal flows (denoted **Theorem 1**), and three other theorems (**Theorems 2, 3, 4**) that are equivalent to it. We can think of these three as the icer, regionally proximal, and group theoretic versions of the Furstenberg theorem. These three theorems have generalizations for distal extensions (denoted **Theorems 2g, 3g, 4g**) which again are equivalent to the Furstenberg structure theorem for distal extensions (**Theorem 1g**). The goal of this section is to clarify the relationship between the different approaches to the structure of distal extensions by giving explicit arguments showing that the four theorems are equivalent. We then give a construction of the so-called Furstenberg tower for a distal extension; a classical consequence of the structure theorem. Finally we comment on various proofs of these theorems valid in the metric, countable group, and general cases. We begin with explicit statements of each of the four theorems along with their generalizations.

Theorem 1 (Furstenberg Structure Theorem) Let:

(i) $R \subset S$ with $R \neq S$ be icers on M, and
(ii) M/R be distal.

Then there exists an icer N on M such that:

(a) $R \subset N \subset S$,
(b) $M/N \to M/S$ is an almost periodic extension, and
(c) $N \neq S$.

In particular every non-trivial distal flow has a non-trivial equicontinuous factor. When $S = M \times M$ this theorem shows that any non-trivial distal flow has a non-trivial equicontinuous factor. This allows an inductive construction of the so-called Furstenberg tower of almost periodic extensions and inverse limits. This tower begins with $\{pt\}$ (when S is taken to be $M \times M$) or with the maximal equicontinuous factor $M/S_{eq}(R)$ of M/R (when S is taken to be $S_{eq}(R)$) and ends with M/R. (See **20.14** at the end of this section.)

Theorem 1g (Furstenberg Structure Theorem for distal extensions) Let:

(i) $R \subset S$ with $R \neq S$ be icers on M, and
(ii) $M/R \to M/S$ be a distal extension.

Then there exists an icer N on M such that:

(a) $R \subset N \subset S$,
(b) $M/N \to M/S$ is an almost periodic extension, and
(c) $N \neq S$.

This theorem is sometimes referred to as the generalized or relative Furstenberg theorem. We can think of it as saying that every non-trivial distal extension "contains" a non-trivial almost periodic extension. This theorem allows an inductive construction of the so-called Furstenberg tower of almost periodic extensions and inverse limits beginning with M/S and ending with M/R. (Again see **20.14** at the end of this section.) Note of course that M/R need not itself be a distal flow in this case.

Our second theorem and it's generalization are stated in terms of icers.

Theorem 2 Let:

(i) (X, T) be a distal minimal flow,
(ii) N_X be an icer on X, and
(iii) N_X be topologically transitive.

Then $N_X = \Delta_X$.

This is corollary **4.26**. One immediate consequence of this theorem (proven in **4.25**) is that the only weak mixing minimal distal flow is the trivial one-point flow.

Theorem 2g Let:

(i) (X, T) be a minimal flow,
(ii) N_X be an icer on X, and
(iii) N_X be topologically transitive and pointwise almost periodic.

Then $N_X = \Delta_X$.

This is theorem **9.13**, proven using the quasi-relative product. It says that the only distal weak mixing extension of minimal flows is the trivial one.

Our third theorem and it's generalization are stated in terms of the generalized regionally proximal relation $Q(N)$ defined for any icer N.

Theorem 3 Let:

(i) (X, T) be a distal flow,
(ii) R be an icer on X, and
(iii) $Q(R) = R$.

Then $X = \{pt\}$.

Theorem 3g Let $R \subset S$ be a distal extension with $Q(\pi_R(S)) = \pi_R(S)$. Then $R = S$.

This theorem appears in [McMahon, D., Wu, T.S., *Distal homomorphisms of nonmetric minimal flows*, (1981)]; the authors use it to deduce the Furstenberg structure theorem for minimal distal extensions.

Finally our fourth version of the theorem is a "group-theoretic" one, stated in terms of the iterated derived group A^∞.

Theorem 4 $G^\infty \subset D$.

This was proven in **14.13**.

Theorem 4g Let $R \subset S$ be a distal extension. Then $G(S)^\infty \subset G(R)$.

We now give explicit arguments showing that these four theorems are equivalent. We will concentrate on showing that the generalized versions are equivalent; the arguments in the "absolute" case are essentially the same. We will prove that **1g** $\Longrightarrow$ **2g** $\Longrightarrow$ **3g** $\Longrightarrow$ **1g** and **2g** $\Longrightarrow$ **4g** $\Longrightarrow$ **3g**.

20.1 THM 1g IMPLIES THM 2g: Let (X, T) be minimal, N_X a topologically transitive pointwise almost periodic icer on X, and assume that Theorem 1g holds. We will show that $N_X = \Delta_X$.

PROOF: 1. Let R be an icer on M with $X = M/R$.
2. Let $S = \pi_R^{-1}(N_X)$.
3. S is an icer on M and $\pi_R(S) = N_X$. (by 2)
4. $R \subset S$ is distal. (by 2, 3, **18.9**, N_X is pointwise almost periodic)
5. $Q(\pi_R(S)) = \pi_R(S)$. (by 3, **19.4**, N_X is topologically transitive)
6. Let N be an icer on M with $R \subset N \subset S$ and $N \subset S$ an almost periodic extension.
7. $\Delta_{M/N} \underset{\mathbf{19.3}}{=} Q(\pi_N(S)) \underset{\mathbf{19.6}}{=} \pi_N^R(Q(\pi_R(S))) \underset{\mathbf{5}}{=} \pi_N^R(\pi_R(S)) = \pi_N(S)$.
8. $N = S$. (by 7)
9. $R = S$. (by 4, 8, and Theorem 1g)
10. $N_X = \pi_R(S) = \Delta_X$. (by 3, 9)

20.2 THM 2g IMPLIES THM 3g: Let $R \subset S$ be a distal extension with $Q(\pi_R(S)) = \pi_R(S)$, and assume that Theorem 2g holds. We will show that $R = S$.

PROOF: 1. $\pi_R(gr(G(S)')) \underset{\mathbf{19.9}}{=} Q(\pi_R(S)) = \pi_R(S) \underset{\mathbf{18.9}}{=} \pi_R(gr(G(S)))$.
($R \subset S$ is distal)

2. $G(S)'G(R) = G(S)$. (by 1)
3. $G(S)^\infty G(R) = G(S)$. (by 2, **11.18**)
4. $\pi_R\left(\overline{gr(G(S)^\infty)}\right) = \pi_R(gr(G(S)^\infty) = \pi_R(gr(G(S))) = \pi_R(S)$.
(by 3, **18.9**, $R \subset S$ is distal)
5. $(\pi_R(S), T)$ is topologically transitive. (by 4, **12.5**, **4.21**)
6. $(\pi_R(S), T)$ is pointwise almost periodic. (by **18.9**, $R \subset S$ is distal)
7. $\pi_R(S) = \Delta$ and hence $R = S$. (by 5, 6, and **Theorem 2g**)

20.3 THM 3g IMPLIES THM 1g: Let $R \subset S$ be a distal extension with $R \neq S$, and assume that Theorem 3g holds. We will show that there exists $R \subset N \subset S$ with $N \neq S$, such that $N \subset S$ is an almost periodic extension.

PROOF: 1. $\pi_R(gr(G(S)') \underset{\mathbf{19.9}}{=} Q(\pi_R(S)) \underset{\textbf{Thm 3g}}{\neq} \pi_R(S) \underset{\mathbf{18.9}}{=} \pi_R(gr(G(S))$.
2. $G(R)G(S)'$ is a proper closed subgroup of $G(S)$. (by 1, **11.7**, and **10.10**)
3. There exists an icer N on M with $R \subset N \subset S$ and $G(N) = G(R)G(S)'$.
(by 2, **18.20**)
4. $N \subset S$ is a distal extension and $N \neq S$. (by 2, 3, **18.8**, $R \subset S$ is distal)
5. $N \subset S$ is an almost periodic extension. (by 3, 4, **19.11**)

20.4 THM 2g IMPLIES THM 4g: Let $R \subset S$ be a distal extension, and assume that theorem 2g holds. We will show that $G(S)^\infty \subset G(R)$.

PROOF: 1. $G(R)G(S)^\infty = G(S)^\infty G(R)$.
($G(S)^\infty$ is normal in $G(S)$ by **11.17**)
2. Let $N = R \circ gr(G(S)^\infty)$.
3. N is an icer on M with $R \subset N \subset S$ and $G(N) = G(R)G(S)^\infty$.
(by 2, **18.20**)
4. $\pi_R(N) = \pi_R(\overline{G(S)^\infty}) = \pi_R(gr(G(S)^\infty)$ is an icer on M/R.
(by 2, 3, **6.11**)
5. $(\pi_R(N), T)$ is topologically transitive and pointwise almost periodic.
(by 4, **4.21**, **12.5**, **18.9**)
6. $\pi_R(N) = \Delta$ and hence $N = R$. (by 5, and **Theorem 2g**)
7. $G(S)^\infty \subset G(R)$. (by 3, 6)

20.5 THM 4g IMPLIES THM 3g: Let $R \subset S$ be a distal extension with $Q(\pi_R(S)) = \pi_R(S)$, and assume that theorem 4g holds. We will show that $R = S$.

PROOF: 1. $\pi_R(gr(G(S)')) \underset{\mathbf{19.9}}{=} Q(\pi_R(S)) = \pi_R(S) \underset{\mathbf{18.9}}{=} \pi_R(gr(G(S)))$.
($R \subset S$ is distal)
2. $G(S)'G(R) = G(S)$. (by 1)
3. $G(R) = G(S)^\infty G(R) = G(S)$. (by 2, **11.18**, and **Theorem 4g**)
4. $R = S$. (by 3, **18.20**, $R \subset S$ is distal)
One important and classical consequence of these theorems is the so-called Furstenberg tower for a distal extension of minimal flows. This is analogous to the PI-tower discussed in the notes to section 19. One can give a proof using that tower (see **19.N.3**); instead we give a direct proof which takes advantage of the properties of the relative product and the subgroups of G developed in the previous sections.

Theorem 20.6 (the Furstenberg tower) Let $R \subset S$ be a distal extension. Then there exists a family $(S_\alpha \mid \alpha \leq \nu)$ of icers such that:

(a) $S_0 = S$ and $S_\nu = R$,
(b) $S_{\alpha+1} \subset S_\alpha$ is an almost periodic extension, and
(c) $S_\beta = \bigcap_{\alpha<\beta} S_\alpha$ for all limit ordinals $\beta \leq \nu$.

PROOF: We proceed by induction.
1. Assume that for every ordinal $\beta < \gamma$, a family $(S_\alpha \mid \alpha \leq \beta)$ has been constructed such that:

(a) $S_0 = S$ and $S_\alpha = R \circ gr(G(S)^\alpha)$, for all $\alpha \leq \beta$,
(b) $S_{\alpha+1} \subset S_\alpha$ is an almost periodic extension, and
(c) $S_\rho = \bigcap_{\alpha<\rho} S_\alpha$ for all limit ordinals $\rho \leq \beta$.

2. If γ is a limit ordinal, set $S_\gamma = \bigcap_{\beta<\gamma} S_\beta$,
3. $G(S_\gamma) = G\left(\bigcap_{\beta<\gamma} S_\beta\right) \underset{\mathbf{7.19}}{=} \bigcap_{\beta<\gamma} G(S_\beta)$

$$= _1 \bigcap_{\beta<\gamma} G(R)G(S)^\beta \underset{\mathbf{10.12}}{=} G(R) \bigcap_{\beta<\gamma} G(S)^\beta \underset{\mathbf{11.17}}{=} G(R)G(S)^\gamma.$$

4. $R \subset S_\gamma \subset S$ are distal extensions.
(by 1, 2, **18.8**, $R \subset S$ is a distal extension)
5. $S_\gamma = R \circ gr(G(S)^\gamma)$. (by 3, 4, **18.9**)
6. If $\gamma = \beta + 1$, set $S_\gamma = R \circ gr\left((G(S)^\beta)'\right) = R \circ gr(G(S)^\gamma)$.
7. S_γ is an icer. (by 6, **18.20**)
8. $S_\gamma \subset S_\beta$ is an almost periodic extension. (by 6, **18.8**, and **19.11**)

9. By induction there exists ν such that $S_\nu = S_{\nu+1}$.
10. $G(S)^\nu = (G(S)^\nu)' = G(S)^\infty$. (by 9, **11.17**)
11. $S_\nu = R \circ gr(G(S)^\infty = R$. (by 9, 10, **Theorem 4g**)

It is an interesting exercise to understand how the theorems we have been discussing can be proven in certain special cases, especially since our proof of Theorem 2g in the most general context (see **9.13**) involves a quite technical construction using the quasi-relative product. In the chart below we outline some alternative proofs when certain assumptions are made about either the space X or the group T; the approach taken here has been to focus on the icer version (Theorems **2** and **2g**).

Comments on the proof of the Theorems 2 and 2g in various cases:

	In order to prove 2	In order to prove 2g
When X is metrizable	**All** metric flows which are distal and top. tr. are minimal (by **4.19**) Thm 2 follows	**All** metric flows which are p.a.p. and top. tr. are minimal (by **4.19**) Thm 2g follows
When T is countable	**All** flows which are distal and top. tr. are minimal (by **4.23**) Thm 2 follows	by **All** flows which are p.a.p. and top. tr. are minimal (by **4.23**) Thm 2g follows
The general case	All flows which are distal and top. tr. are minimal (by **4.24**) Thm 2 follows	Use the proof given in **9.13**

Note that in the literature, the most popular approach to proving the Furstenberg structure theorem in the metric case is to use the outline above to deduce Theorem 2 or 2g, and then deduce Theorems 1 and 1g. For the general case one approach is to prove Theorem 3g (see [Auslander, (1988)]) and use it to deduce Theorem 1g; here the difficult technicalities appear in the proof of Theorem 3g (for which Auslander refers to [McMahon, D., Wu, S.T., *Distal homomorphisms of nonmetric minimal flows*, (1981)]). Our approach has been to emphasize Theorem 2g in the general case, introducing the quasi-relative product as a means of giving a clear proof. In the literature the focus has mainly been on deducing the Furstenberg theorem, so the fact that all of these theorems are equivalent seems not to have been explicitly emphasized. This fact and the metric approach leads us to ask the following question: Theorem 2g says that any **icer** R, on a minimal flow, with (R, T) pointwise almost periodic and topologically transitive must be minimal and hence trivial; what if (R, T)

is any flow? In other words, does **4.23** hold in the uncountable group case? If pointwise almost periodicity is strengthened to distality the answer is yes; in the metric and the countable group case the answer is also yes. In general we do not know the answer to this question.

EXERCISE FOR CHAPTER 20

Exercise 20.1 Let:

(i) $R \subset S$ be a RIC extension, and
(ii) the only factor $N \supset R$ such that $N \subset S$ is an almost periodic extension, be $N = S$.

Show that $R \subset S$ is a weak-mixing extension (that is $(\pi_R(S), T)$ is topologically transitive).

References

Akin, E. *Recurrence in Topological Dynamics* Plenum Press, New York, London (1997).

Auslander, J. *Regular minimal sets I*, Trans. Amer. Math. Soc. 123 1966 469–479.

Auslander, J. *Minimal Flows and Their Extensions*, North Holland, Amsterdam (1988).

Auslander, J; Ellis, D. B.; Ellis, R. *The regionally proximal relation*, Trans. Amer. Math. Soc. 347 (1995), no. 6, 2139–2146.

Auslander, J; Glasner, S. *Distal and highly proximal extensions of minimal flows*, Indiana University Journal vol. 26 (1977) no. 4.

Bourbaki, N. *Elements de Mathematique X #3, Topologie Generale*, Hermann & Cie, (1949).

Dugundji, James *Topology*, Allyn and Bacon Inc., (1966).

Ellis, Robert. *Locally compact transformation groups*, Duke Math Journal vol. 24 (1957) no. 2 119–126.

Ellis, Robert. *A semigroup associated with a transformation group*, Trans. Amer. Math. Soc. 94 (1960) 272–281.

Ellis, Robert. *Group-like extensions of minimal sets*, Trans. Amer. Math. Soc. 127 (1967) 125–135.

Ellis, Robert. *The structure of group-like extensions of minimal sets*, Trans. Amer. Math. Soc. 134 (1968) 261–287.

Ellis, R. *Lectures on Topological Dynamics*, Benjamin, New York (1969).

Ellis, Robert. *The Veech structure theorem*, Trans. Amer. Math. Soc. 186 (1973), 203–218 (1974).

Ellis, Robert. *The Furstenberg structure theorem*, Pacific J. Math. 76 (1978), no. 2, 345–349.

Ellis, Robert; Glasner, S.; Shapiro, L., *Proximal-isometric (PI) flows*, Advances in Math. 17 (1975).

Glasner, Shmuel. *Compressibility properties in topological dynamics*, Amer. J. Math. 97 (1975), 148–171.

Glasner, S, *Proximal Flows*, Lecture Notes in Mathematics, Vol. 517. Springer-Verlag, Berlin-New York, (1976).

Gleason, A., *Projective topological spaces*, Illinois Journal 2 # 4A, (1958), 482–489.

James, I. M. *Topological and Uniform Spaces*, Undergraduate Texts in Mathematics, Springer-Verlag, New York-Heidelberg (1987).

Kelley, John L. *General Topology,* D. Van Nostrand Company, Inc., Toronto-New York-London, (1955).

Massey, W. S. *Algebraic Topology: an Introduction*, Reprint of the 1967 edition. Graduate Texts in Mathematics, Vol. 56. Springer-Verlag, New York-Heidelberg (1977).

McMahon, Douglas C.; Nachman, Louis J. *An intrinsic characterization for PI flows*, Pacific J. Math. 89 (1980), no. 2, 391–403.

McMahon, D.; Wu, T. S. *Distal homomorphisms of nonmetric minimal flows*, Proc. Amer. Math. Soc. 82 (1981), no. 2, 283–287.

Munkres James R., *Topology, a First Course* Prentice-Hall, Inc., Englewood Cliffs, New Jersey (1975).

Rudin, Walter, *Principles of Mathematical Analysis* McGraw-Hill Book Company, Inc., New York-Toronto-London, (1953).

Shapiro, Leonard, *Proximality in minimal transformation groups*, Proc. Amer. Math. Soc. 26 (1970) 521–525.

Veech, William A. *The equicontinuous structure relation for minimal abelian transformation groups*, Amer. J. Math. 90 (1968) 723–732.

Veech, William A. *Point-distal flows*, Amer. J. Math. 92 (1970) 205–242.

Veech, William A. *Topological dynamics*, Bull. Amer. Math. Soc. 83 (1977), no. 5, 775–830.

Index

Printed in the United States
by Baker & Taylor Publisher Services